Lecture Notes
in Control and Information Sciences 230

Editor: M. Thoma

AF676789

Springer
London
Berlin
Heidelberg
New York
Barcelona
Budapest
Hong Kong
Milan
Paris
Santa Clara
Singapore
Tokyo

B. Siciliano and K.P. Valavanis (Eds)

Control Problems in Robotics and Automation

Springer

Series Advisory Board

A. Bensoussan · M.J. Grimble · P. Kokotovic · H. Kwakernaak
J.L. Massey · Y.Z. Tsypkin

Editors

Professor Bruno Siciliano
Dipartimento di Informatica e Sistemistica,
Università degli Studi di Napoli Federico II,
Via Claudio 21, 80125 Napoli, Italy

Professor Kimon P. Valavanis
Robotics and Automation Laboratory,
Center for Advanced Computer Studies,
University of Southwestern Louisiana,
Lafayette, LA 70505-4330, USA

ISBN 3-540-76220-5 Springer-Verlag Berlin Heidelberg New York

British Library Cataloguing in Publication Data
A catalogue record for this book is available from the British Library

Library of Congress Cataloging-in-Publication Data
Control problems in robotics and automation / B. Siciliano and K.P. Valavanis, eds.
p. cm. - - (Lecture notes in control and information sciences : 230)
Includes bibliographical references (p.).
ISBN 3-540-76220-5 (alk. paper)
1. Automatic control. 2. Robots- -Control systems. 3. Automation.
I. Siciliano, Bruno, 1959- . II. Valavanis, K. (Kimon) III. Series
TJ213.C5725 1998
629.8- -dc21 97-31960

Apart from any fair dealing for the purposes of research or private study, or criticism or review, as permitted under the Copyright, Designs and Patents Act 1988, this publication may only be reproduced, stored or transmitted, in any form or by any means, with the prior permission in writing of the publishers, or in the case of reprographic reproduction in accordance with the terms of licences issued by the Copyright Licensing Agency. Enquiries concerning reproduction outside those terms should be sent to the publishers.

© Springer-Verlag London Limited 1998
Printed in Great Britain

The use of registered names, trademarks, etc. in this publication does not imply, even in the absence of a specific statement, that such names are exempt from the relevant laws and regulations and therefore free for general use.

The publisher makes no representation, express or implied, with regard to the accuracy of the information contained in this book and cannot accept any legal responsibility or liability for any errors or omissions that may be made.

Typesetting: Camera ready by editors
Printed and bound at the Athenæum Press Ltd, Gateshead
69/3830-543210 Printed on acid-free paper

Foreword

It is rather evident that if we are to address successfully the control needs of our society in the 21st century, we need to develop new methods to meet the new challenges, as these needs are imposing ever increasing demands for better, faster, cheaper and more reliable control systems. There are challenging control needs all around us, in manufacturing and process industries, in transportation and in communications, to mention but a few of the application areas. Advanced sensors, actuators, computers, and communication networks offer unprecedented opportunities to implement highly ambitious control and decision strategies. There are many interesting control problems out there which urgently need good solutions. These are exciting times for control, full of opportunities. We should identify these new problems and challenges and help the development and publication of fundamental results in new areas, areas that show early promise that will be able to help address the control needs of industry and society well into the next century. We need to enhance our traditional control methods, we need new ideas, new concepts, new methodologies and new results to address the new problems. Can we do this? This is the challenge and the opportunity.

Among the technology areas which demand new and creative approaches are complex control problems in robotics and automation. As automation becomes more prevalent in industry and traditional slow robot manipulators are replaced by new systems which are smaller, faster, more flexible, and more intelligent, it is also evident that the traditional PID controller is no longer a satisfactory method of control in many situations. Optimum performance of industrial automation systems, especially if they include robots, will demand the use of such approaches as adaptive control methods, intelligent control, "soft computing" methods (involving neural networks, fuzzy logic and evolutionary algorithms). New control systems will also require the ability to handle uncertainty in models and parameters and to control lightweight, highly flexible structures. We believe complex problems such as these, which are facing us today, can only be solved by cooperation among groups across traditional disciplines and over international borders, exchanging ideas and sharing their particular points of view.

In order to address some of the needs outlined above, the IEEE Control Systems Society (CSS) and the IEEE Robotics and Automation Society (RAS) sponsored an *International Workshop on Control Problems in Robotics and Automation: Future Directions* to help identify problems and promising solutions in that area. The CSS and the RAS are leading the effort to identify future and challenging control problems that must be addressed to meet future needs and demands, as well as the effort to provide solutions to these problems. The Workshop marks ten years of fruitful collaboration between the sponsoring Societies.

On behalf of the CSS and RAS, we would like to express our sincere thanks to Kimon Valavanis and Bruno Siciliano, the General and Program Chairs of the Workshop for their dedication, ideas and hard work. They have brought together a truly distinguished group of robotics, automation, and control experts and have made this meeting certainly memorable and we hope also useful, with the ideas that have been brought forward being influential and direction setting for years to come. Thank you.

We would like also to thank the past CSS President Mike Masten and the past RAS President T.-J.Tarn for actively supporting this Workshop in the spirit of cooperation among the societies. It all started as an idea at an IEEE meeting, also in San Diego, in early 1996. We hope that it will lead to future workshops and other forms of cooperation between our societies.

Panos J. Antsaklis
President, IEEE Control Systems Society

George A. Bekey
President, IEEE Robotics and Automation Society

Preface

The purpose of the book is to focus on the state-of-the-art of control problems in robotics and automation. Beyond its tutorial value, the book aims at identifying challenging control problems that must be addressed to meet future needs and demands, as well as at providing solutions to the identified problems.

The book contains a selection of invited and submitted papers presented at the *International Workshop on Control Problems in Robotics and Automation: Future Directions*, held in San Diego, California, on December 9, 1997, in conjunction with the 36th IEEE Conference on Decision and Control. The Workshop has been jointly sponsored by the IEEE Control Systems Society and the IEEE Robotics and Automation Society.

The key feature of the book is its wide coverage of relevant problems in the field, discussed by world-recognized leading experts, who contributed chapters for the book. From the vast majority of control aspects related to robotics and automation, the Editors have tried to opt for those "hot" topics which are expected to lead to significant achievements and breakthroughs in the years to come.

The sequence of the topics (corresponding to the chapters in the book) has been arranged in a progressive way, starting from the closest issues related to industrial robotics, such as force control, multirobots and dexterous hands, to the farthest advanced issues related to underactuated and nonholonomic systems, as well as to sensors and fusion. An important part of the book has been dedicated to automation by focusing on interesting issues ranging from the classical area of flexible manufacturing systems to the emerging area of distributed multi-agent control systems.

A reading track along the various contributions of the sixteen chapters of the book is outlined in the following.

Robotic systems have captured the attention of control researchers since the early 70's. In this respect, it can be said that the motion control problem for rigid robot manipulators is now completely understood and solved. Nonetheless, practical robotic tasks often require interaction between the manipulator and the environment, and thus a *force control* problem arises. The chapter by *De Schutter et al.* provides a comprehensive classification of different approaches where force control is broadened to a differential-geometric context.

Whenever a manipulation task exceeds the capability of a single robot, a *multirobot cooperative system* is needed. A number of issues concerning the modelling and control of such a kind of system are surveyed in the chapter by *Uchiyama*, where the problem of robust holding of the manipulated object is emphasized.

Multifingered *robot hands* can be regarded as a special class of multirobot systems. The chapter by *Bicchi et al.* supports a minimalist approach to design of *dexterous* end effectors, where nonholonomy plays a key role.

Force feedback becomes an essential requirement for *teleoperation* of robot manipulators, and *haptic interfaces* have been devised to alleviate the task of remote system operation by a computer user. The chapter by *Salcudean* points out those control features that need to be addressed for the manipulation of virtual environments.

A radically different approach to the design control problem for complex systems is offered by *fuzzy control*. The potential of such approach is discussed in the chapter by *Hsu and Fu*, in the light of a performance enhancement obtained by either a learning or a suitable approximation procedure. The application to mechanical systems, including robot manipulators, is developed.

Modelling robot manipulators as rigid mechanical systems is an idealization that becomes unrealistic when higher performance is sought. *Flexible manipulators* are covered in the chapter by *De Luca*, where both joint elasticity and link flexibility are considered with special regard to the demanding problem of trajectory control.

Another interesting type of mechanical systems is represented by walking machines. The chapter by *Hurmuzlu* concentrates on the locomotion of *bipedal robots*. Active vs. passive control strategies are discussed where the goal is to generate stable gait patterns.

Unlike the typical applications on ground, *free-floating robotic systems* do not have a fixed base, e.g. in the space or undersea environment. The derivation of effective models becomes more involved, as treated in the chapter by *Egeland and Pettersen*. Control aspects related to motion coordination of vehicle and manipulator, or else to system underactuation, are brought up.

The more general class of *underactuated mechanical systems* is surveyed in the chapter by *Spong*. These include flexible manipulators, walking robots, space and undersea robots. The dynamics of such systems place them at the forefront of research in advanced control. Geometric nonlinear control and passivity-based control methods are invoked for stabilization and tracking control purposes.

The chapter by *Canudas de Wit* concerns the problem of controlling mobile robots and multibody vehicles. An application-oriented overview of some actual trends in control design for these systems is presented which also touches on the realization of transportation systems and intelligent highways.

Control techniques for mechanical systems such as robots typically rely on the feedback information provided by proprioceptive sensors, e.g. position, velocity, force. On the other hand, heteroceptive sensors, e.g. tactile, proximity, range, provide a useful tool to enrich the knowledge about the operational environment. In this respect, vision-based robotic systems have represented a source of active research in the field. The fundamentals of the various proposed approaches are described in the chapter by *Corke and Hager*, where the interdependence of vision and control is emphasized and the closure of a visual-feedback control loop (*visual servoing*) is shown as a powerful means to ensure better accuracy.

The employment of multiple sensors in a control system calls for effective techniques to handle disparate and redundant sensory data. In this respect, *sensor fusion* plays a crucial role as evidenced in the chapter by *Henderson et al.*, where architectural techniques for developing wide area sensor network systems are described.

Articulated robot control tasks, e.g. assembly, navigation, perception, human-robot shared control, can be effectively abstracted by resorting to the theory of *discrete event systems*. This is the subject of the chapter by *McCarragher*, where constrained motion systems are examined to demonstrate the advantages of discrete event theory in regarding robots as part of a complete automation system. Process monitoring techniques based on the detection and identification of discrete events are also dealt with.

Flexible manufacturing systems have traditionally constituted the ultimate challenge for automation in industry. The chapter by *Luh* is aimed at presenting the basic job scheduling problem formulation and a relevant solution methodology. A practical case study is taken to discuss the resolution and the implications of the scheduling problem.

Integration of sensing, planning and control in a manufacturing work-cell represents an attractive problem in intelligent control. A unified framework for *task synchronization* based on a Max-Plus algebra model is proposed in the chapter by *Tarn et al.* where the interaction between discrete and continuous events is treated in a systematic fashion.

The final chapter by *Sastry et al.* is devoted to a different type of automation other than the industrial scenario; namely, air traffic management. This is an important example of control of *distributed multi-agent systems*. Owing to technological advances, new levels of system efficiency and safety can be reached. A decentralized architecture is proposed where air traffic control functionality is moved on board aircraft. Conflict resolution strategies are illustrated along with verification methods based on Hamilton-Jacobi, automata, and game theories.

The book is intended for graduate students, researchers, scientists and scholars who wish to broaden and strengthen their knowledge in robotics and automation and prepare themselves to address and solve control problems in the next century.

We hope that this Workshop may serve as a milestone for closer collaboration between the IEEE Control Systems Society and the IEEE Robotics and Automation Society, and that many more will follow in the years to come.

We wish to thank the Presidents Panos Antsaklis and George Bekey, the Executive and Administrative Committees of the Control Systems Society and Robotics and Automation Society for their support and encouragement, the Members of the International Steering Committee for their suggestions, as well as the Contributors to this book for their thorough and timely preparation of the book chapters. The Editors would also like to thank Maja Matijašević and Cathy Pomier for helping them throughout the Workshop, and a special note of mention goes to Denis Gračanin for his assistance during the critical stage of the editorial process. A final word of thanks is for Nicholas Pinfield, Engineering Editor, and his assistant Michael Jones of Springer-Verlag, London, for their collaboration and patience.

September 1997

Bruno Siciliano
Kimon P. Valavanis

Table of Contents

List of Contributors xvii

Force Control: A Bird's Eye View
Joris De Schutter, Herman Bruyninckx, Wen-Hong Zhu, and
Mark W. Spong 1

1. Introduction 1
2. Basics of Force Control 2
   2.1 Basic Approaches 2
   2.2 Examples 3
   2.3 Basic Implementations 4
   2.4 Properties and Performance of Force Control 6
3. Multi-Degree-of-Freedom Force Control 8
   3.1 Geometric Properties 8
   3.2 Constrained Robot Motion 9
   3.3 Multi-Dimensional Force Control Concepts 10
   3.4 Task Specification and Control Design 11
4. Robust and Adaptive Force Control 13
   4.1 Geometric Errors 13
   4.2 Dynamics Errors 14
5. Future Research 15

Multirobots and Cooperative Systems
Masaru Uchiyama 19

1. Introduction 19
2. Dynamics of Multirobots and Cooperative Systems 21
3. Derivation of Task Vectors 24
   3.1 External and Internal Forces/Moments 24
   3.2 External and Internal Velocities 25
   3.3 External and Internal Positions/Orientations 26
4. Cooperative Control 27
   4.1 Hybrid Position/Force Control 27
   4.2 Load Sharing 28
5. Recent Research and Future Directions 30

6. Conclusions ... 31

Robotic Dexterity via Nonholonomy
Antonio Bicchi, Alessia Marigo, and Domenico Prattichizzo ... 35

1. Introduction ... 35
2. Nonholonomy on Purpose ... 37
3. Systems of Rolling Bodies ... 42
   3.1 Regular Surfaces ... 42
   3.2 Polyhedral Objects ... 44
4. Discussion and Open Problems ... 46

Control for Teleoperation and Haptic Interfaces
Septimiu E. Salcudean ... 51

1. Teleoperation and Haptic Interfaces ... 51
2. Teleoperator Controller Design ... 52
   2.1 Modeling Teleoperation Systems ... 52
   2.2 Robust Stability Conditions ... 54
   2.3 Performance Specifications ... 54
   2.4 Four-Channel Controller Architecture ... 55
   2.5 Controller Design via Standard Loop Shaping Tools ... 56
   2.6 Parametric Optimization-based Controller Design ... 57
   2.7 Nonlinear Transparent Control ... 58
   2.8 Passivation for Delays and Interconnectivity ... 58
   2.9 Adaptive Teleoperation Control ... 59
   2.10 Dual Hybrid Teleoperation ... 60
   2.11 Velocity Control with Force Feedback ... 61
3. Teleoperation Control Design Challenges ... 61
4. Teleoperation in Virtual Environments ... 62
5. Conclusion ... 63

Recent Progress in Fuzzy Control
Feng-Yih Hsu and Li-Chen Fu ... 67

1. Introduction ... 67
2. Mathematical Foundations ... 68
3. Enhanced Fuzzy Control ... 69
   3.1 Learning-based Fuzzy Control ... 69
   3.2 Approximation-based Fuzzy Control ... 72
4. Conclusion ... 80

Trajectory Control of Flexible Manipulators
Alessandro De Luca ... 83

1. Introduction ... 83
2. Robots with Elastic Joints ... 84

2.1 Dynamic Modeling ... 85
2.2 Generalized Inversion Algorithm ... 86
3. Robots with Flexible Links ... 92
3.1 Dynamic Modeling ... 92
3.2 Stable Inversion Control ... 94
3.3 Experimental Results ... 99
4. Conclusions ... 102

Dynamics and Control of Bipedal Robots
Yildirim Hurmuzlu ... 105

1. How Does a Multi-link System Achieve Locomotion? ... 105
1.1 Inverted Pendulum Models ... 106
1.2 Impact and Switching ... 107
2. Equations of Motion and Stability ... 108
2.1 Equations of Motion During the Continuous Phase of Motion ... 108
2.2 Impact and Switching Equations ... 109
2.3 Stability of the Locomotion ... 110
3. Control of Bipedal Robots ... 113
3.1 Active Control ... 113
3.2 Passive Control ... 114
4. Open Problems and Challenges in the Control of Bipedal Robots ... 114

Free-Floating Robotic Systems
Olav Egeland and Kristin Y. Pettersen ... 119

1. Kinematics ... 119
2. Equation of Motion ... 121
3. Total System Momentum ... 125
4. Velocity Kinematics and Jacobians ... 125
5. Control Deviation in Rotation ... 126
6. Euler Parameters ... 127
7. Passivity Properties ... 127
8. Coordination of Motion ... 128
9. Nonholonomic Issues ... 128

Underactuated Mechanical Systems
Mark W. Spong ... 135

1. Introduction ... 135
2. Lagrangian Dynamics ... 136
2.1 Equilibrium Solutions and Controllability ... 139
3. Partial Feedback Linearization ... 140
3.1 Collocated Linearization ... 140
3.2 Non-collocated Linearization ... 140
4. Cascade Systems ... 141

4.1 Passivity and Energy Control 142
4.2 Lyapunov Functions and Forwarding 143
4.3 Hybrid and Switching Control 145
4.4 Nonholonomic Systems 145
5. Conclusions 147

Trends in Mobile Robot and Vehicle Control
Carlos Canudas de Wit 151

1. Introduction 151
2. Preliminaries 152
3. Automatic Parking 153
4. Path Following 157
5. Visual-based Control System 162
6. Multibody Vehicle Control 164
6.1 Multibody Train Vehicles 164
6.2 Car Platooning in Highways and Transportation Systems 168
7. Conclusions 172

Vision-based Robot Control
Peter I. Corke and Gregory D. Hager 177

1. Introduction 177
2. Fundamentals 178
2.1 Camera Imaging and Geometry 178
2.2 Image Features and the Image Feature Parameter Space 179
2.3 Camera Sensor 180
3. Vision in Control 181
3.1 Position-based Approach 182
3.2 Image-based Approach 182
3.3 Dynamics 185
4. Control and Estimation in Vision 186
4.1 Image Feature Parameter Extraction 186
4.2 Image Jacobian Estimation 188
4.3 Other 188
5. The Future 189
5.1 Benefits from Technology Trends 189
5.2 Research Challenges 189
6. Conclusion 190

Sensor Fusion
Thomas C. Henderson, Mohamed Dekhil, Robert R. Kessler, and Martin L. Griss 193

1. Introduction 193
2. State of the Art Issues in Sensor Fusion 194

2.1 Theory 195
2.2 Architecture 195
2.3 Agents 195
2.4 Robotics 195
2.5 Navigation 195
3. Wide Area Sensor Networks 196
3.1 Component Frameworks 197
4. Robustness 199
4.1 Instrumented Sensor Systems 201
4.2 Adaptive Control 202
5. Conclusions 205

Discrete Event Theory for the Monitoring and Control of Robotic Systems
Brenan J. McCarragher 209

1. Introduction and Motivation 209
2. Discrete Event Modelling 210
2.1 Modelling using Constraints 210
2.2 An Assembly Example 212
2.3 Research Challenges 213
3. Discrete Event Control Synthesis 215
3.1 Controller Constraints 215
3.2 Command Synthesis 216
3.3 Event-level Adaptive Control 217
3.4 Research Challenges 218
4. Process Monitoring 220
4.1 Monitoring Techniques 220
4.2 Control of Sensory Perception 221
4.3 Research Challenges 222

Scheduling of Flexible Manufacturing Systems
Peter B. Luh 227

1. Introduction 227
1.1 Classification of FMS 228
1.2 Key Issues in Operating an FMS 228
1.3 Scope of This Chapter 229
2. Problem Formulation 229
2.1 Formulation of a Job Shop Scheduling Problem 229
2.2 Differences between FMS and Job Shop Scheduling 230
3. Solution Methodology 232
3.1 Approaches for Job Shop Scheduling 232
3.2 Methods for FMS Scheduling 233
4. A Case Study of the Apparel Production 233
4.1 Description of the FMS for Apparel Production 234

4.2 Mathematical Problem Formulation 235
4.3 Solution Methodology 237
4.4 Numerical Results 239
5. New Promising Research Approaches 240

Task Synchronization via Integration of Sensing, Planning, and Control in a Manufacturing Work-cell
Tzyh-Jong Tarn, Mumin Song, and Ning Xi 245

1. Introduction 245
2. A Max-Plus Algebra Model 248
3. Centralized Multi-Sensor Data Fusion 252
4. Event-based Planning and Control 254
5. Experimental Results 257
6. Conclusions 259

Advanced Air Traffic Automation: A Case Study in Distributed Decentralized Control
Claire J. Tomlin, George J. Pappas, Jana Košecká, John Lygeros, and Shankar S. Sastry 261

1. New Challenges: Intelligent Multi-agent Systems 261
   1.1 Analysis and Design of Multi-agent Hybrid Control Systems ... 263
2. Introduction to Air Traffic Management 264
3. A Distributed Decentralized ATM 266
4. Advanced Air Transportation Architectures 267
   4.1 Automation on the Ground 268
   4.2 Automation in the Air 268
5. Conflict Resolution 271
   5.1 Noncooperative Conflict Resolution 272
   5.2 Resolution by Angular Velocity 276
   5.3 Resolution by Linear Velocity 280
   5.4 Cooperative Conflict Resolution 282
   5.5 Verification of the Maneuvers 292
6. Conclusions 292

List of Contributors

Antonio Bicchi
Centro "E. Piaggio"
Università degli Studi di Pisa
Via Diotisalvi 2
56126 Pisa, Italy
bicchi@piaggio.ccii.unipi.it

Herman Bruyninckx
Department of Mechanical Engineering
Katholieke Universiteit Leuven
Celestijnenlaan 300 B
3001 Heverlee–Leuven, Belgium
Herman.Bruyninckx@mech.kuleuven.ac.be

Carlos Canudas de Wit
Laboratoire d'Automatique de Grenoble
ENSIEG–INPG
38402 Saint-Martin-d'Hères, France
canudas@lag.ensieg.inpg.fr

Peter I. Corke
CSIRO Manufacturing Science and Technology
Kenmore, QLD 4069, Australia
pic@cat.csiro.au

Mohamed Dekhil
Department of Computer Science
University of Utah
Salt Lake City, UT 84112, USA
dekhil@cs.utah.edu

Alessandro De Luca
Dipartimento di Informatica e Sistemistica
Università degli Studi di Roma "La Sapienza"
Via Eudossiana 18
00184 Roma, Italy
adeluca@giannutri.caspur.it

Joris De Schutter
Department of Mechanical Engineering
Katholieke Universiteit Leuven
Celestijnenlaan 300 B
3001 Heverlee–Leuven, Belgium
Joris.DeSchutter@mech.kuleuven.ac.be

Olav Egeland
Department of Engineering Cybernetics
Norwegian University of Science and Technology
7034 Trondheim, Norway
Olav.Egeland@itk.ntnu.no

Li-Chen Fu
Department of Electrical Engineering
National Taiwan University
Taipei, Taiwan 10764, ROC
lichen@csie.ntu.edu.tw

Martin L. Griss
Hewlett Packard Labs
Palo Alto, CA 94301, USA
griss@hplsrd.hpl.hp.com

Gregory D. Hager
Department of Computer Science
Yale University
New Haven, CT 06520, USA
hager-greg@cs.yale.edu

Thomas C. Henderson
Department of Computer Science
University of Utah
Salt Lake City, UT 84112, USA
tch@cs.utah.edu

Feng-Yih Hsu
Department of Electrical Engineering
National Taiwan University
Taipei, Taiwan 10764, ROC
fyhsu@smart.csie.ntu.edu.tw

Yildirim Hurmuzlu
Mechanical Engineering Department
Southern Methodist University
Dallas, TX 75275, USA
hurmuzlu@seas.smu.edu

Robert R. Kessler
Department of Computer Science
University of Utah
Salt Lake City, UT 84112, USA
kessler@cs.utah.edu

Jana Košecká
Department of Electrical Engineering and Computer Science
University of California at Berkeley
Berkeley, CA 94720, USA
janka@robotics.eecs.berkeley.edu

Peter B. Luh
Department of Electrical and Systems Engineering
University of Connecticut
Storrs, CT 06269, USA
luh@brc.uconn.edu

John Lygeros
Department of Electrical Engineering and Computer Science
University of California at Berkeley
Berkeley, CA 94720, USA
lygeros@robotics.eecs.berkeley.edu

Alessia Marigo
Centro "E. Piaggio"
Università degli Studi di Pisa
Via Diotisalvi 2
56126 Pisa, Italy
marigo@piaggio.ccii.unipi.it

Brenan J. McCarragher
Department of Engineering, Faculties
Australian National University
Canberra, ACT 0200, Australia
Brenan.McCarragher@anu.edu.au

George J. Pappas
Department of Electrical Engineering and Computer Science
University of California at Berkeley
Berkeley, CA 94720, USA
gpappas@robotics.eecs.berkeley.edu

Kristin Y. Pettersen
Department of Engineering Cybernetics
Norwegian University of Science and Technology
7034 Trondheim, Norway
Kristin.Ytterstad.Pettersen@itk.ntnu.no

Domenico Prattichizzo
Centro "E. Piaggio"
Università degli Studi di Pisa
Via Diotisalvi 2
56126 Pisa, Italy
domenico@piaggio.ccii.unipi.it

Septimiu E. Salcudean
Department of Electrical and Computer Engineering
University of British Columbia
2356 Main Mall
Vancouver, BC, Canada V6T 1Z4
tims@ee.ubc.ca

Shankar S. Sastry
Department of Electrical Engineering and Computer Science
University of California at Berkeley
Berkeley, CA 94720, USA
sastry@robotics.eecs.berkeley.edu

Mumin Song
Department of Systems Science and Mathematics
Washington University
One Brookings Drive
St. Louis, MO 63130, USA
song@wuauto.wustl.edu

Mark W. Spong
Coordinated Science Laboratory
University of Illinois at Urbana–Champaign
1308 W. Main St.
Urbana, IL 61801, USA
m-spong@uiuc.edu

Tzyh-Jong Tarn
Department of Systems Science and Mathematics
Washington University
One Brookings Drive
St. Louis, MO 63130, USA
tarn@wuauto.wustl.edu

Claire J. Tomlin
Department of Electrical Engineering and Computer Science
University of California at Berkeley
Berkeley, CA 94720, USA
clairet@robotics.eecs.berkeley.edu

Masaru Uchiyama
Department of Aeronautics and Space Engineering
Tohoku University
Aramaki aza-Aoba, Aoba-ku
Sendai 980, Japan
uchiyama@space.mech.tohoku.ac.jp

Ning Xi
Department of Electrical Engineering
Michigan State University
East Lansing, MI 48824, USA
xi@wuauto.wustl.edu

Wen-Hong Zhu
Department of Mechanical Engineering
Katholieke Universiteit Leuven
Celestijnenlaan 300 B
3001 Heverlee–Leuven, Belgium
Wen-Hong.Zhu@mech.kuleuven.ac.be

Force Control: A Bird's Eye View

Joris De Schutter[1], Herman Bruyninckx[1], Wen-Hong Zhu[1], and Mark W. Spong[2]

[1] Department of Mechanical Engineering, Katholieke Universiteit Leuven, Belgium
[2] Coordinated Science Laboratory, University of Illinois at Urbana–Champaign, USA

This chapter summarizes the major conclusions of twenty years of research in robot force control, points out remaining problems, and introduces issues that need more attention. By looking at force control from a distance, a lot of common features among different control approaches are revealed; this allows us to put force control into a broader (e.g. differential-geometric) context. The chapter starts with the basics of force control at an introductory level, by focusing at one or two degrees of freedom. Then the problems associated with the extension to the multidimensional case are described in a differential-geometric context. Finally, robustness and adaptive control are discussed.

1. Introduction

The purpose of force control could be quite diverse, such as applying a controlled force needed for a manufacturing process (e.g. deburring or grinding), pushing an external object using a controlled force, or dealing with geometric uncertainty by establishing controlled contacts (e.g. in assembly). This chapter summarizes the major conclusions of twenty years of research in robot force control, points out remaining problems, and introduces issues that, in the authors' opinions, need more attention.

Rather than discussing details of individual force control implementations, the idea is to step back a little bit, and look at force control from a distance. This reveals a lot of similarities among different control approaches, and allows us to put force control into a broader (e.g. differential-geometric) context. In order to achieve a high information density this text works with short, explicit statements which are briefly commented, but not proven. Some of these statements are well known and sometimes even trivial, some others reflect the personal opinion and experience of the authors; they may not be generally accepted, or at least require further investigation. Nevertheless we believe this collection of statements represents a useful background for future research in force control.

This chapter is organized as follows: Section 2. presents the basics of force control at an introductory level, by focusing at one or two degrees of freedom. Section 3. describes in a general differential-geometric context the problems associated with the extension to the multi-dimensional case.

Section 4. discusses robustness and adaptive control. Finally, Section 5. points at future research directions.

2. Basics of Force Control

2.1 Basic Approaches

The two most common basic approaches to force control are Hybrid force/position control (hereafter called Hybrid control), and Impedance control. Both approaches can be implemented in many different ways, as discussed later in this section. Hybrid control [16, 12] is based on the decomposition of the workspace into purely motion controlled directions and purely force controlled directions. Many tasks, such as inserting a peg into a hole, are naturally described in the ideal case by such task decomposition. Impedance control [11], on the other hand, does not regulate motion or force directly, but instead regulates the ratio of force to motion, which is the mechanical impedance.

Both Hybrid control and Impedance control are highly idealized control architectures. To start with, the decomposition into purely motion controlled and purely force controlled directions is based on the assumption of ideal constraints, i.e. rigid and frictionless contacts with perfectly known geometry. In practice, however, the environment is characterized by its impedance, which could be inertial (as in pushing), resistive (as in sliding, polishing, drilling, etc.) or capacitive (spring-like, e.g. compliant wall). In general the environment dynamics are less known than the robot dynamics. In addition there could be errors in the modeled contact geometry (or contact kinematics)[1], e.g. the precise location of a constraint, or a bad orientation of a tangent plane. Both environment dynamics and geometric errors result in motion in the force controlled directions, and contact forces in the position controlled directions.[2] Hence, the impedance behavior of the robot in response to these imperfections, which is usually neglected in Hybrid control designs, is of paramount importance. Impedance control provides only a partial answer, since, in order to obtain an acceptable task execution, the robot impedance should be tuned to the environment dynamics and contact geometry. In addition, both Hybrid control and Impedance control have to cope with other imperfections, such as unknown robot dynamics (e.g. joint friction, joint and link flexibility, backlash, inaccurately known inertia parameters, etc.), measurement noise, and other external disturbances.

In order to overcome some of the fundamental limitations of the basic approaches, the following improvements have been proposed. The combina-

[1] As stated in the introduction dealing with geometric uncertainty is an important motivation for the use of force control!

[2] In some cases there is even an explicit need to combine force and motion in a single direction, e.g. when applying a contact force on an object which lies on a moving conveyor belt.

tion of force and motion control in a single direction has been introduced in the Hybrid control approach, first in [10, 8], where it is termed feedforward motion in a force controlled direction, and more recently in [5, 18], where it is termed parallel force/position control (hereafter called Parallel control). In each case force control dominates over motion control, i.e. in case of conflict the force setpoint is regulated at the expense of a position error. On the other hand, Hybrid control and Impedance control can be combined into Hybrid impedance control [1], which allows us to simultaneously regulate impedance and either force or motion.

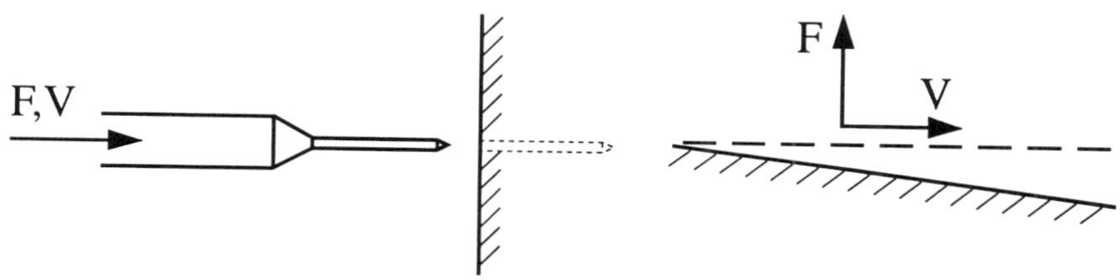

Fig. 2.1. Left: One-dimensional drilling. Right: Following a planar contour

2.2 Examples

In the first example, Fig. 2.1 (left), one needs to control the position of a tool (drill) along a straight line in order to drill a hole. This is an example of a (highly) resistive environment. The speed of the motion depends on the environment (hardness of the material), the properties of the tool (maximum allowable force), as well as the robot dynamics (actuator limits, friction, etc.). Hence it is natural to regulate both force and motion in the same direction.

Several strategies might be considered:

1. **Pure force control:** A constant force is commanded. The tool moves as material is removed so that position control is not required. The desired force level is determined by the maximum allowable force (so as not to break the drill) and by the maximum allowable speed (so as not to damage the material being drilled). Successful task execution then requires knowledge of the environment dynamics.
2. **Pure position control:** A desired velocity trajectory is commanded. This strategy would work in a highly compliant environment where excessive forces are unlikely to build up and damage the tool. In a stiff or highly resistive environment, the properties of the tool and environment

must be known with a high degree of precision. Even then, a pure position control strategy would be unlikely to work well since even small position errors result in excessively large forces.

3. **Pure impedance control:** This approach is similar to the pure position control strategy, except that the impedance of the robot is regulated to avoid excessive force buildup. However, in this approach there is no guarantee of performance and successful task execution would require that the dynamics of the robot and environment be known with a high accuracy in order to determine the commanded reference velocity and the desired closed loop impedance parameters.
4. **Force control with feedforward motion, or parallel control:** In this approach both a motion controller and a force controller would be implemented (by superposition). The force controller would be given precedence over the motion controller so that an error in velocity would be tolerated in order to regulate the force level. Again, this approach would require accurate knowledge of the environment dynamics in order to determine the reference velocity and the desired force level. In a more advanced approach the required reference velocity is estimated and adapted on-line [8].
5. **Hybrid impedance control:** In this approach the nature of the environment would dictate that a force controller be applied as in 1. This guarantees force tracking while simultaneously regulating the manipulator impedance. Impedance regulation, in addition to force control, is important if there are external disturbances (a knot in wood, for example) which could cause the force to become excessive.

In the second example, Fig. 2.1 (right), the purpose is to follow a planar surface with a constant contact force and a constant tangential velocity. In the Hybrid control approach it is natural to apply pure force control in the normal direction and pure position control in the tangential direction. However, if the surface is misaligned, the task execution results in motion in the force controlled direction, and contact forces (other than friction) in the position controlled direction. In terms of impedance, the environment is resistive (in case of surface friction) in tangential direction, and capacitive in normal direction. Hence it is natural in the Hybrid impedance control to regulate the robot impedance to be noncapacitive in the normal direction, and capacitive in the tangential directions, in combination with force control in normal direction and position control in tangential direction [1]. Hence, a successful task execution would require accurate knowledge of both the environment dynamics and the contact geometry.

2.3 Basic Implementations

There are numerous implementations of both Hybrid control and Impedance control. We only present a brief typology here. For more detailed reviews the reader is referred to [22, 15, 9].

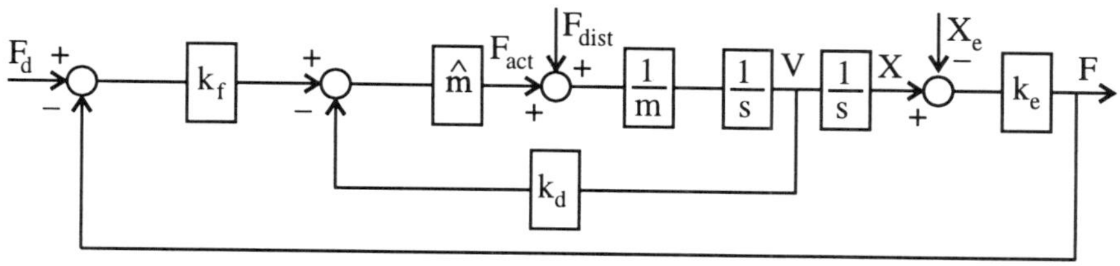

Fig. 2.2. Direct force control

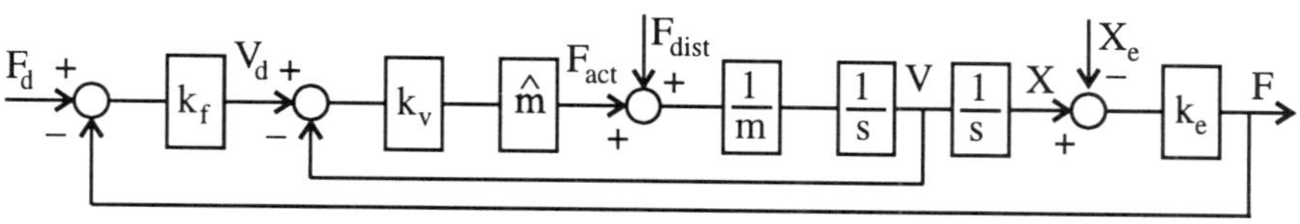

Fig. 2.3. Force control with inner position/velocity control loop

In Hybrid control we focus on the implementation of pure force control. As a first option, measured force errors are directly converted to actuator forces or torques to be applied at the robot joints. This is called direct force control hereafter. Fig. 2.2 depicts this for the 1 d.o.f. case. The robot has mass m, and is in contact with a compliant environment with stiffness k_e. F_d is the desired contact force; F is the actual contact force which is measured using a force sensor at the robot wrist; k_f is a proportional force control gain; damping is provided by adding velocity feedback[3], using feedback constant k_d; F_{act} is the actuator force; F_{dist} is an external disturbance force; x and v represent the position and the velocity of the robot; x_e represents the position of the environment. Notice that an estimate of the robot mass, $\hat{m}$, is included in the controller in order to account for the robot dynamics. In the multiple d.o.f. case this is replaced by a full dynamic model of the robot. As a second option, measured force errors are converted to desired motion, either desired position, or desired velocity, which is executed by a position or velocity control loop. This implementation is called inner position (or velocity)/outer force control. Fig. 2.3 depicts this for the case of an inner velocity loop. The velocity controller includes a feedback gain, k_v, and again a dynamic model of the robot. In many practical implementations, however, the velocity controller merely consists of a PI feedback controller, without dynamic model. Feedforward motion can be introduced by adding an extra desired velocity (not shown in figure) to the velocity resulting from the force feedback control. Comparison of Figs. 2.2 and 2.3 reveals that both

[3] Instead of taking the derivative of measured force signals, which are usually too noisy.

implementations are very similar. However, the advantage of the inner/outer implementation is that the bandwidth of the inner motion control loop can be made faster than the bandwidth of the outer force control loop.[4] Hence, if the inner and outer loops are tuned consecutively, force disturbances are rejected more efficiently in the inner/outer implementation.[5] Since errors in the dynamic model can be modeled as force disturbances, this explains why the inner/outer implementation is more robust with respect to errors in the robot dynamic model (or even absence of such model).

As for Impedance control, the relationship between motion and force can be imposed in two ways, either as an *impedance* or as an *admittance*. In *impedance* control the robot reacts to deviations from its planned position and velocity trajectory by generating forces. Special cases are a stiffness or damping controller. In essence they consist of a PD position controller, with position and velocity feedback gains adjusted in order to obtain the desired compliance.[6] No force sensor is needed. In *admittance* control, the measured contact force is used to modify the robot trajectory. This trajectory can be imposed as a desired acceleration or a desired velocity, which is executed by a motion controller which may involve a dynamic model of the robot.

2.4 Properties and Performance of Force Control

Properties of force control have been analysed in a systematic way in [7] for the Hybrid control approach, and in [1] for the Impedance control approach. The statements presented below are inspired by a detailed study and comparison of both papers, and by our long experimental experience. Due to space limitations detailed discussions are omitted.

Statement 2.1. An equivalence exists between pure force control, as applied in Hybrid control, and Impedance control. Both types of controllers can be converted to each other.

Statement 2.2. All force control implementations, when optimized, are expected to have similar bandwidths.

---

[4] Force control involves noncollocation between actuator and sensor, while this is not the case for motion control. In case of noncollocation the control bandwidth should be 5 to 10 times lower than the first mechanical resonance frequency of the robot in order to preserve stability; otherwise bandwidths up to the first mechanical resonance frequency are possible, see e.g. [17] for a detailed analysis.

[5] Of course, the same effect can be achieved by choosing highly overdamped closed-loop dynamics in the direct force control case, i.e. by taking a large k_d. However, this requires a high sampling rate for the direct force controller. (Note that velocity controllers are usually implemented as analog controllers.)

[6] In the multiple d.o.f. case the position and velocity feedback gain matrices are position dependent in order to achieve constant stiffness and damping matrices in the operational space.

This is because the bandwidth is mainly limited due to system imperfections such as backlash, flexibility, actuator saturation, nonlinear friction, etc., which are independent of the control law. As a result:

Statement 2.3. The apparent advantage of impedance control over pure force is its freedom to regulate impedance. However, this freedom can only be exercised within a limited bandwidth.

In order to evaluate the robustness of a force controller, one should study: (i) its capability to reject force disturbances, e.g. due to imperfect cancellation of the robot dynamics (cfr. Sect. 2.3); (ii) its capability to reject motion disturbances, e.g. due to motion or misalignment of the environment (cfr. Sect. 2.1); (iii) its behaviour out of contact and at impact (this is important for the transition phase, or approach phase, between motion in free space and motion in contact).

Statement 2.4. The capability to reject force disturbances is proportional to the contact compliance.

Statement 2.5. The capability to reject motion disturbances is proportional to the contact compliance.

Statement 2.6. The force overshoot at impact is proportional to the contact stiffness.

A larger approach speed can be allowed if the environment is more compliant. Then, combining Statements 2.5 and 2.6:

Statement 2.7. For a given uncertainty in the task geometry a larger task execution speed can be allowed if the environment is more compliant.

Statement 2.8. The capability to reject force disturbances is larger in the inner/outer implementations.

This is explained in Sect. 2.3

When controlling motion in free space, the use of a dynamic model of the robot is especially useful when moving at high speeds. At very low speeds, traditional joint PID controllers perform better, because they can better deal with nonlinear friction. Now, the speed of motion in contact is often limited due to the nature of the task. Hence:

Statement 2.9. In case of a compliant environment, the performance of inner/outer control is better than or equal to direct force control.

However, due to the small signal to noise ratios and resolution problems of position and velocity sensors at very low speeds:

Statement 2.10. The capability to establish stable contact with a hard environment is better for direct force control than for inner/outer control. (A low-pass filter should be used in the loop.)

3. Multi-Degree-of-Freedom Force Control

All concepts discussed in the previous section generalize to the multi-degree-of-freedom case. However, this generalization is not always straightforward. This section describes the fundamental physical differences between the one-dimensional and multi-dimensional cases, which *every* force control algorithm should take into account. As before, most facts are stated without proofs.

3.1 Geometric Properties

(The necessary background for this section can be found in [14] and references therein.) The first major distinctions are between *joint space* and *Cartesian space* (or "operational space"):

Statement 3.1. Joint space and Cartesian space models are equivalent coordinate representations of the same physical reality. However, the equivalence breaks down at the robot's singularities.

(This text uses the term "configuration space" if joint space or Cartesian space is meant.)

Statement 3.2 (Kinematic coupling). Changing position, velocity, force, torque, ... in one degree of freedom in joint space induces changes in all degrees of freedom in Cartesian space, and vice versa.

The majority of publications use linear algebra (vectors and matrices) to model a constrained robot, as well as to describe controllers and prove their properties. This often results in neglecting that:

Statement 3.3. The geometry of operational space is not that of a vector space.

The fundamental reason is that rotations do not commute, either with other rotations or with translations. Also, there is not a set of globally valid coordinates to represent orientation of a rigid body whose time derivative gives the body's instantaneous angular velocity.

Statement 3.4. Differences and magnitudes of rigid body positions, velocities and forces are not uniquely defined; neither are the "shortest paths" between two configurations. Hence, position, velocity and force errors are not uniquely determined by subtracting the coordinate vectors of desired and measured position, velocity and force.

Statement 3.3 is well-known, in the sense that the literature (often implicitly) uses two different Jacobian matrices for a general robot: the first is the matrix of partial derivatives (with respect to the joint angles) of the forward position kinematics of the robot; in the second, every column represents the instantaneous velocity of the end-effector due to a unit velocity at the corresponding joint and zero velocities at the other joints. Both Jacobians differ. But force control papers almost always choose one of both, without explicitly mentioning which one, and using the same notation "J."

Statement 3.4 is much less known. It implies that the basic concepts of velocity and/or force errors are not as trivial as one might think at first sight: if the desired and actual position of the robot differ, velocity and force errors involve the comparison of quantities at different configurations of the system. Since the system model is not a vector space, this comparison requires a definition of how to "transport" quantities defined at different configurations to the same configuration in order to be compared. This is called *identification* of the force and velocity spaces at different configurations. A practical consequence of Statement 3.4 is that these errors are different if different coordinate representations are chosen. However, this usually has no significant influence in practice, since a good controller succeeds in making these errors small, and hence also the mentioned differences among different coordinate representations.

3.2 Constrained Robot Motion

The difference between controlling a robot in free space and a robot in contact with the environment is due to the *constraints* that the environment imposes on the robot. Hence, the large body of theories and results in constrained systems in principle applies to force-controlled robots. Roughly speaking, the difference among the major force control approaches is their (implicit, default) *constraint model*:

Statement 3.5. Hybrid/Parallel control works with geometric constraints. Impedance control works with dynamic constraints.

Geometric ("holonomic") constraints are constraints on the *configuration* of the robot. In principle, they allow us to *eliminate* a number of degrees of freedom from the system, and hence to work with a lower-dimensional controller. ("In principle" is usually not exactly the same as "in practice"...) Geometric constraints are the conceptual model of *infinitely stiff* constraints.

Dynamic constraints are relationships among the configuration variables, their time derivatives and the constraint forces. Dynamic constraints represent *compliant/damped/inertial* interactions. They do not allow us to work in a lower-dimensional configuration space. An exact dynamic *model* of the robot/environment interaction is difficult to obtain in practice, especially if the contact between robot and environment changes continuously.

Most theoretical papers on modeling (and control) of constrained robots use a Lagrangian approach: the constrained system's dynamics are described by a Lagrangian function (combining kinetic and potential energy) with external inputs (joint torques, contact forces, friction, ...). The contact forces can theoretically be found via *d'Alembert's principle*, using Lagrangian multipliers. In this context it is good to know that:

Statement 3.6. Lagrange multipliers are well-defined for all systems with constraints that are linear in the velocities; constraints that are non-linear in the velocities give problems [4];

and

Statement 3.7. (Geometric) contact constraints are linear in the velocities.

The above-mentioned Lagrange-d'Alembert models have practical problems when the geometry and/or dynamics of the interaction robot-environment are not accurately known.

3.3 Multi-Dimensional Force Control Concepts

The major implication of Statement 3.4 for robot force control is that there is no *natural* way to identify the spaces of positions (and orientations), velocities, and forces. It seems mere common sense that quantities of completely different nature cannot simply be added, but nevertheless:

Statement 3.8. Every force control law adds position, velocity and/or force errors together in some way or another, and uses the result to generate setpoints for the joint actuators.

The way errors of different physical nature are combined forms the basic distinction among the three major force control approaches:

1. **Hybrid control.** This approach [13, 16] idealizes any interaction with the environment as *geometric* constraints. Hence, a number of motion degrees of freedom ("velocity-controlled directions") are eliminated, and replaced by "force-controlled directions." This means that a hybrid force controller *selects* n position or velocity components and $6 - n$ force components, subtracts the measured values from the desired values in the lower-dimensional motion and force subspaces, multiplies with a weighting factor ("dynamic control gains") and finally adds the results from the two subspaces. Hence, hybrid control makes a conceptual difference between (i) taking into account the geometry of the constraint, and (ii) determining the dynamics of the controls in the motion and force subspaces.
2. **Impedance/Admittance control.** This approach does not distinguish between constraint geometry and control dynamics: it *weighs* the (complete) contributions from contact force errors or positions and velocities errors, respectively, with *user-defined* (hence arbitrary) weighting matrices. These (shall) have the physical dimensions of impedance or admittance: stiffness, damping, inertia, or their inverses.

3. **Parallel control.** This approach combines some advantages of both other methods: it keeps the geometric constraint approach as model paradigm to think about environment interaction (and to specify the desired behavior of the constrained system), but it weighs the complete contributions from position, velocity and/or force errors in a user-defined (hence arbitrary) way, giving priority to force errors. The motivation behind this approach is to increase the robustness; Section 4. gives more details.

In summary, all three methods do exactly the same thing (as they should do). They only differ in (i) the motion constraint paradigm, (ii) the place in the control loop where the gains are applied, and (iii) which (partial) control gains are *by default* set to one or zero. "Partial control gains" refers to the fact that control errors are multiplied by control gains in different stages, e.g. at the sensing stage, the stage of combining errors from different sources, or the transformation from joint position/velocity/force set-points into joint torques/currents/voltages.

Invariance under coordinate changes is a desirable property of any controller. It means that the dynamic behavior of the controlled system (i.e. a robot in contact with its environment) is not changed if one changes (i) the reference frame(s) in which the control law is expressed, and (ii) the physical units (e.g. changing centimeters in inches changes the moment component of a generalized force differently than the linear force component). Making a force control law invariant is not very difficult:

Statement 3.9. The weighting matrices used in all three force control approaches represent the geometric concept of a *metric* on the configuration space. A metric allows to measure distances, to transport vectors over configuration spaces that are not vector spaces, and to determine shortest paths in configuration space. A metric is the standard geometric way to identify different spaces, i.e. motions, velocities, forces. The coordinate expressions of a metric transform according to well-known formulas. Applying these transformation formulas is sufficient to make a force control law invariant.

3.4 Task Specification and Control Design

As in any control application, a force controller has many complementary faces. The following paragraphs describe only those aspects which are particular to force control:

1. **Model paradigm.** The major paradigms (Hybrid, Impedance, Parallel) all make several (implicit) assumptions, and hence it is not advisable to transport a force control law blindly from one robot system to another. Force controllers are more sensitive than motion controllers to the system they work with, because the interaction with a changing environment is much more difficult to model and identify correctly than the dynamic and

kinematic model of the robot itself, especially in the multiple degree-of-freedom case.

2. **Choice of coordinates.** This is not much of a problem for free-space motion, but it does become an important topic if the robot has to control several contacts in parallel on the same manipulated object. For multiple degree-of-freedom systems, it is not straightforward to describe the contact kinematics and/or dynamics at each separate contact on the one hand, and the resulting kinematics and dynamics of the robot's endpoint on the other hand. Again, this problem increases when the contacts are time-varying and the environment is (partially) unknown. See [3] for kinematic models of multiple contacts in parallel.
3. **Task specification.** In addition to the physical constraints imposed by the interaction with the environment, the user must specify his own extra constraints on the robot's behavior. In the Hybrid/Parallel paradigms, the task specification is "geometric": the user must define the *natural constraints* (which degrees of freedom are "force-controlled" and which are "velocity controlled") and the *artificial constraints* (the control setpoints in all degrees of freedom). The Impedance/Admittance paradigm requires a "dynamic" specification, i.e. a set of impedances/admittances. This is a more indirect specification method, since the real behavior of the robot depends on how these specified impedances *interact* with the environment. In practice, there is little difference between the task specification in both paradigms: where the user expects motion constraints, he specifies a more compliant behavior; where no constraints are expected, the robot can react stiffer.
4. **Feedforward calculation.** The ideal case of perfect knowledge is the only way to make all errors zero: the models with which the force controller works provide perfect knowledge of the future, and hence perfect feedforward signals can be calculated. Of course, a general contact situation is far from completely predictable, not only quantitatively, but, which is worse, also qualitatively: the contact configuration can change abruptly, or be of a different type than expected. This case is again not exceptional, but by definition rather standard for force-controlled systems with multiple degrees of freedom.
5. **On-line adaptation.** Coping with the above-mentioned quantitative and qualitative changes is a major and actual challenge for force control research. Section 4. discusses this topic in some more detail.
6. **Feedback calculation.** Every force controller wants to make (a combination of) motion, velocity and/or force errors "as small as possible." The different control paradigms differ in what combinations they emphasize. Anyway, the goal of feedback control is to dissipate the "energy" in the error function. Force control is more sensitive than free-space motion control since, due to the contacts, this energy can change drastically under small motions of the robot.

The design of a force controller involves the choice of the arbitrary weights among all input variables, and the arbitrary gains to the output variables, in such a way that the following (conflicting) control design goals are met: stability, bandwidth, accuracy, robustness. The performance of a controller is difficult to prove, and as should be clear from the previous sections, any such proof depends heavily on the model paradigm.

4. Robust and Adaptive Force Control

Robustness of a controller is its capability to keep controlling the system (albeit with degraded performance), even when confronted with quantitative and qualitative model errors. Model errors can be geometric or dynamic, as described in the following subsections.

4.1 Geometric Errors

As explained in Sect. 2.1 geometric errors in the contact model result in motion in the force controlled directions, and contact forces in the position controlled directions. Statements 2.4–2.8 in Sect. 2.4 already dealt with robustness issues in this respect.

The Impedance/Admittance paradigm starts with this robustness issue as primary motivation; Hybrid controllers should be made robust explicitly. If this is the case Hybrid controllers perform better than Impedance controllers. For example:

1. **Making contact with an unknown surface.** Impedance control is designed to be robust against this uncertainty, i.e. the impact force will remain limited. A Hybrid controller could work with two different constraint models, one for free space motion and one for impact transition. Alternatively, one could use only the model describing the robot in contact, and make sure the controller is robust against the fact that initially the expected contact force does not yet exist. In this case the advantage of the Hybrid controller over the Impedance controller is that, after impact, the contact force can be regulated accurately.
2. **Moving along a surface with unknown orientation.** Again, Impedance control is designed to be robust against this uncertainty in the contact model; Hybrid control uses a more explicit contact model (higher in the above-mentioned hierarchy) to describe the geometry of the constraint, but the controller should be able to cope with forces in "velocity-controlled directions" and motions in the "force-controlled directions." If so, contact force regulation will be more accurate in the Hybrid control case.

Hence, Hybrid control and Impedance control are complementary, and:

Statement 4.1. The purpose of combining Hybrid Control and Impedance Control, such as in Hybrid impedance control or Parallel control, is to improve robustness.

Another way to improve robustness is to adapt on-line the geometric models that determine the paradigm in which the controller works. Compared to the "pure" force control research, on-line adaptation has received little attention in the literature, despite its importance.

The goal is to make a local model of the contact geometry, i.e. roughly speaking, to estimate (i) the tangent planes at each of the individual contacts, and (ii) the type of each contact (vertex-face, edge-edge, etc.). Most papers limit their presentation to the simplest cases of single, vertex-face contacts; the on-line adaptation then simplifies to nothing more than the estimation of the axis of the measured contact force. The most general case (multiple time-varying contact configurations) is treated in [3]. The theory covers all possible cases (with contacts that fall within the "geometric constraints" class of the Hybrid paradigm!). In practice the estimation or identification of uncertainties in the geometric contact models often requires "active sensing": the motion of the manipulated object resulting from the nominal task specification does not persistently excite all uncertainties and hence extra identification subtasks have to be superimposed on the nominal task. Adaptive control based on an explicit contact model has a potential danger in the sense that interpreting the measurements in the wrong model type leads to undesired behavior; it only increases the robustness *if* the controller is able to (i) recognize (robustly!) transitions between different contact types, and (ii) reason about the probability of different contact hypotheses. Especially this last type of "intelligence" is currently beyond the state of the art, as well as completely automatic active sensing procedures.

4.2 Dynamics Errors

Most force control approaches assume that the robot dynamics are perfectly known and can be conquered exactly by servo control. In practice, however, uncertainties exist. This motivates the use of either robust control or model based control to improve force control accuracy.

Robust control [6] involves a simple control law, which treats the robot dynamics as a disturbance. However, right now robust control can only ensure stability in the sense of uniformly ultimate boundedness, not asymptotic stability.

On the other hand, model-based control is used to achieve asymptotic stability. Briefly speaking, model-based control can be classified into two categories: linearization via nonlinear feedback [20, 21] and passivity-based control [2, 19, 23]. Linearization approaches usually have two calculation steps. In the first step, a nonlinear mapping is designed so that an equivalent linear system is formed by connecting this mapping to the robot dynamics. In the second step, linear control theory is applied to the overall system. Most linearization approaches assume that the robot dynamics are perfectly known so that nonlinear feedback can be applied to cancel the robot dynamics. Nonlinear feedback linearization approaches can be used to carry out a robustness analysis against parameter uncertainty, as in [20], but they cannot deal with parameter adaptation.

Parameter adaptation can be addressed by passivity-based approaches. These are developed using the inherent passivity between robot joint velocities and joint torques [2]. Most model-based control approaches are using a Lagrangian robot model, which is computationally inefficient. This has motivated the *virtual decomposition* approach [23], an adaptive Hybrid approach based on passivity. In this approach the original system is virtually decomposed into subsystems (rigid links and joints) so that the control problem of the complete system is converted into the control problem of each subsystem independently, plus the issue of dealing with the dynamic interactions among the subsystems. In the control design, only the dynamics of the subsystems instead of the dynamics of the complete system are required. Each subsystem can be treated independently in view of control design, parameter adaptation and stability analysis. The approach can accomplish a variety of control objectives (position control, internal force control, constraints, and optimizations) for generalized high-dimensional robotic systems. Also, it can include actuator dynamics, joint flexibility, and has potential to be extended to environment dynamics. Each dynamic parameter can be adjusted within its lower and upper bounds independently. Asymptotic stability of the complete system is guaranteed in the sense of Lyapunov.

5. Future Research

Most of the "low-level" (i.e. set-point) force control performance goals are met in a satisfactory way: many people have succeeded in making stable and accurate force controllers, with acceptable bandwidth. However, force control remains a challenging research area.

A unified theoretical framework is still lacking, describing the different control paradigms as special limit cases of a general theory. This area is slowly but steadily progressing, by looking at force control as a specific example of a nonlinear mechanical system to which differential-geometric concepts and tools can be applied. Singular perturbation is another nonlinear control concept that might be useful to bridge the gap between geometric and dynamic constraints.

Robustness means different things to different people. Hence, refinement of the robustness concept (similar to what happened with the stability concept) is another worthwhile theoretical challenge.

From a more practical point of view, future research should produce systems with improved intermediate and high-level performance and user-friendliness:

1. **Intermediate-level performance.** This is the control level at which system models are given, which however have to be adapted on line in order to compensate for quantitative errors. Further progress is needed on how to identify the errors both in the geometric and dynamic robot and environment models (and how to compensate for them), and especially on how to integrate geometric and dynamic adaptation.
2. **High-level performance.** This level is (too) slowly getting more attention. It should make a force-controlled system robust against unmodeled events, using "intelligent" force/motion signal processing and reasoning tools to decide (semi)autonomously and robustly when to perform control model switches, when to re-plan (parts of) the user-specified task, when to add active sensing, etc. The required intelligence could be model-based or not (e.g. neural networks, etc.).
3. **User-friendliness.** Current task specification tools are not really worth that name since they are rather control-oriented and not application-oriented. Force control systems should be able to use domain-specific knowledge bases, allowing the user to concentrate on the semantics of his tasks and not on *how* they are to be executed by the control system: the model and sensor information needed to execute the task is extracted automatically from knowledge and data bases, and vice versa. How to optimize the human interaction with an intelligent high-level force controller is another open question.

All these developments have strong parallels in other robotic systems under, for example, ultrasonic and/or visual guidance. Whether force-controlled systems (or sensor-based systems in general) will ever be used outside of academic or strictly controlled industrial environments will be determined in the first place by the progress achieved in these higher-level control challenges, more than by simply continuing the last two decades' research on low-level control aspects.

References

[1] Anderson R J, Spong M W 1988 Hybrid impedance control of robotic manipulators. *IEEE J Robot Automat.* 4:549–556

[2] Arimoto S 1995 Fundamental problems of robot control: Parts I and II. *Robotica.* 13:19–27, 111–122

[3] Bruyninckx H, Demey S, Dutré S, De Schutter J 1995 Kinematic models for model based compliant motion in the presence of uncertainty. *Int J Robot Res.* 14:465–482

[4] Cariñena J F, Rañada M F 1993 Lagrangian systems with constraints. *J Physics A.* 26:1335–1351

[5] Chiaverini S, Sciavicco L 1993 The parallel approach to force/position control of robotic manipulators. *IEEE Trans Robot Automat.* 9:361–373

[6] Dawson D M, Qu Z, Carrol J J 1992 Tracking control of rigid-link electrically-driven robot manipulators. *Int J Contr.* 56

[7] De Schutter J 1987 A study of active compliant motion control methods for rigid manipulators using a generic scheme. In: *Proc 1987 IEEE Int Conf Robot Automat.* Raleigh, NC, pp 1060–1065

[8] De Schutter J 1988 Improved force control laws for advanced tracking applications. In: *Proc 1988 IEEE Int Conf Robot Automat.* Philadelphia, PA, pp 1497–1502

[9] De Schutter J, Bruyninckx H 1995 Force control of robot manipulators. In: Levine W S (ed) *The Control Handbook* CRC Press, Boca Raton, FL, pp 1351–1358

[10] De Schutter J, Van Brussel H 1988 Compliant robot motion II. A control approach based on external control loops. *Int J Robot Res.* 7:18–33

[11] Hogan N 1985 Impedance control: An approach to manipulation. *ASME J Dyn Syst Meas Contr.* 107:1–7

[12] Khatib O 1987 A unified approach for motion and force control of robot manipulators: The operational space formulation. *IEEE J Robot Automat.* 3:43–53

[13] Mason M 1981 Compliance and force control for computer controlled manipulators. *IEEE Trans Syst Man Cyber.* 11:418–432

[14] Murray R M, Li Z, Sastry S S 1994 *A Mathematical Introduction to Robotic Manipulation* CRC Press, Boca Raton, FL

[15] Patarinski S, Botev R 1993 Robot force control, a review. *Mechatronics.* 3:377–398

[16] Raibert M H, Craig J J 1981 Hybrid position/force control of manipulators. *ASME J Dyn Syst Meas Contr.* 103:126–133

[17] Rankers A M 1997 Machine dynamics in mechatronic systems. An engineering approach. PhD thesis, Twente University, The Netherlands

[18] Siciliano B 1995 Parallel force/position control of robot manipulators. In: Giralt G, Hirzinger G (eds) *Robotics Research: The Seventh International Symposium.* Springer-Verlag, London, UK, pp 78–89

[19] Slotine J-J E, Li W 1988 Adaptive manipulator control: A case study. *IEEE Trans Automat Contr.* 33:995–1003

[20] Spong M W, Vidyasagar M 1989 *Robot Dynamics and Control.* Wiley, New York

[21] Tarn T J, Wu Y, Xi N, Isidori A 1996 Force regulation and contact transition control. *IEEE Contr Syst Mag.* 16(1):32–40

[22] Whitney D E 1987 Historic perspective and state of the art in robot force control. *Int J Robot Res.* 6(1):3–14

[23] Zhu W H, Xi Y G, Zhang Z J, Bien Z, De Schutter J 1997 Virtual decomposition based control for generalized high dimensional robotic systems with complicated structure. *IEEE Trans Robot Automat.* 13:411–436

Multirobots and Cooperative Systems

Masaru Uchiyama

Department of Aeronautics and Space Engineering, Tohoku University, Japan

Multiple robots executing a task on an object form a complex mechanical system that has been a target of enthusiastic research in the field of robotics and control for a decade. The chapter presents the state of the art of multirobots and cooperative systems and discusses control issues related to the topic. Kinematics and dynamics of the system is to clarify a framework for control and will give an answer to the question: what is the cooperation of the multiple robots? Different control schemes such as hybrid position/force control, load-sharing control, etc., may be designed in the framework. The chapter presents and discusses those control schemes, and briefs examples of real systems that are being studied in the author's laboratory. The examples include a couple of advanced systems such as a robot with two flexible-arms and a system consisting of many simple cooperative robots.

1. Introduction

In the early 1970's, not late after the emergence of robotics technologies, multirobots and cooperative systems began to be interested in by some robotics researchers. Examples of their research include that by Fujii and Kurono [4], Nakano et al. [12], and Takase et al. [16]. Those pieces of work discussed important key issues in the control of multirobots and cooperative systems, such as master/slave control, force/compliance control, and task space control. Nakano et al. [12] proposed master/slave force control for the coordination of the two robots to carry an object cooperatively. They pointed out the necessity of force control for the cooperation. Fujii and Kurono's proposal in [4], on the other hand, is compliance control for the coordination; they defined a task vector with respect to the object frame and controlled the compliance expressed in the frame. Interesting features in the work by Fujii and Kurono [4] and also by Takase et al. [16], by the way, are that both of the work implemented force/compliance control without using any force/torque sensors; they exploited the back-drivability of the actuators. The importance of this technique in practical applications, however, was not recognized at that time. More complicated techniques to use precise force/torque sensors lured people in robotics.

In the 1980's, with growing research in robotics, research on the multirobots and cooperative systems attracted more researchers [7]. Definition of task vectors with respect to the object to be handled [3], dynamics and control of the closed-loop system formed by the multiple robots and the object

[10, 17], and force control issues such as hybrid position/force control [5, 22] were explored. Through the research work, strong theoretical background for the control of the multirobots and cooperative systems is being formed, as is described below, and giving basis for research on more advanced topics.

How to parameterize the constraint forces/moments on the object, based on the dynamic model for the closed-loop system, is an important issue to be studied; the parameterization gives a task vector for the control and, hence, an answer to one of the most frequently asked questions in the field of multirobots and cooperative systems, that is, how to control simultaneously the trajectory of the object, the contact forces/moments on the object, the load sharing among the robots, and even the external forces/moments on the object.

Many researchers have challenged solving the problem; force/moment decomposition may be a key to solving the problem and has been studied by Uchiyama and Dauchez [19, 20], Walker et al. [29], and Bonitz and Hsia [1]. Parameterization of the internal forces/moments on the object to be intuitively understood is important. Williams and Khatib have given a solution to this [31]. Cooperative control schemes based on the parameterization are then designed; they include hybrid control of position/motion and forces [19, 20], [30, 13], and impedance control [8].

Load sharing among the robots is also an interesting issue on which many papers have been published [18, 26, 23, 21, 27, 28]. The load sharing is for optimal distribution of the load among the robots. Also, it may be exploited for robust holding of the object when the object is held by the robots without being grasped rigidly. In both cases, anyhow, it becomes a problem of optimization and can be solved by either heuristic methods [26] or mathematical methods [23, 21].

Recent research is focused on more advanced topics such as handling of flexible objects [34, 15, 33, 14] and cooperative control of flexible robots [6, 32]. Once modeling and control problem is solved, the flexible robot is a robot with many merits [25]: it is light-weight, compliant, and hence safe, etc. The topics of recent days also include slip detection and compensation in non-grasped manipulation [11], elaboration of kinematics for more sophisticated tasks [2], and decentralized control [9].

Another important issue that should be studied, by the way, is practical implementation of the proposed schemes. From practical points of view, sophisticated equipments such as force/torque sensors had better be avoided because they make the system complicated and, hence, unreliable and more expensive. Rebirth of the early method by Fujii and Kurono [4] should be attractive for people in industry. Hybrid position/force control without using any force/torque sensors but using the motor currents only is being successfully implemented in [24].

The rest of this chapter is organized as follows: In Sect. 2. dynamics formulation of closed-loop systems consisting of multiple robots and an object is presented. In Sect. 3. the constraint forces/moments on the object derived in Sect. 2. are elaborated; they are parameterized by external and internal forces/moments. In Sect. 4. a hybrid position/force control scheme that is based on the results in the previous section, is presented, before load-sharing control being discussed. Advanced topics in Sect. 5. are mainly those of research in the author's laboratory. This chapter is concluded in Sect. 6.

2. Dynamics of Multirobots and Cooperative Systems

Consider the situation depicted in Fig. 2.1 where two robots hold a single object. The robots and the object form a closed kinematic chain and, therefore, equations of motion for the system is easily obtained. A point here is that the system is an over-actuated system where the number of actuators to drive the system is more than the number of degrees of freedom of the system. Therefore, how to deal with the constraint forces/moments acting on the system becomes crucial. Here, we formulate those as the forces/moments that the robots impart to the object.

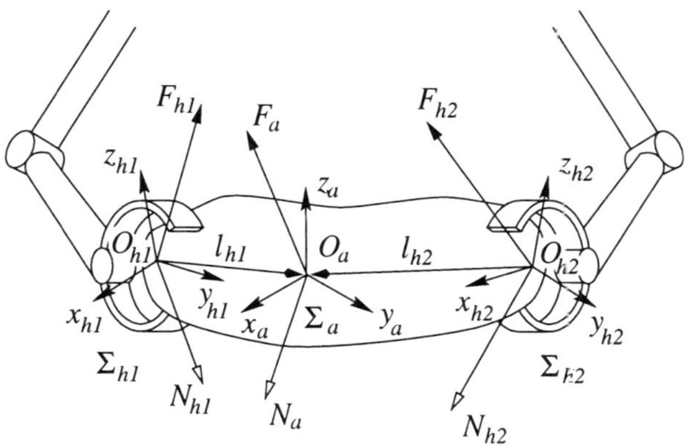

Fig. 2.1. Two robots holding an object

A model for the analysis that we introduce here is a lumped-mass model and a concept of virtual stick. The virtual stick concept was originally presented in kinematics formulation [19, 20]. The object is modeled as a point with mass and moment of inertia, and the two robots holds the point through the virtual sticks. The point has the same mass and moment of inertia as the object and is located on the center of mass. The model is illustrated in Fig. 2.2 with definitions of the frames Σ_a and Σ_i $(i = 1,\ 2)$ that will be used later in this chapter. With this modeling the formulation becomes straightforward.

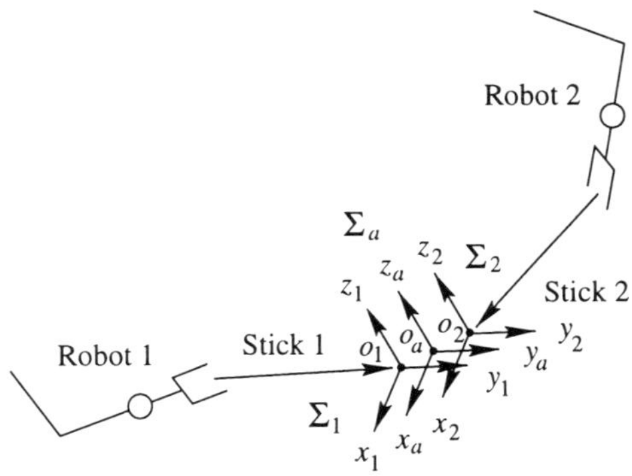

Fig. 2.2. A lumped-mass model with virtual sticks

Let denote the forces and moments at the point acting on the object through the robot i as $\boldsymbol{f}_i$, then, the forces and moments reacting on the robot through the object is $-\boldsymbol{f}_i$, and the equations of motion of the robot i is given by

$$\boldsymbol{M}_i(\boldsymbol{\theta}_i)\,\ddot{\boldsymbol{\theta}}_i + \boldsymbol{G}_i(\boldsymbol{\theta}_i, \dot{\boldsymbol{\theta}}_i) = \boldsymbol{\tau}_i + \boldsymbol{J}_i^T(\boldsymbol{\theta}_i)\,(-\boldsymbol{f}_i) \tag{2.1}$$

where $\boldsymbol{\theta}_i$ is a vector of the joint variables, $\boldsymbol{\tau}_i$ is a vector of the joint torques or forces, $\boldsymbol{M}_i(\boldsymbol{\theta}_i)$ is an inertia matrix, $\boldsymbol{G}_i(\boldsymbol{\theta}_i, \dot{\boldsymbol{\theta}}_i)$ represents the joint torques or forces due to the centrifugal, Coriolis, gravity, and friction torques or forces at the joints. $\boldsymbol{J}_i(\boldsymbol{\theta}_i)$ is the Jacobian matrix to transform the velocity of the joint variables $\dot{\boldsymbol{\theta}}_i$ into the velocity of the frame Σ_i at the tip of the virtual stick.

Another factor to influence the dynamics of the system is that of the object which in this case is obtained as one for a rigid body. Supposing the position and orientation of the object be represented by a vector $\boldsymbol{p}_a$, we have the following equation of motion:

$$\boldsymbol{M}_o(\boldsymbol{\phi})\,\ddot{\boldsymbol{p}}_a + \boldsymbol{G}_o(\boldsymbol{\phi}, \dot{\boldsymbol{\phi}}) = \boldsymbol{f}_1 + \boldsymbol{f}_2 \tag{2.2}$$

where $\boldsymbol{\phi}$ is a vector to represent orientation angles of the object, $\boldsymbol{M}_o(\boldsymbol{\phi})$ is an inertia matrix of the object, and $\boldsymbol{G}_o(\boldsymbol{\phi}, \dot{\boldsymbol{\phi}})$ represents nonlinear components of the inertial forces such as gravity, centrifugal, and Coriolis forces.

The geometrical constraints imposed on the system come from the fact that the two robots hold the object. Denote the position and orientation of the object calculated from the joint variables of the robot i as $\boldsymbol{p}_i$, and suppose that the vector is given by

$$\boldsymbol{p}_i = \boldsymbol{H}_i(\boldsymbol{\theta}_i)\,. \tag{2.3}$$

Since the object is rigid, the constraints are represented by

$$\boldsymbol{p}_a = \boldsymbol{H}_1(\boldsymbol{\theta}_1) = \boldsymbol{H}_2(\boldsymbol{\theta}_2)\,. \tag{2.4}$$

Now, we have a set of fundamental equations to describe the dynamics of the closed-loop system, that consists of the differential equations (2.1) and (2.2) to describe the dynamics of the robots and the object, respectively, and the algebraic equation (2.4) to represent the constraint condition.

The system of equations forms a singular system and the solution is obtained as follows [10]: The differential equations (2.1) and (2.2) are written by one equation as

$$\boldsymbol{M}(\boldsymbol{q})\,\ddot{q} + \boldsymbol{G}(\boldsymbol{q},\dot{\boldsymbol{q}}) = \boldsymbol{\tau} + \boldsymbol{J}^T(\boldsymbol{q})\,\boldsymbol{\lambda} \tag{2.5}$$

where $\boldsymbol{M}(\boldsymbol{q})$ is the inertia matrix of the whole system, $\boldsymbol{G}(\boldsymbol{q},\dot{\boldsymbol{q}})$ represents the nonlinear components of the whole system, $\boldsymbol{q}$ is a vector of generalized coordinates that consist of the joint variables of the robots and the position and orientation of the object, $\boldsymbol{\tau}$ represents the generalized forces, and $\boldsymbol{J}(\boldsymbol{q})$ is a Jacobian matrix. $\boldsymbol{\lambda}$ represents constraint forces/moments. The constraint condition (2.4) is written in a compact form as

$$\boldsymbol{H}(\boldsymbol{q}) = \boldsymbol{0}. \tag{2.6}$$

Combining Eqs. (2.5) and (2.6), we have

$$\begin{bmatrix} \boldsymbol{M}(\boldsymbol{q}) & \boldsymbol{0} \\ \boldsymbol{0} & \boldsymbol{0} \end{bmatrix} \begin{bmatrix} \ddot{q} \\ \dot{\lambda} \end{bmatrix} = \begin{bmatrix} \boldsymbol{\tau} - \boldsymbol{G}(\boldsymbol{q},\dot{\boldsymbol{q}}) + \boldsymbol{J}^T(\boldsymbol{q})\,\boldsymbol{\lambda} \\ \boldsymbol{H}(\boldsymbol{q}) \end{bmatrix}. \tag{2.7}$$

It is noted that the matrix in the left-hand side of the equation is singular and hence direct integration of Eq. (2.7) is impossible, of course.

The solution of Eq. (2.7) is obtained after the reduction transformation as follows [10]: Differentiating the constraint condition twice w.r.t. time, we have

$$\ddot{\boldsymbol{H}}(\boldsymbol{q}) = \boldsymbol{J}(\boldsymbol{q})\,\ddot{q} + \dot{\boldsymbol{J}}(\boldsymbol{q})\,\dot{q} = \boldsymbol{0}. \tag{2.8}$$

Since $\boldsymbol{M}(\boldsymbol{q})$ in Eq. (2.5) is positive definite, its inverse exists and we have

$$\ddot{q} = \boldsymbol{M}(\boldsymbol{q})^{-1}\left\{\boldsymbol{\tau} + \boldsymbol{J}^T(\boldsymbol{q})\,\boldsymbol{\lambda} - \boldsymbol{G}(\boldsymbol{q},\dot{\boldsymbol{q}})\right\}. \tag{2.9}$$

Substituting Eq. (2.9) into Eq. (2.8), we have

$$\boldsymbol{J}(\boldsymbol{q})\,\boldsymbol{M}(\boldsymbol{q})^{-1}\boldsymbol{J}^T(\boldsymbol{q})\,\boldsymbol{\lambda} = \boldsymbol{J}(\boldsymbol{q})\left[\boldsymbol{M}(\boldsymbol{q})^{-1}\left\{\boldsymbol{G}(\boldsymbol{q},\dot{\boldsymbol{q}}) - \boldsymbol{\tau}\right\}\right] - \dot{\boldsymbol{J}}(\boldsymbol{q})\,\dot{q}. \tag{2.10}$$

Therefore,

$$\boldsymbol{\lambda} = \left\{\boldsymbol{J}(\boldsymbol{q})\,\boldsymbol{M}(\boldsymbol{q})^{-1}\boldsymbol{J}^T(\boldsymbol{q})\right\}^{-1}\left\{\boldsymbol{J}(\boldsymbol{q})\left[\boldsymbol{M}(\boldsymbol{q})^{-1}\left\{\boldsymbol{G}(\boldsymbol{q},\dot{\boldsymbol{q}}) - \boldsymbol{\tau}\right\}\right] - \dot{\boldsymbol{J}}(\boldsymbol{q})\,\dot{\boldsymbol{q}}\right\}. \tag{2.11}$$

From Eqs. (2.9) and (2.11), we obtain $\boldsymbol{q}$ and $\boldsymbol{\lambda}$, that is the solution for a given $\boldsymbol{\tau}$.

3. Derivation of Task Vectors

The task vector consists of a set of variables that is convenient for describing a given task. A set of Cartesian coordinates in the workspace forms a task vector for a task of carrying an object in the workspace, for example. For more complicated tasks that include constrained motion, it has to be defined not only as position/orientation of the object but also as forces/moments acting on the object. In this section, we derive task vectors to describe a task to be executed by multirobots and cooperative systems.

The constraint forces/moments $\boldsymbol{f}_i$ are those applied to the object by the robot i and are obtained from Eq. (2.11) when the joint torques or forces $\boldsymbol{\tau}_i$ are given. Since $\boldsymbol{f}_i$ is 6-dimensional, the forces/moments applied to the object by the two robots are altogether 12-dimensional, six of which are for driving the object, and the rest of which do not contribute to the motion of the object but yield internal forces/moments on the object. Noting this intuition, we derive the task vector for the cooperating two robots [19, 20, 18].

3.1 External and Internal Forces/Moments

First, the external forces/moments on the object are defined as those to drive the object. That is,

$$\begin{aligned} \boldsymbol{f}_a &= \boldsymbol{f}_1 + \boldsymbol{f}_2 \\ &= \begin{bmatrix} \boldsymbol{I}_6 & \boldsymbol{I}_6 \end{bmatrix} \begin{bmatrix} \boldsymbol{f}_1^T & \boldsymbol{f}_2^T \end{bmatrix}^T \\ &= \boldsymbol{W}\boldsymbol{\lambda} \end{aligned} \tag{3.1}$$

where $\boldsymbol{W}$ is a 6×12 matrix with range of 6-dimension and null space of 6-dimension. $\boldsymbol{I}_n$ is the unit matrix of n-dimension. This relation is shown in Fig. 3.1 (a). A solution $\boldsymbol{\lambda}$ for a given $\boldsymbol{f}_a$ is

$$\begin{aligned} \boldsymbol{\lambda} &= \boldsymbol{W}^+\boldsymbol{f}_a + \left(\boldsymbol{I}_{12} - \boldsymbol{W}^+\boldsymbol{W}\right)\boldsymbol{z} \\ &= \boldsymbol{W}^+\boldsymbol{f}_a + \begin{bmatrix} \boldsymbol{I}_6 & -\boldsymbol{I}_6 \end{bmatrix}^T \boldsymbol{f}_r \\ &= \boldsymbol{W}^+\boldsymbol{f}_a + \boldsymbol{V}\boldsymbol{f}_r \end{aligned} \tag{3.2}$$

where $\boldsymbol{W}^+$ is the Moore-Penrose inverse of $\boldsymbol{W}$ given by

$$\boldsymbol{W}^+ = \begin{bmatrix} \frac{1}{2}\boldsymbol{I}_6 \\ \frac{1}{2}\boldsymbol{I}_6 \end{bmatrix}, \tag{3.3}$$

and $\boldsymbol{z}$ is an arbitrary vector of 12-dimension. The second term of the right hand side of Eq. (3.2) represents the null space of $\boldsymbol{W}$, and $\boldsymbol{V}$ represents its bases by which the vector $\boldsymbol{f}_r$ is represented. The relation is shown in Fig. 3.1 (b). It is apparent when viewing $\boldsymbol{V}$ that $\boldsymbol{f}_r$ represents forces/moments being applied by the two robots in opposite directions. We call the forces/moments represented by $\boldsymbol{f}_r$ internal forces/moments. Solving Eq. (3.2) for $\boldsymbol{f}_a$ and $\boldsymbol{f}_r$, we have

$$\boldsymbol{f}_a = \boldsymbol{f}_1 + \boldsymbol{f}_2 \tag{3.4}$$

$$\boldsymbol{f}_r = \frac{1}{2}(\boldsymbol{f}_1 - \boldsymbol{f}_2). \tag{3.5}$$

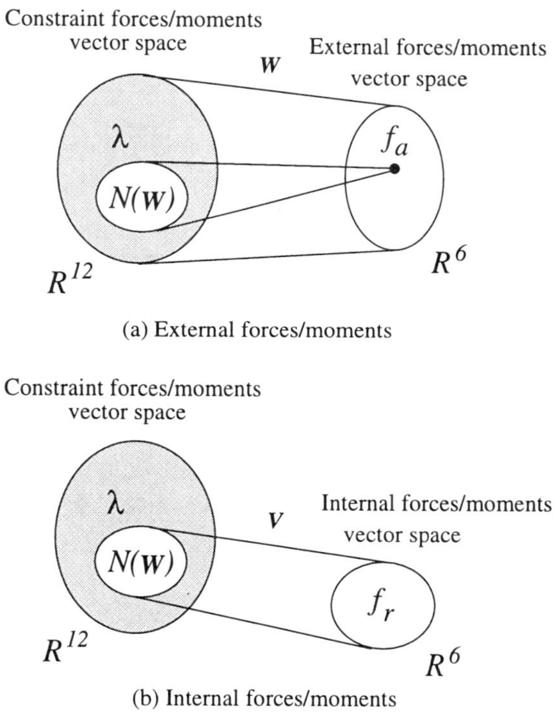

Fig. 3.1. External and internal forces/moments

3.2 External and Internal Velocities

The velocities corresponding to the external and internal forces/moments are derived using the principle of virtual work, as follows:

$$\boldsymbol{s}_a = \frac{1}{2}(\boldsymbol{s}_1 + \boldsymbol{s}_2) \tag{3.6}$$

$$\Delta\boldsymbol{s}_r = \boldsymbol{s}_1 - \boldsymbol{s}_2 \tag{3.7}$$

where $\boldsymbol{s}_a$, $\Delta\boldsymbol{s}_r$, $\boldsymbol{s}_1$ and $\boldsymbol{s}_2$ are velocity vectors corresponding to $\boldsymbol{f}_a$, $\boldsymbol{f}_r$, $\boldsymbol{f}_1$ and $\boldsymbol{f}_2$, respectively. The velocities $\boldsymbol{s}_a$, $\boldsymbol{s}_1$ and $\boldsymbol{s}_2$ are those of Σ_a, Σ_1 and Σ_2 in Fig. 2.2, respectively.

3.3 External and Internal Positions/Orientations

The positions/orientations corresponding to the external and internal forces/moments are derived by integrating the relation in Eqs. (3.6) and (3.7), as follows:

$$\boldsymbol{p}_a = \frac{1}{2}(\boldsymbol{p}_1 + \boldsymbol{p}_2) \tag{3.8}$$

$$\Delta\boldsymbol{p}_r = \boldsymbol{p}_1 - \boldsymbol{p}_2 \tag{3.9}$$

where $\boldsymbol{p}_a$, $\Delta\boldsymbol{p}_r$, $\boldsymbol{p}_1$ and $\boldsymbol{p}_2$ are position/orientation vectors corresponding to $\boldsymbol{s}_a$, $\boldsymbol{s}_r$, $\boldsymbol{s}_1$ and $\boldsymbol{s}_2$, respectively. The positions/orientations $\boldsymbol{p}_a$, $\boldsymbol{p}_1$ and $\boldsymbol{p}_2$ are those of Σ_a, Σ_1 and Σ_2 in Fig. 2.2, respectively.

An alternative way of representing the positions/orientations is to use the homogeneous transformation matrix [18]: The positions and orientations of the frames Σ_1 Σ_2 in Fig. 2.2 is represented by

$$\boldsymbol{H}_i = \begin{bmatrix} \boldsymbol{n}_i & \boldsymbol{o}_i & \boldsymbol{a}_i & \boldsymbol{x}_i \\ 0 & 0 & 0 & 1 \end{bmatrix}. \tag{3.10}$$

Corresponding to the positions/orientations $\boldsymbol{p}_a$ and $\Delta\boldsymbol{p}_r$, the homogeneous transformation matrix to represent the position/orientation of the frame Σ_a:

$$\boldsymbol{H}_a = \begin{bmatrix} \boldsymbol{n}_a & \boldsymbol{o}_a & \boldsymbol{a}_a & \boldsymbol{x}_a \\ 0 & 0 & 0 & 1 \end{bmatrix} \tag{3.11}$$

and the vectors $\Delta\boldsymbol{x}_r$, $\Delta\boldsymbol{\Omega}_r$ to represent the small (virtual) deformation of the object are derived as follows:

$$\boldsymbol{n}_a = \frac{1}{2}(\boldsymbol{n}_1 + \boldsymbol{n}_2) \tag{3.12}$$

$$\boldsymbol{o}_a = \frac{1}{2}(\boldsymbol{o}_1 + \boldsymbol{o}_2) \tag{3.13}$$

$$\boldsymbol{a}_a = \frac{1}{2}(\boldsymbol{a}_1 + \boldsymbol{a}_2) \tag{3.14}$$

$$\boldsymbol{x}_a = \frac{1}{2}(\boldsymbol{x}_1 + \boldsymbol{x}_2) \tag{3.15}$$

$$\Delta\boldsymbol{x}_r = \boldsymbol{x}_1 - \boldsymbol{x}_2 \tag{3.16}$$

$$\Delta\boldsymbol{\Omega}_r = \frac{1}{2}(\boldsymbol{n}_2 \times \boldsymbol{n}_1 + \boldsymbol{o}_2 \times \boldsymbol{o}_1 + \boldsymbol{a}_2 \times \boldsymbol{a}_1). \tag{3.17}$$

4. Cooperative Control

In the previous section we have seen that the task vectors for the cooperating two robots are the external and internal forces/moments, velocities, and positions/orientations. The internal positions/orientations are constrained in the task of carrying a rigidly held object. Therefore, a certain force-related control scheme should be applied to the cooperative control.

There have been proposed various schemes regarding the force-related control. They include compliance control [4], hybrid control of position/motion and force [5, 22, 19, 20, 30, 13], and impedance control [8]. Any of those control schemes will be successfully applied to the cooperative control if the task vector is properly chosen. For those systems that this chapter deals with and in which constraint conditions are clearly stated, however, hybrid position/force control will be most suitably used. Section 4.1, therefore, describes the hybrid position/force control [19, 20].

Load sharing is also an important issue to be addressed in the cooperative control. The problem is how to distribute the load to each robot; a strong robot may share the load more than a weak one, for instance. This is possible because the cooperative system has redundant actuators; if the system has only sufficient number of actuators for supporting the load, no optimization of load distribution is possible. Section 4.2 elaborates this problem according to our previous work [18, 26, 23, 21]. Also, it should be noted that the work by Unseren [27, 28] is more comprehensive.

4.1 Hybrid Position/Force Control

Using the equations derived in Sect. 3. the task vectors for the hybrid position/force control are defined as

$$\boldsymbol{z} = \begin{bmatrix} \boldsymbol{p}_a^T & \Delta\boldsymbol{p}_r^T \end{bmatrix}^T \tag{4.1}$$

$$\boldsymbol{u} = \begin{bmatrix} \boldsymbol{s}_a^T & \Delta\boldsymbol{s}_r^T \end{bmatrix}^T \tag{4.2}$$

$$\boldsymbol{h} = \begin{bmatrix} \boldsymbol{f}_a^T & \boldsymbol{f}_r^T \end{bmatrix}^T \tag{4.3}$$

where $\boldsymbol{z}$, $\boldsymbol{u}$, and $\boldsymbol{h}$ are the task position, velocity, and force vectors, respectively. The organization of the control scheme is shown in Fig. 4.1, diagrammatically. The suffixes r, c and m represent the reference value, current value and control command, respectively. The command vector $\boldsymbol{e}_r$ to the actuators of the two robots is calculated by

$$\boldsymbol{e}_r = \boldsymbol{e}_z + \boldsymbol{e}_h \tag{4.4}$$

where $\boldsymbol{e}_z$ is the command vector for the position control and is calculated by

$$\boldsymbol{e}_z = \boldsymbol{K}_z\, \boldsymbol{J}_\theta^{-1} \boldsymbol{G}_z(s)\, \boldsymbol{S}\boldsymbol{B}_a\, (\boldsymbol{z}_r - \boldsymbol{z}_c) \tag{4.5}$$

and e_h is the command vector for the force control and is calculated by

$$e_h = K_h J_\theta^T G_h(s)(I - S)(h_r - h_c). \tag{4.6}$$

B_a in Eq. (4.5) is a matrix to transform the errors of orientation angles into a rotation vector. J_θ is the Jacobian matrix to transform the vector $\dot{\theta} = [\dot{\theta}_1^T \ \dot{\theta}_2^T]^T$ into the task vector of velocity u. $G_z(s)$ and $G_h(s)$ are operator matrices representing position and force control laws, respectively. The matrices K_z and K_h are assumed to be diagonal. Their diagonal elements convert velocity and force commands into actuator commands, respectively. S is a matrix to switch the control modes from position to force or vice versa. S is diagonal and its diagonal elements take the values of 1 or 0. The ith workspace coordinate is position-controlled if the ith diagonal element of S is 1, and force-controlled if 0. I is the unit matrix with the same dimension as S. θ_c and λ_c are vectors of the measured joint-variables and the measured forces/moments, respectively.

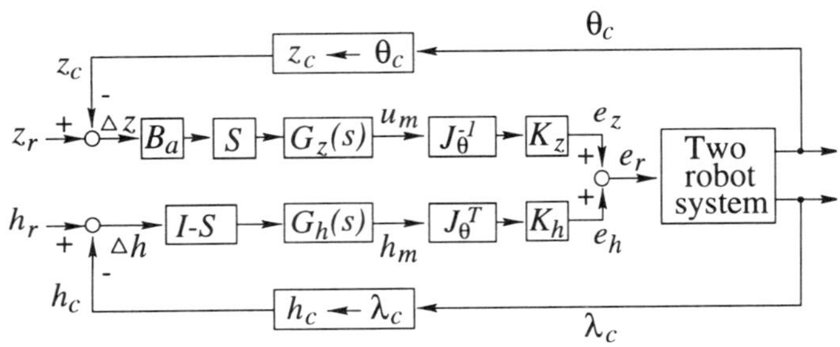

Fig. 4.1. A hybrid position/force control scheme

In the above control scheme, without distinguishing a master nor a slave, the two robots are controlled cooperatively. It is not necessary to assign master nor slave modes to each robot. Also, in the control of internal forces/moments, since the references to the external positions/orientations are sent to the both robots, the disturbance from the position-control loop to the force-control loop is decreased. This enables the above scheme to attain more precise force control than the master/slave scheme [22].

4.2 Load Sharing

We can introduce a load-sharing matrix in the framework presented in Sect. 3. By replacing the Moore-Penrose inverse in Eq. (3.2) by a generalized inverse, we obtain:

$$\lambda = W^- f_a + V f_r' \tag{4.7}$$

where

$$\boldsymbol{W}^- = \begin{bmatrix} \boldsymbol{K}^T & (\boldsymbol{I}_6 - \boldsymbol{K})^T \end{bmatrix}^T . \tag{4.8}$$

The matrix $\boldsymbol{K}$ is the load-sharing matrix. We can prove easily that the non-diagonal elements of $\boldsymbol{K}$ only yield vectors in the null space of $\boldsymbol{W}$, that is, the space of internal forces/moments. Therefore, without losing generality, let us choose $\boldsymbol{K}$ such that:

$$\boldsymbol{K} = \text{diag} \begin{bmatrix} \alpha_i \end{bmatrix} \tag{4.9}$$

where we call α_i a load-sharing coefficient.

Now, the problem we have to deal with is that of how to tune the load sharing coefficient α_i to ensure correct manipulation of the object by the two robots. To answer this question, we have to notice first that by mixing Eqs. (3.2) and (4.7), we obtain:

$$\boldsymbol{f}_r = \boldsymbol{V}^{-1} \left(\boldsymbol{W}^- - \boldsymbol{W}^+ \right) \boldsymbol{f}_a + \boldsymbol{f}'_r \tag{4.10}$$

which, keeping in mind that only $\boldsymbol{f}_a$ and $\boldsymbol{\lambda}$ are really existing forces/moments, notifies that:

- $\boldsymbol{f}_r$, $\boldsymbol{f}'_r$ and α_i are "artificial" parameters introduced for better understanding of the manipulation process, and
- $\boldsymbol{f}'_r$ and α_i are not independent; the concept of internal forces/moments and the concept of load sharing are mathematically mixed with each other.

Therefore, we can conclude that to tune the load sharing coefficients or to choose suitable internal forces/moments is strictly equivalent from the mathematical and also from the performance point of view. One of $\boldsymbol{f}_r$, $\boldsymbol{f}'_r$ and α_i constitute the independent parameters, that are redundant parameters to be optimized for load sharing. This is more generally stated in [27, 28]. We have proposed to tune the internal forces/moments $\boldsymbol{f}_r$ for simplicity of equations and also for consistency with control [23, 21].

One interesting problem regarding the load sharing is that of robust holding: a problem to determine the forces/moments $\boldsymbol{\lambda}$, which the two robots apply to the object, in order not to drop it even when disturbing external forces/moments are applied. Tasks to illustrate the problem are shown in Fig. 4.2. This problem can be solved by tuning the internal forces/moments (or the load-sharing coefficients, of course).

This problem is addressed in [26], where conditions to keep holding are expressed by the forces/moments at the end-effectors, and Eq. (4.7) being substituted into the conditions, a set of linear inequalities for both $\boldsymbol{f}'_r$ and α_i are obtained as:

$$\boldsymbol{A}\boldsymbol{f}'_r + \boldsymbol{B}\boldsymbol{\alpha} < \boldsymbol{c} \tag{4.11}$$

where $\boldsymbol{A}$ and $\boldsymbol{B}$ are 6×6 matrices, $\boldsymbol{c}$ a 6-dimensional vector, and

$$\boldsymbol{\alpha} = \begin{bmatrix} \alpha_1, & \alpha_2, & \cdots, & \alpha_6 \end{bmatrix}^T . \tag{4.12}$$

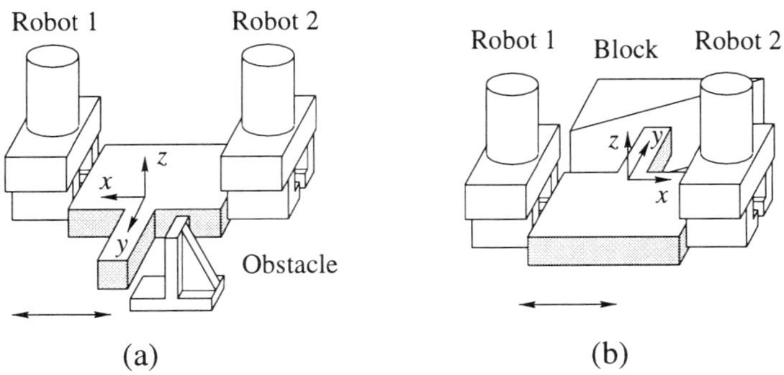

Fig. 4.2. Tasks of robust holding

In [26], a solution of $\boldsymbol{\alpha}$ for the inequality is obtained, heuristically. The above inequality can be transformed into that with respect to $\boldsymbol{f}_r$, of course, but the parameter α_i is fitter to such heuristic algorithm because α_i can be understood intuitively.

The same problem may be solved mathematically: introducing an objective function to be optimized, we can formulate the problem as that of mathematical programming. For that purpose, we choose a quadratic function of $\boldsymbol{f}_r$ as

$$\min \boldsymbol{f}_r^T \boldsymbol{Q} \boldsymbol{f}_r \tag{4.13}$$

where $\boldsymbol{Q}$ is a 6×6 positive definite matrix. The objective function represents a kind of energy to be consumed by the joint actuators; the robots consume electric energy at the actuators in order to yield the internal forces/moments $\boldsymbol{f}_r$. The problem to minimize the objective function under the constraints is a quadratic programming problem. A solution can be found in [23, 21].

5. Recent Research and Future Directions

Recent research regarding the multirobots and cooperative systems is focused on more advanced topics such as handling of flexible objects [34, 15, 33, 14] and control of multiple flexible-robots [6, 32]. The former enhances the capability of manipulation and the latter the robot itself. Both have the same dynamic model in the sense that elastic bodies are included in both of the closed-loop structures of the cooperative systems.

The flexible robot is a robot with light weight and structural compliance [25]. Due to the compliance, demerits such as positioning errors and structural vibrations take place. Nevertheless, merits with it such as light weight, compliance, and safety, are worth being paid for by the disadvantages. The flexible robot is certainly a robot of future; powerful computational means in the future will make it possible to implement even sophisticated control algorithms. Kinematics, dynamics and compliance of the flexible robot should be studied systematically before the implementation, and exploitation of the compliance should be a key to successful implementation.

The advanced topics of recent days also include slip detection and compensation in non-grasped manipulation. Cooperating multiple robots experience slip when grasps on the object are materialized only by the internal forces developed due to each robot. Such manipulations without physical grasps have got many constraints like friction between a robot's finger-tip and the object, and the friction cone defined due to it. A contact-point slip is evident if any of the constraints is overlooked. This slip causes not only manipulation errors but also a failure of system control. However, if this slip or its effects are compensated just after its occurrence, then successful manipulation is possible even in an enhanced workspace. The research in the author's laboratory [11] is regarding on this topic and concentrates on slip detection and its compensation for robust holding with using position information of each robot only. This kind of control with massive sensory-information will make the cooperation robuster and will be a research target in the future.

Other advanced topics for future research will include elaboration of kinematics for more sophisticated tasks [2] and decentralized control [9].

Another important issue that should be studied, by the way, is practical implementation of the proposed control schemes. The results regarding the cooperative control that the researchers have yielded so far are of value, of course, but not being used in industry. Why are they not being used? A reason will be that the schemes require sophisticated force/torque sensors and special control software that is incompatible to current industrial robots. From practical points of view, sophisticated equipments such as force/torque sensors had better be avoided; they make the system complicated and, hence, unreliable and more expensive. Rebirth of the early method by Fujii and Kurono [4] should be attractive for people in industry. To see if this solution is feasible, we are implementing the hybrid position/force control scheme in Sect. 4.1 by a two-arm robot developed for experimental research on application to shipbuilding work and are yielding successful results [24]. We have found a key to this implementation is compensation of the friction at the robot joints.

6. Conclusions

This chapter has presented a general perspective of the state of the art of multirobots and cooperative systems. First, it presented a historical perspective and, then, gave fundamentals of the kinematics, statics, and dynamics of such systems. Definition of task vectors highlighted the results and gave a basis on which cooperative control schemes such as hybrid position/force control, load-sharing control, etc. were designed systematically. Then, it presented application of the load-sharing control to robust holding. It also presented a couple of advanced topics of recent days and future directions of research; the topics include cooperative control of multiple flexible-robots, robust holding with slip detection, and practical implementation of the hybrid position/force control without using any force/torque sensors but with exploiting the motor currents. In concluding this chapter, we should note that application of theoretical results to real robot systems is of prime importance, and that efforts in future research will be directed in this direction to yield stronger results.

References

[1] Bonitz R G, Hsia T C 1994 Force decomposition in cooperating manipulators using the theory of metric spaces and generalized inverses. In: *Proc 1994 IEEE Int Conf Robot Automat.* San Diego, CA, pp 1521–1527

[2] Chiacchio P, Chiaverini S, Siciliano B 1995 Redundancy resolution for two cooperative spatial manipulators with a sliding contact. In: *Theory and Practice of Robots and Manipulators, Proc RoManSy 10.* Springer-Verlag, Vienna, Austria, pp 119–124

[3] Dauchez P, Zapata R 1985 Co-ordinated control of two cooperative manipulators: The use of a kinematic model. In: *Proc 15th Int Symp Industr Robot.* Tokyo, Japan, pp 641–648

[4] Fujii S, Kurono S 1975 Coordinated computer control of a pair of manipulators. In: *Proc 4th IFToMM World Congr.* Newcastle-upon-Tyne, UK, pp 411–417

[5] Hayati S 1986 Hybrid position/force control of multi-arm cooperating robots. In: *Proc 1986 IEEE Int Conf Robot Automat.* San Francisco, CA, pp 82–89

[6] Kim J-S, Yamano M, Uchiyama M 1997 Lumped-parameter modeling for cooperative control of two flexible manipulators. *1997 Asia-Pacific Vibr Conf.* Kyongju, Korea

[7] Koivo A J, Bekey G A 1988 Report of workshop on coordinated multiple robot manipulators: planning, control, and applications. *IEEE J Robot Automat.* 4:91–93

[8] Kosuge K, Koga M, Nosaki K 1989 Coordinated motion control of robot arm based on virtual internal model. In: *Proc 1989 IEEE Int Conf Robot Automat.* Scottsdale, AZ, pp 1097–1102

[9] Kosuge K, Oosumi T 1996 Decentralized control of multiple robots handling an object. In: *Proc 1996 IEEE/RSJ Int Conf Intel Robot Syst.* Osaka, Japan, pp 318–323

[10] McClamroch N H 1986 Singular systems of differential equations as dynamic models for constrained robot systems. In: *Proc 1986 IEEE Int Conf Robot Automat.* San Francisco, CA, pp 21–28

[11] Munawar K, Uchiyama M 1997 Slip compensated manipulation with cooperating multiple robots. *36th IEEE Conf Decision Contr.* San Diego, CA

[12] Nakano E, Ozaki S, Ishida T, Kato I 1974 Cooperational control of the anthropomorphous manipulator 'MELARM'. In: *Proc 4th Int Symp Industr Robot.* Tokyo, Japan, pp 251–260

[13] Perdereau P, Drouin M 1996 Hybrid external control for two robot coordinated motion. *Robotica.* 14:141–153

[14] Sun D, Shi X, Liu Y 1996 Modeling and cooperation of two-arm robotic system manipulating a deformable object. In: *Proc 1996 IEEE Int Conf Robot Automat.* Minneapolis, MN, pp 2346–2351

[15] Svinin M M, Uchiyama M 1994 Coordinated dynamic control of a system of manipulators coupled via a flexible object. In: *Prepr 4th IFAC Symp Robot Contr.* Capri, Italy, pp 1005–1010

[16] Takase K, Inoue H, Sato K, Hagiwara S 1974 The design of an articulated manipulator with torque control ability. In: *Proc 4th Int Symp Industr Robot.* Tokyo, Japan, pp 261–270

[17] Tarn T J, Bejczy A K, Yun X 1988 New nonlinear control algorithms for multiple robot arms. *IEEE Trans Aerosp Electron Syst.* 24:571–583

[18] Uchiyama M 1990 A unified approach to load sharing, motion decomposing, and force sensing of dual arm robots. In: Miura H, Arimoto S (eds) *Robotics Research: The Fifth International Symposium.* MIT Press, Cambridge, MA, pp 225–232

[19] Uchiyama M, Dauchez P 1988 A symmetric hybrid position/force control scheme for the coordination of two robots. In: *Proc 1988 IEEE Int Conf Robot Automat.* Philadelphia, PA, pp 350–356

[20] Uchiyama M, Dauchez P 1993 Symmetric kinematic formulation and non-master/slave coordinated control of two-arm robots. *Advanc Robot.* 7:361–383

[21] Uchiyama M, Delebarre X, Amada H, Kitano T 1994 Optimum internal force control for two cooperative robots to carry an object. In: *Proc 1st World Automat Congr.* Maui, HI, vol 2, pp 111–116

[22] Uchiyama M, Iwasawa N, Hakomori K 1987 Hybrid position/force control for coordination of a two-arm robot. In: *Proc 1987 IEEE Int Conf Robot Automat.* Raleigh, NC, pp 1242–1247

[23] Uchiyama M, Kanamori Y 1993 Quadratic programming for dextrous dual-arm manipulation. In: *Robotics, Mechatronics and Manufacturing Systems, Trans IMACS/SICE Int Symp Robot Mechatron Manufact Syst.* Elsevier, Amsterdam, The Netherlands, pp 367–372

[24] Uchiyama M, Kitano T, Tanno Y, Miyawaki K 1996 Cooperative multiple robots to be applied to industries. In: *Proc 2nd World Automat Congr.* Montpellier, France, vol 3, pp 759–764

[25] Uchiyama M, Konno A 1996 Modeling, controllability and vibration suppression of 3D flexible robots. In: Giralt G, Hirzinger G (eds) *Robotics Research, The Seventh International Symposium.* Springer-Verlag, London, UK, pp 90–99

[26] Uchiyama M, Yamashita T 1991 Adaptive load sharing for hybrid controlled two cooperative manipulators. In: *Proc 1991 IEEE Int Conf Robot Automat.* Sacramento, CA, pp 986–991

[27] Unseren M A 1994 A new technique for dynamic load distribution when two manipulators mutually lift a rigid object. Part 1: The proposed technique. In: *Proc 1st World Automat Congr.* Maui, HI, vol 2, pp 359–365

[28] Unseren M A 1994 A new technique for dynamic load distribution when two manipulators mutually lift a rigid object. Part 2: Derivation of entire system model and control architecture. In: *Proc 1st World Automat Congr.* Maui, HI, vol 2, pp 367–372

[29] Walker I D, Freeman R A, Marcus S I 1991 Analysis of motion and internal force loading of objects grasped by multiple cooperating manipulators. *Int J Robot Res.* 10:396–409

[30] Wen J T, Kreutz-Delgado K 1992 Motion and force control of multiple robotic manipulators. *Automatica.* 28:729–743

[31] Williams D, Khatib O 1993 The virtual linkage: A model for internal forces in multi-grasp manipulation. In: *Proc 1993 IEEE Int Conf Robot Automat.* Atlanta, GA, pp 1025–1030
[32] Yamano M, Kim J-S, Uchiyama M 1997 Experiments on cooperative control of two flexible manipulators working in 3D space. *1997 Asia-Pacific Vibr Conf.* Kyongju, Korea
[33] Yukawa T, Uchiyama M, Nenchev D N, Inooka H 1996 Stability of control system in handling of a flexible object by rigid arm robots. In: *Proc 1996 IEEE Int Conf Robot Automat.* Minneapolis, MN, pp 2332–2339
[34] Zheng Y F, Chen M Z 1993 Trajectory planning for two manipulators to deform flexible beams. In: *Proc 1993 IEEE Int Conf Robot Automat.* Atlanta, GA, vol 1, pp 1019–1024

Robotic Dexterity via Nonholonomy

Antonio Bicchi, Alessia Marigo, and Domenico Prattichizzo

Centro "E. Piaggio", Università degli Studi di Pisa, Italy

In this paper we consider some new avenues that the design and control of versatile robotic end–effectors, or "hands", are taking to tackle the stringent requirements of both industrial and servicing applications. A point is made in favour of the so–called minimalist approach to design, consisting in the reduction of the hardware complexity to the bare minimum necessary to fulfill the specifications. It will be shown that to serve this purpose best, more advanced understanding of the mechanics and control of the hand–object system is necessary. Some advancements in this direction are reported, while few of the many problems still open are pointed out.

1. Introduction

The development of mechanical hands for grasping and fine manipulation of objects has been an important part of robotics research since its beginnings. Comparison of the amazing dexterity of the human hand with the extremely elementary functions performed by industrial grippers, compelled many robotics researchers to try and bring some of the versatility of the anthropomorphic model in robotic devices. From the relatively large effort spent by the research community towards this goal, several robot hands sprung out in laboratories all over the world. The reader is referred to detailed surveys such as e.g. [15, 34, 13, 27, 2].

Multifingered, "dextrous" robot hands often featured very advanced mechanical design, sensing and actuating systems, and also proposed interesting analysis and control problems, concerning e.g. the distribution of control action among several agents (fingers) subject to complex nonlinear bounds. Notwithstanding the fact that hands designed in that phase of research were often superb engineering projects, the community had to face a very poor penetration to the factory floor, or to any other scale application. Among the various reasons for this, there is undoubtedly the fact that dextrous robot hands were too mechanically complex to be industrially viable in terms of cost, weight, and reliability.

Reacting to this observation, several researchers started to reconsider the problem of obtaining good grasping and manipulation performance by using mechanically simpler devices. This approach can be seen as an embodiment of a more general, "minimalist" attitude at robotics design (see e.g. works reported in [3]). It often turns out that this is indeed possible, provided that more sophisticated analysis, programming and control tools are employed.

The challenge is to make available theoretical tools which allow to reduce the hardware cost at little incremental cost of basic research.

One instance of this process of hardware reduction without sacrificing performance can be seen in devices for "power grasping", or "whole–arm manipulation", i.e. devices that exploit all their parts to contact and constrain the manipulated part, and not just their end–effectors (or fingertips, in the case of hands). From the example of human grasp, it is evident that power grasps using also the palm and inner phalanges are more robust than fingertip grasping, for a given level of actuator strength. However, using inner parts of the kinematic chain, which have reduced mobility in their operational space, introduces important limitations in terms of controllability of forces and motions of the manipulated part, and ensue non–trivial complications in control. Such considerations are dealt with at some length in references [37, 36], and will not be reported here.

In this paper, we will focus on the achievement of dexterity with simplified hardware. By dexterity we mean here (in a somewhat restrictive sense) the ability of a hand to relocate and reorient an object being manipulated among its fingers, without loosing the grasp on it. Salisbury [23] showed first that the minimum theoretical number of d.o.f.'s to achieve dexterity in a hand with rigid, hard–finger, non–rolling and non–sliding contacts, is 9. As a simple explanation of this fact, consider that at least three hard–fingers are necessary to completely restrain an object. On the other hand, as no rolling nor sliding is allowed, fingers must move so as to track with the contact point on their fingertip the trajectory generated by the corresponding contact point on the object, while this moves in 3D space. Hence, 3 d.o.f.'s per finger are strictly necessary. If the non–rolling assumption is lifted, however, the situation changes dramatically, as nonholonomy enters the picture. The analysis of manipulation in the presence of rolling has been pioneered by Montana [25], Cai and Roth [9], Cole, Hauser, and Sastry [11], Li and Canny [20].

In this paper we report on some results that have been obtained in the study of rolling objects, in view of the realization of a robot gripper that exploits rolling to achieve dexterity. A first prototype of such device, achieving dexterity with only four actuators, was presented by Bicchi and Sorrentino [5]. Further developments have been described in [4, 22].

Although nonholonomy seems to be a promising approach to reducing the complexity, cost, weight, and unreliability of the hardware used in robotic hands, it is true in general that planning and controlling nonholonomic systems is more difficult than holonomic ones. Indeed, notwithstanding the efforts spent by applied mathematicians, control engineers, and roboticists on the subject, many open problems remain unsolved at the theoretical level, as well as at the computational and implementation level.

The rest of the paper is organized as follows. In Sect. 2. we overview applications of nonholonomic mechanical systems to robotics, and provide a rather broad definition of nonholonomy that allows to treat in a uniform way phenomena with a rather different appearance. In Sect. 3. we make the point on the state–of–the–art in manipulation by rolling, with regard to both regular and irregular surfaces. We conclude the paper in Sect. 4. with a discussion of the open problems in planning and controlling such devices.

2. Nonholonomy on Purpose

A knife–edge cutting a sheet of paper and a cat falling onto its feet are common examples of natural nonholonomic systems. On the other hand, bicycles and cars (possibly with trailers) are familiar examples of artificially designed nonholonomic devices. While nonholonomy in a system is often regarded as an annoying side–effect of other design considerations (this is how most people consider e.g. maneuvering their car for parking in parallel), it is possible that nonholonomy is introduced on purpose in the design in order to achieve specific goals. The Abdank–Abakanowicz's integraph and the Henrici–Corradi harmonic analyzer reported by Neimark and Fufaev [30] are nineteenth–century, very ingenuous examples in this sense, where the nonholonomy of rolling of wheels and spheres are exploited to mechanically construct the primitive and the Fourier series expansion of a plotted function, respectively.

Another positive aspect of nonholonomy, and actually the one that motivates the perspective on robotic design considered in this paper, is the reduction in the number of actuators it may allow. In order to make the idea evident, consider the standard definition of a nonholonomic system as given in most mechanics textbooks:

Definition 2.1. *A mechanical system described by its generalized coordinates* $\mathbf{q} = (q_1, q_2, \ldots, q_n)^T$ *is called nonholonomic if it is subject to constraints of the type*

$$\mathbf{c}(\mathbf{q}(t), \dot{\mathbf{q}}(t)) = 0, \tag{2.1}$$

and if there is no equation of the form $\hat{\mathbf{c}}(\mathbf{q}(t)) = 0$ *such that* $\frac{d\hat{\mathbf{c}}(\mathbf{q}(t))}{dt} = \mathbf{c}(\mathbf{q}(t), \dot{\mathbf{q}}(t))$. *If linear in* $\dot{\mathbf{q}}$, *i.e. if it can be written as*

$$\mathbf{c}(\mathbf{q}, \dot{\mathbf{q}}) = \mathbf{A}(\mathbf{q})\dot{\mathbf{q}} = 0,$$

a constraint is called Pfaffian.

A Pfaffian set of constraints can be rewritten in terms of a basis $\mathbf{G}(q)$ of the kernel of $\mathbf{A}(q)$, as[1]

[1] in more precise geometrical terms, the rows of $A(q)$ are the covector fields of the active constraints forming a codistribution, and the columns of $\mathbf{G}(q)$ are a set of vector fields spanning the annihilator of the constraint codistribution. If the constraints are smooth and independent, both the codistribution and distribution are nonsingular.

$$\dot{\mathbf{q}} = \mathbf{G}(\mathbf{q})\mathbf{u} \tag{2.2}$$

This is the standard form of a nonlinear, driftless control system. In the related vocabulary, components of **u** are "inputs". The non–integrability of the original constraint has its control–theoretic counterpart in Frobenius Theorem, stating that a nonsingular distribution is integrable if and only if it is involutive. In other words, if the distribution spanned by $\mathbf{G}(q)$ is not involutive, motions along directions that are not in the span of the original vector fields are possible for the system.

From this fact follows the most notable characteristic of nonholonomic systems with respect to minimalist robotic design, i.e. that they can be driven to a desired equilibrium configuration in a d–dimensional configuration manifold using less than d inputs. In a kinematic bicycle, for instance, two inputs (the forward velocity and the steering rate) are enough to steer the system to any desired configuration in its 4–dimensional state space. Notice that these "savings" are unique to nonlinear systems, as a linear system always requires as many inputs as states to be steered to arbitrary equilibrium states (this property being in fact equivalent to functional controllability of outputs for linear systems).

Since "inputs" in engineering terms translates into "actuators", devices designed by intentionally introducing nonholonomic mechanisms can spare hardware costs without sacrificing dexterity. Few recent works in mechanism design and robotics reported on the possibility of exploiting nonholonomic mechanical phenomena in order to design devices that achieve complex tasks with a reduced number of actuators (see e.g. [39, 5, 12, 35]).

It is worthwhile mentioning at this point that nonholonomy occurs not only because of rolling, but also in systems of different types, such as for instance:

- Systems subject to conservation of angular momentum, as is the case of the falling cat. This type of nonholonomy can be exploited for instance for orienting a satellite with only two torque actuators [26], or reconfiguring a satellite–manipulator system [29, 17].
- Underactuated mechanical systems, such as robot arms with some free joints, usually result in dynamic, second–order nonintegrable, nonholonomic constraints [32]. This may allow reconfiguration of the whole system by controlling only actuated joints, as e.g. in [1, 12].
- Nonholonomy may be exhibited by piecewise holonomic systems, such as switching electrical systems [19], or mechanical systems with discontinuous phenomena due to intermittent contacts, Coulomb friction, etc.. Brockett [8] discussed some deep mathematical aspects of the rectification of vibratory motion in connection with the problem of realizing miniature piezoelectric motors (see Fig. 2.1). He stated in that context that "from the point of view of classical mechanics, rectifiers are necessarily non–holonomic systems". Lynch and Mason [21] used controlled slippage to build a 1–joint "manipulator" that can reorient and displace arbitrarily

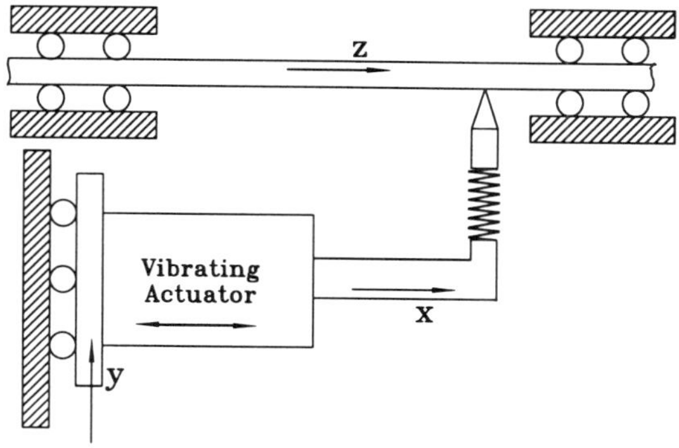

Fig. 2.1. Illustrating the principle of a mechanical rectifier after Brockett. The tip of the vibrating element oscillates in the x direction, while a variable pressure against the rod is controlled in the y direction. When the contact pressure is larger than a threshold y_0, dry friction forces the rod to translate in the z direction

most planar mechanical parts on a a conveyor belt, thus achieving control on a 3 dimensional configuration space by using one controlled input (the manipulator's actuator) and one constant drift vector (the belt velocity). Ostrowski and Burdick [33] gave a rather general mathematical model of locomotion in natural and artificial systems, showing how basically any locomotion system is a nonholonomic system. In these examples, however, a more general definition of nonholonomy has to be considered to account for the discontinuous nature of the phenomena occurring.

- Nonholonomy can be exhibited by inherently discrete systems. The simple experiment of rolling a die onto a plane without slipping, and bringing it back after any sufficiently rich path, shows that its orientation has changed in general (see Fig. 2.2). The fact that almost all polyhedra can be brought close to a desired position and orientation by rolling on a plate, to be discussed shortly, can be used to build dextrous hands for manipulation of general (non–smooth) mechanical parts. Once again, these nonholonomic phenomena can not be described and studied based on classical differential geometric tools.

A more general definition than (2.1) is given below for time–invariant systems:

Definition 2.2. *Consider a system evolving in a configuration space $\mathcal{Q}$, a time set (continuous or discrete) $\mathcal{T}$, and a bundle of input sets $\mathcal{A}$, such that for each input set $\mathcal{A}(\mathbf{q},\mathbf{t})$ defined at $\mathbf{q} \in \mathcal{Q}$, $\mathbf{t} \in \mathcal{T}$, it holds $\mathbf{a} : (\mathbf{q},\mathbf{t}) \rightarrow \mathbf{q}'$, $\mathbf{q}' \in \mathcal{Q}, \forall \mathbf{a} \in \mathcal{A}(\mathbf{q},\mathbf{t})$. If it is possible to decompose $\mathcal{Q}$ in a projection or* base *space $\mathcal{B} = \Pi(\mathcal{Q})$ and a* fiber bundle *$\mathcal{F}$, such that $\mathcal{B} \times \mathcal{F} = \mathcal{Q}$ and there exists a sequence of inputs in $\mathcal{A}$ starting at $\mathbf{q}_0$ and steering the* *system to* $\mathbf{q}^\star = \mathbf{a}_n(\mathbf{q}_{n-1}, \mathbf{t}_{n-1}) \circ \cdots \circ \mathbf{a}_1(\mathbf{q}_0, \mathbf{t}_0)$, *such that* $\Pi(\mathbf{q}_0) = \Pi(\mathbf{q}^\star)$ *but* $\mathbf{q}_0 \neq \mathbf{q}^\star$, *then the system is nonholonomic at* $\mathbf{q}_0$.

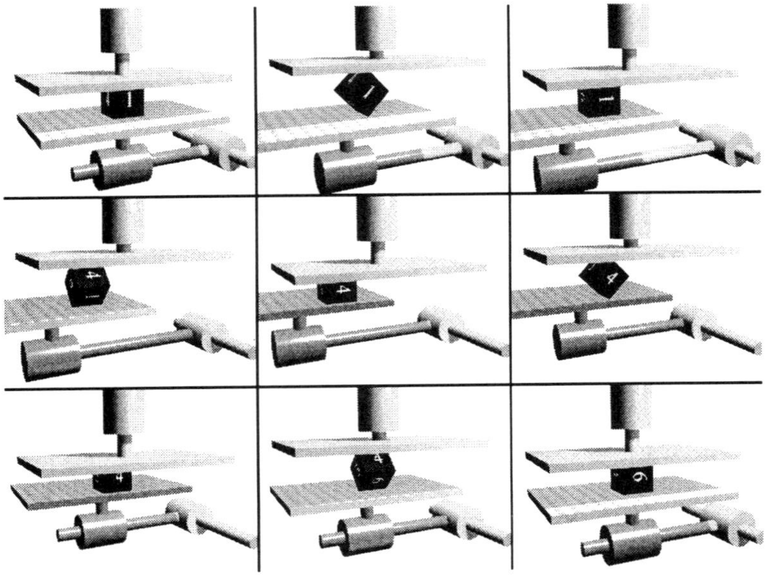

Fig. 2.2. A die being rolled between two parallel plates. After four tumbles over its edges, the center of the die comes back to its initial position, while its orientation has changed

According to this definition, a system is nonholonomic if there exist controls that make some configurations go through closed cycles, while the rest of configurations undergo net changes per cycle (see Fig. 2.3).

For instance, in the continuous, nonholonomic Heisenberg system

$$\begin{bmatrix} \dot{x}_1 \\ \dot{x}_2 \\ \dot{x}_3 \end{bmatrix} = \begin{bmatrix} 1 \\ 0 \\ -x_2 \end{bmatrix} u_1 + \begin{bmatrix} 0 \\ 1 \\ x_1 \end{bmatrix}, \tag{2.3}$$

it is well known (see e.g. [8]) that "Lie–bracket motions" in the direction of

$$[\mathbf{G}_1(\mathbf{x}), \mathbf{G}_2(\mathbf{x})] = \begin{bmatrix} 0 \\ 0 \\ -2 \end{bmatrix}$$

are generated by any pair of simultaneous periodic zero–average functions $u_1(\cdot), u_2(\cdot)$. Definition 1 specializes in this case with $\mathcal{Q} = \mathbb{R}^3$, $\mathcal{T} = \mathbb{R}_+$, and

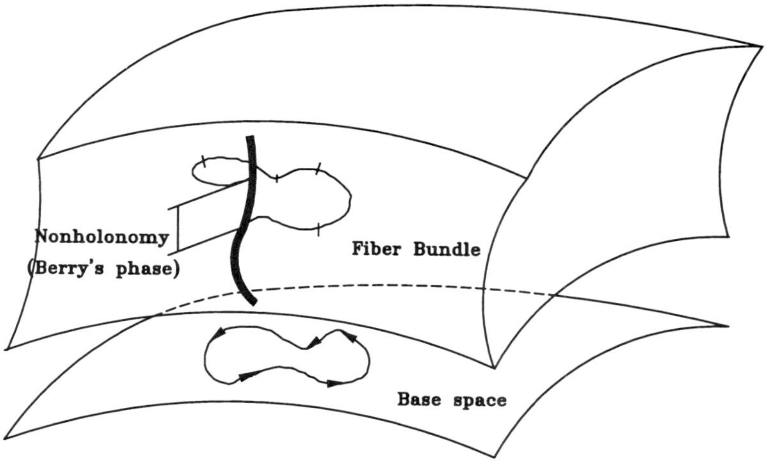

Fig. 2.3. Illustrating the definition of nonholonomic systems

$\mathcal{A}(\mathbf{x}, t) = \{\exp(t(\mathbf{G}_1 u_1 + \mathbf{G}_2 u_2))\,\mathbf{x}, \forall$ piecewise continuous $u_i(\cdot) : [0, t] \to \mathbb{R}, i = 1, 2.\}$. The base space is simply the x_1, x_2 plane, and the fibers are in the x_3 direction. Periodic inputs generate closed paths in the base space, corresponding to a fiber motion of twice the (signed) area enclosed on the base by the path.

As an instance of embodiment of the above definition in a piecewise holonomic system, consider the simplified version of one of Brockett's rectifiers in Fig. 2.1. The two regimes of motion, without and with friction, are

$$\begin{bmatrix} \dot{x} \\ \dot{y} \\ \dot{z} \end{bmatrix} = \begin{bmatrix} u_1 \\ u_2 \\ 0 \end{bmatrix} = \begin{bmatrix} 1 \\ 0 \\ 0 \end{bmatrix} u_1 + \begin{bmatrix} 0 \\ 1 \\ 0 \end{bmatrix} u_2, \quad y < y_0;$$

and

$$\begin{bmatrix} \dot{x} \\ \dot{y} \\ \dot{z} \end{bmatrix} = \begin{bmatrix} u_1 \\ u_2 \\ u_1 \end{bmatrix} = \begin{bmatrix} 1 \\ 0 \\ 1 \end{bmatrix} u_1 + \begin{bmatrix} 0 \\ 1 \\ 0 \end{bmatrix} u_2, \quad y \geq y_0,$$

respectively. In this case, base variables can be identified as x and y, while the fiber variable is z. Time is continuous, but the input bundle is discontinuous:

$$\mathcal{A}(x,y,z,t)=\begin{cases} y<y_0: & \begin{cases} x & \rightarrow & x+\int_0^t u_1(\sigma)d\sigma; \\ y & \rightarrow & y+\int_0^t u_2(\sigma)d\sigma; \\ z & \rightarrow & z; \end{cases} \\ \\ y\geq y_0: & \begin{cases} x & \rightarrow & x+\int_0^t u_1(\sigma)d\sigma; \\ y & \rightarrow & y+\int_0^t u_2(\sigma)d\sigma; \\ z & \rightarrow & z+\int_0^t u_1(\sigma)d\sigma. \end{cases} \end{cases}$$

By changing frequency and phase of the two inputs, different directions and velocities of the rod motion can be achieved. Note in particular that input u_1 need not actually to be tuned finely, as long as it is periodic, and can be fixed e.g. as a resonant mode of the vibrating actuator. Fixing a periodic $u_1(\cdot)$ and tuning only u_2 still guarantees in this case the (non-local) controllability of the nonholonomic system: notice here the interesting connection with results on controllability of systems with drift reported by Brockett ([6], Theorem 4 and Hirschorn's Theorem 5).

Finally, consider how the above definition of nonholonomic system specializes to the case of rolling a polyhedron. Considering only configurations with one face of the polyhedron sitting on the plate, these can be described by fixing a point and a line on the polyhedron (excluding lines that are perpendicular to any face), taking their normal projections to the plate, and affixing coordinates x, y to the projected point, and θ to the angle of the projected line, with respect to some reference frame fixed to the plate. Therefore, $\mathcal{Q} = \mathbb{R}^2 \times S^1 \times F$, where F is the finite set of m face of the polyhedron. As the only actions that can be taken on the polyhedron are assumed to be "tumbles", i.e. rigid rotations about one of edges of the face currently lying on the plate that take the corresponding adjacent faces down to the plate, we take $\mathcal{T} = \mathbb{N}_+$ and $\mathcal{A}$ the bundle of m different, finite sets of neighbouring configurations just described. Figure 2.2 shows how a closed path in the base variables (x, y) generates a $\pi/2$ counterclockwise rotation and a change of contact face.

3. Systems of Rolling Bodies

For the reader's convenience, we report here some preliminaries that help in fixing the notation and resume the background. For more details, see e.g. [28, 5, 4, 10].

3.1 Regular Surfaces

The kinematic equations of motion of the contact points between two bodies with regular surface (i.e. with no edges or cusps) rolling on top of each other describe the evolution of the (local) coordinates of the contact point on the finger surface, $\alpha_f \in \mathbb{R}^2$, and on the object surface, $\alpha_o \in \mathbb{R}^2$, along with the holonomy angle ψ between the x–axes of two gauss frames fixed on the surfaces at the contact points, as they change according to the rigid relative motion of the finger and the object described by the relative velocity $\mathbf{v}$ and angular velocity ω. According to the derivation of Montana [25], in the presence of friction one has

$$\begin{aligned} \dot{\alpha}_f &= \mathbf{M}_f^{-1}\mathbf{K}_r^{-1}\begin{bmatrix} -\omega_y \\ \omega_x \end{bmatrix}; \\ \dot{\alpha}_o &= \mathbf{M}_o^{-1}\mathbf{R}_\psi\mathbf{K}_r^{-1}\begin{bmatrix} -\omega_y \\ \omega_x \end{bmatrix}; \\ \dot{\psi} &= \mathbf{T}_f\mathbf{M}_f\dot{\alpha}_f + \mathbf{T}_o\mathbf{M}_o\dot{\alpha}_o; \end{aligned} \tag{3.1}$$

where $\mathbf{K}_r = \mathbf{K}_f + \mathbf{R}_\psi\mathbf{K}_o\mathbf{R}_\psi$ is the relative curvature form, $\mathbf{M}_o, \mathbf{M}_f, \mathbf{T}_o, \mathbf{T}_f$ are the object and finger metric and torsion forms, respectively, and

$$\mathbf{R}_\psi = \begin{bmatrix} \cos\psi & -\sin\psi \\ -\sin\psi & -\cos\psi \end{bmatrix}.$$

The rolling kinematics (3.1) can readily be written, upon specialization of the object surfaces, in the standard control form

$$\dot{\xi} = \mathbf{g}_1(\xi)v_1 + \mathbf{g}_2(\xi)v_2, \tag{3.2}$$

where the state vector $\xi \in \mathbb{R}^5$ represents a local parameterization of the configuration manifold, and the system inputs are taken as the relative angular velocities $v_1 = \omega_x$ and $v_2 = \omega_y$. Applying known results from nonlinear system theory, some interesting properties of rolling pairs have been shown. The first two concern controllability of the system:

Theorem 3.1. *(from [20]) A kinematic system comprised of a sphere rolling on a plane is completely controllable. The same holds for a sphere rolling on another sphere, provided that the radii are different and neither is zero.*

Theorem 3.2. *(from [4]) A kinematic system comprised of any smooth, strictly convex surface of revolution rolling on a plane is completely controllable.*

Remark 3.1. Motivated by the above results, it seems reasonable to conjecture that a kinematic system comprised of *almost* any pair of surfaces is controllable. Such fact is indeed important in order to guarantee the possibility of building a dextrous hand manipulating arbitrary (up to practical constraints) objects.

The following propositions concern the possibility of finding coordinate transforms and static state feedback laws which put the plate–ball system in special forms, which are of interest for designing planning and control algorithms:

Proposition 3.1. *The plate–ball system can not be put in* chained form *[27]; it is not* differentially flat *[38]; it is not* nilpotent *[14].*

These results prevent the few powerful planning and control algorithms known in the literature to be applied to kinematic rolling systems (of which the plate–ball system is a prototype). The following positive result however holds:

Theorem 3.3. *(from [5]). Assuming that either surface in contact is (locally) a plane, there exist a state diffeomorphism and a regular static state feedback law such that the kinematic equations of contact (3.1) assume a strictly triangular structure.*

The relevance of the strictly triangular form to planning stems from the fact that the flow of the describing ODE can be integrated directly by quadratures. Whenever it is possible to compute the integrals symbolically, the planning problem is reduced to the solution of a set of nonlinear algebraic equations, to which problem many well–known numerical methods apply.

3.2 Polyhedral Objects

The above mentioned simple experiment of rolling a die onto a plane without slipping hints to the fact that manipulation of parts with non–smooth (e.g. polyhedral) surface can be advantageously performed by rolling. However, while for analysing rolling of regular surfaces the powerful tools of differential geometry and nonlinear control theory are readily available, the surface regularity assumption is rarely verified with industrial parts, which often have edges and vertices.

Although some aspects of graspless manipulation of polyhedral objects by rolling have been already considered in the robotics literature, a complete study on the analysis, planning, and control of rolling manipulation for polyhedral parts is far from being available, and indeed it comprehends many aspects, some of which appear to be non–trivial. In particular, the lack of a differentiable structure on the configuration space of a rolling polyhedron deprives us of most techniques used with regular surfaces. Moreover, peculiar phenomena may happen with polyhedra, which have no direct counterpart with regular objects. For instance, in the examples reported in Figs. 3.1 and 3.2, it is shown that two apparently similar objects can reach configurations belonging to a very fine and to a coarse grid, respectively. In the second case, the mesh of the grid can actually be made arbitrarily small by manipulating the object long enough; in such case, the reachable set is said to be dense.

In fact, considering the description of the configuration set of a rolling polyhedron provided in Sect. 2. it can be observed that the state space $\mathcal{Q}$ is the union of l copies of $\mathbb{R}^2 \times S^1$. The subset of reachable configurations from some initial configuration $\mathcal{R}$ is given by the set of points reached by applying

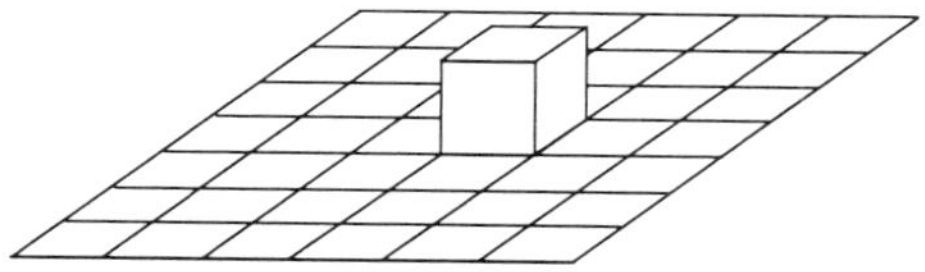

Fig. 3.1. A polyhedron whose reachable set is nowhere dense

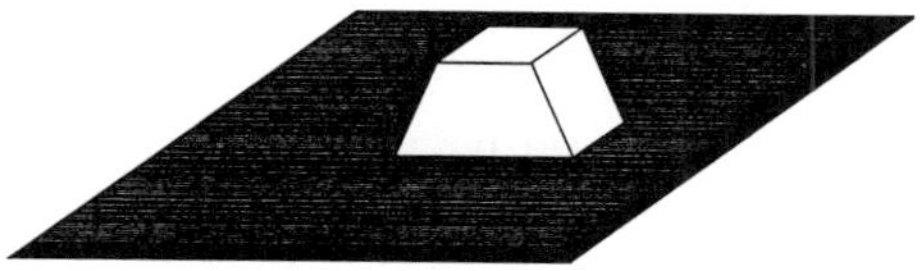

Fig. 3.2. A polyhedron whose reachable set is everywhere dense

all admissible sequences of tumbles to the initial configuration. Notice that the set of all sequences is an infinite but countable set while the configuration space is a finite disjoint union of copies of a 3–dimensional variety. Thus, the set of reachable points is itself countable. Therefore, instead of the more familiar concept of "complete reachability" (corresponding to $\mathcal{R} = \mathcal{Q}$), it will only make sense to investigate a property of "dense reachability" defined as $\mathsf{closure}(\mathcal{R}) = \mathcal{Q}$. In other words, rolling a polyhedron on a plane has the dense reachability property if, for any configuration of the polyhedron and every $\epsilon \in \mathbb{R}_+$, there exists a finite sequence of tumbles that brings the polyhedron closer to the desired configuration than ϵ. We refer in particular to a distance on $\mathcal{Q}$ defined as

$$\|(x_1, y_1, \theta_1, F_i) - (x_2, y_2, \theta_2, F_j)\| = \\ \max\left\{\sqrt{(x_1 - x_2)^2 + (y_1 - y_2)^2},\ |\theta_1 - \theta_2|,\ 1 - \delta(F_i, F_j)\right\}.$$

The term *discrete* will be used for the negation of *dense*. On this regard, the following results were reported in [22] (we recall that the defect angle is 2π minus the sum of the planar angles of all faces concurring at that vertex, and equals the gaussian curvature that can be thought to be concentrated at the vertex):

Theorem 3.4. *The set of configurations reachable by a polyhedron is dense in $\mathcal{Q}$ if and only if there exists a vertex V_i whose defect angle is irrational with π.*

Theorem 3.5. *The reachable set is discrete in both positions and orientations if and only if either of these conditions hold:*

i) all angles of all faces (hence all defect angles) are integer multiples of $\pi/3$, and all lengths of the edges are rational w.r.t. each other;

ii) all angles of all faces (hence all defect angles) are $\pi/2$, and all lengths of the edges are rational w.r.t. each other;
iii) all defect angles are π.

Theorem 3.6. *The reachable set is dense in positions and discrete in orientations if and only if the defect angles are all rational w.r.t. π, and neither conditions i), ii), or iii) of Theorem 3.5 apply.*

Remark 3.2. Parts with a discrete reachable set are very special. Polyhedra satisfying condition i) of Theorem 3.2 are rectangular parallelepipeds, as e.g. a cube or a sum of cubes which is convex. Polyhedra as in condition ii) are those whose surface can be covered by a tessellation of equilateral triangles, as e.g. any Platonic solid except the dodecahedron. Condition iii) is only verified by tetrahedra with all faces equal.

Remark 3.3. Observe that in the above reachability theorems the conditions upon which the density or discreteness of the reachable set depends are in terms of rationality of certain parameters and their ratios. This entails that two very similar polyhedra may have qualitatively different reachable sets. This is for instance the case of the cube and truncated pyramid reported above in Figs. 3.1 and 3.2, respectively, where the latter can be regarded as obtained from the cube by slightly shrinking its upper face. In fact, for any polyhedron whose reachable set has a discrete structure, there exists an arbitrarily small perturbation of some of its geometric parameters that achieves density.

In view of these remarks, and considering that in applications the geometric parameters of the parts will only be known to within some *tolerance*, i.e. a bounded neighborhood of their nominal value, a formulation of the planning problems ignoring robustness of results w.r.t. modeling errors will make little sense in applications.

4. Discussion and Open Problems

One way of reducing what is probably the single highest cost source in robotic devices, i.e. their actuators, is offered by nonholonomy. It has been shown in this paper how nonholonomic phenomena are actually much more pervasive in practical applications than usually recognized. However, the real challenge posed by nonholonomic systems is their effective control, including analysis of their structural properties, planning, and stabilization. The situation of research in these fields is briefly reviewed below.

Controllability. A nonholonomic system according to the above definition may not be *completely* nonholonomic, i.e. not completely controllable in some or all of the various senses that are defined in the nonlinear control literature. Detecting controllability is a much easier task for continuous driftless systems, such as e.g. the case of two bodies rolling on top of each other (see Eq. 3.1), because of the tools made available by nonlinear geometric control theory [16, 31]. Even in this case, though, there remains an open question to prove the conjecture that almost any pair of rolling bodies are controllable, or in other words, to characterize precisely the class of bodies which are not controllable, and to show that this subset is meager. Another question, practically a most important one, is to define a viable (i.e. computable and accurate) definition of a "controllability function" for nonholonomic systems, capable of conveying a sense of how intense the control activity has to be to achieve the manipulation goals, in a similar way as "manipulability" indices are defined in holonomic robots.

The controllability question is much harder for discontinuous systems or for systems with discrete input sets. As discussed above, relatively novel problems appear in the study of the reachable set, such as density or lattice structures. Very few tools are available from systems and automata theory to deal with such systems: consider to this regard that even the apparently simple problem of deciding the density of the reachable set of a 1–dimensional, linear problem

$$x_{k+1} = \lambda x_k + u_k, \quad u_k \in \text{ U, a finite set}$$

is unsolved to the best of our knowledge, and apparently not trivial in general. It is often useful in these problems to notice a possible group structure in the fiber motions induced by closed base space motions (see Fig. 2.3): such group analysis was actually instrumental to the results obtained for the polyhedron rolling problem.

Planning. The planning problem (i.e. the open–loop control) for some particular classes of nonholonomic systems is rather well understood. For instance, two–inputs nilpotentiable systems that can be put, by feedback transformation, in the so–called "chained" form, can be steered using sinusoids [28]; systems that are "differentially flat" can be planned looking at their (flat) outputs only [38]; systems that admit an exact sampled model (and maintain controllability under sampling) can be steered using "multirate control" [24]; nilpotent systems can be steered using the "constructive method" of [18]. However, as already pointed out, systems of rolling bodies do not fall into any of these classes. At present, planning motions of a spherical object onto a planar finger can be done in closed form, while for general objects only iterative solutions are available (e.g. the one proposed in [40]).

Stabilization. The control problem is particularly challenging for nonholonomic systems, due to a theorem of Brockett [7] that bars the possibility of stabilizing a nonholonomic vehicle about a nonsingular configuration by any continuous time–invariant static feedback. Non–smooth, time–varying, and dynamic extension algorithms have been proposed to face the point–stabilization problem for some classes of systems (e.g. chained–form). A stabilization method for a system of rolling bodies, or even for a sphere rolling on a planar finger, is not known to the authors.

References

[1] Arai H, Tanie K, Tachi S 1993 Dynamic control of a manipulator with passive joints in operational space. *IEEE Trans Robot Automat.* 9:85–93

[2] Bicchi A 1995 Hands for dextrous manipulation and powerful grasping: a difficult road towards simplicity. In: Giralt G, Hirzinger G (eds) *Robotics Research: The Seventh International Symposium.* Springer-Verlag, London, UK, pp 2–15

[3] Bicchi A, Goldberg K (eds.) 1996 Proc 1996 Work Minimalism in Robotic Manipulation *1996 IEEE Int Conf Robot Automat.* Minneapolis, MA

[4] Bicchi A, Prattichizzo D, Sastry S S 1995 Planning motions of rolling surfaces. In: *Proc 1995 IEEE Conf Decision Contr.* New Orleans, LA

[5] Bicchi A, Sorrentino R 1995 Dextrous manipulation through tolling. In: *Proc 1995 IEEE Int Conf Robot Automat.* Nagoya, Japan, pp 452–457

[6] Brockett R W 1976 Nonlinear systems and differential geometry. *Proc IEEE.* 64(1):61–72

[7] Brockett R W 1983 Asymptotic stability and feedback stabilization. In: Brockett R W, Millmann R S, Sussman H J (eds) *Differential Geometric Control Theory.* Birkhäuser, Boston, MA, pp 181–208

[8] Brockett R W 1989 On the rectification of vibratory motion. *Sensors and Actuators.* 20:91–96

[9] Cai C, Roth B 1987 On the spatial motion of a rigid body with point contact. In: *Proc 1987 IEEE Int Conf Robot Automat.* Raleigh, NC, pp 686–695

[10] Chitour Y, Marigo A, Prattichizzo D, Bicchi A 1996 Reachability of rolling parts. In: Bonivento C, Melchiorri C, Tolle H (eds) *Advances in Robotics: The ERNET Perspective.* World Scientific, Singapore, pp 51–60

[11] Cole A, Hauser J, Sastry S S 1989 Kinematics and control of a multifingered robot hand with rolling contact. *IEEE Trans Robot Automat.* 34(4)

[12] De Luca A, Mattone R, Oriolo G 1996 Dynamic mobility of redundant robots using end–effector commands. In: *Proc 1996 IEEE Int Conf Robot Automat.* Minneapolis, MA, pp 1760–1767

[13] Grupen R A, Henderson T C, McCammon I D 1989 A survey of general–purpose manipulation. *Int J Robot Res.* 8(1):38–62

[14] Guyon C, Petitot M 1995 Flatness and nilpotency. In: *Proc 3rd Euro Contr Conf.* Rome, Italy

[15] Hollerbach J M 1987 Robot hands and tactile sensing. In: Grimson W E L, Patil R S (eds) *AI in the 1980's and beyond.* MIT Press, Cambridge, MA, pp 317–343

[16] Isidori A 1995 *Nonlinear Control Systems.* (3rd ed) Springer-Verlag, London, UK

[17] Kolmanovsky I V, McClamroch N H, Coppola V T 1995 New results on control of multibody systems which conserve angular momentum. *J Dyn Contr Syst.* 1:447-462

[18] Lafferriere G, Sussmann H 1991 Motion planning for controllable systems without drift. In: *Proc 1991 IEEE Int Conf Robot Automat.* Sacramento, CA, pp 1148–1153

[19] Leonard N E, Krishnaprasad P S 1995 Motion control of drift free, left invariant systems on Lie groups. *IEEE Trans Automat Contr.* 40(9)

[20] Li Z, Canny J 1990 Motion of two rigid bodies with rolling constraint. *IEEE Trans Robot Automat.* 6:62–72

[21] Lynch K M, Mason M T 1995 Controllability of pushing. In: *Proc 1995 IEEE Int Conf Robot Automat.* Nagoya, Japan, pp 112-119

[22] Marigo A, Chitour Y, Bicchi A 1997 Manipulation of polyhedral parts by rolling. In: *Proc 1997 IEEE Int Conf Robot Automat.* Albuquerque, NM

[23] Mason M T, Salisbury J K 1985 *Robot Hands and the Mechanics of Manipulation.* MIT Press, Cambridge, MA

[24] Monaco S, Normand–Cyrot, D 1992 An introduction to motion planning under multirate digital control. In: *Proc 31st IEEE Conf Decision Contr.* Tucson, AZ, pp 1780–1785

[25] Montana D J 1988 The kinematics of contact and grasp. *Int J Robot Res.* 7(3):17–32

[26] Morin P, Samson C 1997 Time varying exponential stabilization of a rigid spacecraft with two control torques. *IEEE Trans Automat Contr.* 42:528–533

[27] Murray R M 1994 Nilpotent bases for a class of non-integrable distributions with applications to trajectory generation for nonholonomic systems. *Math Contr Sign Syst.* 7:58–75

[28] Murray R M, Li Z, Sastry S S 1994 *A Mathematical Introduction to Robotic Manipulation.* CRC Press, Boca Raton, FL

[29] Nakamura Y, Mukherjee R 1993 Exploiting nonholonomic redundancy of free flying space robots. *IEEE Trans Robot Automat.* 9:499–506

[30] Neimark J I, Fufaev N A 1972 *Dynamics of Nonholonomic Systems*, American Mathematical Society Translations of Mathematical Monographs, 38

[31] Nijemeijer H, van der Schaft A J 1990 *Nonlinear Dynamical Control Systems.* Springer-Verlag, Berlin, Germany

[32] Oriolo G, Nakamura Y 1991 Free joint manipulators: Motion control under second–order nonholonomic constraints. In: *Proc IEEE/RSJ Int Work Intel Robot Syst.* Osaka, Japan, pp 1248–1253

[33] Ostrowski J, Burdick J 1995 Geometric perspectives on the mechanics and control of robotic locomotion. In: Giralt G, Hirzinger G (eds) *Robotics Research: The Seventh International Symposium.* Springer-Verlag, London, UK

[34] Pertin–Troccaz J 1989 Grasping: A state of the art. In: *The Robotics Review I*, MIT Press, Cambridge, MA, pp 71–98

[35] Peshkin M, Colgate J E, Moore C 1996 Passive robots and haptic displays based on nonholonomic elements. In: *Proc 1996 IEEE Int Conf Robot Automat.* Minneapolis, MA, pp 551–556

[36] Prattichizzo D, Bicchi A 1997 Consistent specification of manipulation tasks for defective mechanical systems. *ASME J Dyn Syst Meas Contr.* 119

[37] Prattichizzo D, Bicchi A 1997 Dynamic analysis of mobility and graspability of general manipulation systems. *IEEE Trans Robot Automat.* 13

[38] Rouchon P, Fliess M, Lèvine J, Martin P 1993 Flatness, motion planning, and trailer systems. In: *32nd IEEE Conf Decision Contr.* San Antonio, TX, pp 2700–2705

[39] Sørdalen O J, Nakamura Y 1994 Design of a nonholonomic manipulator. In: *Proc 1994 IEEE Int Conf Robot Automat.* San Diego, CA, pp 8–13

[40] Sussmann H, Chitour Y 1993 A continuation method for nonholonomic path–finding problems. In: *Proc IMA Work Robot.*

Control for Teleoperation and Haptic Interfaces

Septimiu E. Salcudean

Department of Electrical and Computer Engineering, University of British Columbia, Canada

The concept of teleoperation has evolved to accommodate not only manipulation at a distance but manipulation across barriers of scale and in virtual environments, with applications in many areas. Furthermore, the design of high-performance force-feedback teleoperation masters has been a significant driving force in the development of novel electromechanical or "haptic" computer-user interfaces that provide kinesthetic and tactile feedback to the computer user. Since haptic interfaces/teleoperator masters must interact with an operator and a real or virtual dynamic slave that exhibits significant dynamic uncertainty, including sometimes large and unknown delays, the control of such devices poses significant challenges. This chapter presents a survey of teleoperation control work and discusses issues of simulation and control that arise in the manipulation of virtual environments.

1. Teleoperation and Haptic Interfaces

Since its introduction in the 1940s, the field of teleoperation has expanded its scope to include manipulation at different scales and in virtual worlds. Teleoperation has been used in the handling of radioactive materials, in subsea exploration and servicing. Its use has been demonstrated in space [20], in the control of construction/forestry machines of the excavator type [36], in microsurgery and micro-manipulation experiments [33, 39] and other areas.

The goal of teleoperation is to achieve "transparency" by mimicking human motor and sensory functions. Within the relatively narrow scope of manipulating a tool, transparency is achieved if the operator cannot distinguish between maneuvering the master controller and maneuvering the actual tool. The ability of a teleoperation system to provide transparency depends largely upon the performance of the master and the performance of its bilateral controller. Ideally, the master should be able to emulate any environment encountered by the tool, from free-space to infinitely stiff obstacles.

The need for force-feedback in general purpose computer-user interfaces has been pointed out in the 1970s [5]. The demand is even higher today, as performance improvements in computer systems have enabled complex applications requiring significant interaction between the user and the computer. Examples include continuous and discrete simulation and optimization, searching through large databases, mechanical and electrical design, education and training.

Thus actuated devices with several degrees of freedom can serve as sophisticated input devices that also provide kinesthetic and tactile feedback to the user. Such devices, called "haptic interfaces", have been demonstrated in a number of applications such as molecular docking [35], surgical training [42], and graphical and force-feedback user interfaces [25].

The design of haptic interfaces is quite challenging, as the outstanding motion and sensing capabilities of the human arm are difficult to match. The performance specifications that haptic interfaces must meet are still being developed, based on a number of psychophysical studies and constraints such as manageable size and cost [19]. Peak acceleration, isotropy and dynamic range of achievable impedances are considered to be very important. A design approach that considers the haptic interface controller has been presented in [29].

This chapter is concerned with the design of controllers for teleoperation and haptic interfaces. A survey of control approaches is presented in Sect. 2. Interesting teleoperation control research topics, at least from this author's perspective, are being discussed in Sect. 3. Issues of haptic interface control for manipulation in virtual environments are discussed in a limited manner in Sect.4. mainly stressing aspects common to teleoperation.

2. Teleoperator Controller Design

The teleoperation controller should be designed with the goal of ensuring stability for an appropriate class of operator and environment models and satisfying an appropriately defined measure of performance, usually termed as *transparency*.

2.1 Modeling Teleoperation Systems

A commonly used teleoperation system model, e.g. [3, 17], is one with five interacting subsystems as shown in Fig. 2.1. The master manipulator, controller, and slave manipulator can be grouped into a single block representing the teleoperator as shown by the dashed line. For n-degree-of-freedom manipulation, the teleoperator can be viewed as a $2n$-port master-controller-slave (MCS) network terminated at one side by an n-port operator block and at the other by an n-port environment network, as shown in Fig. 2.2. The force-voltage analogy is more often used than its dual to describe such systems [3, 17]. It assigns equivalent voltages to forces and currents to velocities. With this analogy, masses, dampers and stiffness correspond to inductances, resistances and capacitances, respectively.

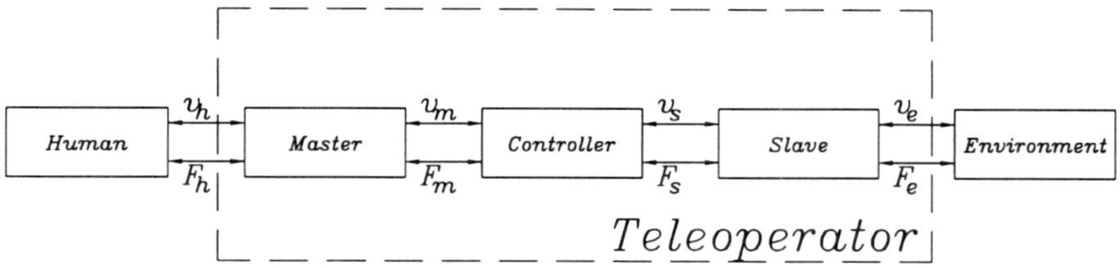

Fig. 2.1. Teleoperation system model

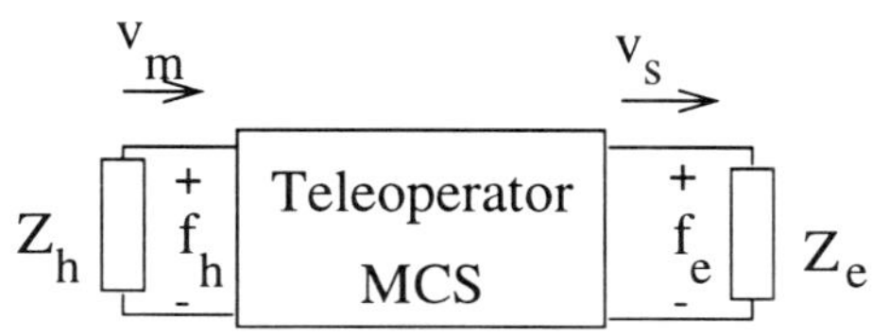

Fig. 2.2. $2n$-port network model of teleoperation system

A realistic assumption about the environment is that it is passive or strictly passive as defined in [12]. Following research showing that the operator hand behaves very much like a passive system [21], the operator block is usually also assumed to be passive.

For simplicity, the above network equivalents are almost always modeled as linear, time-invariant systems in which the dynamics of the force transmission and position responses can be mathematically characterized by a set of network functions. Even though this limits the type of environment and operator interaction that can be characterized, it allows the control system designer to draw upon well-developed network theory for synthesis and analysis of the controller.

The MCS block can be described in terms of hybrid network parameters as follows:

$$\begin{bmatrix} f_h \\ v_s \end{bmatrix} = \begin{bmatrix} Z_{m0} & G_f \\ G_p^{-1} & Y_{s0} \end{bmatrix} \begin{bmatrix} v_m \\ f_e \end{bmatrix}, \tag{2.1}$$

where Z_{m0} is the master impedance and G_p is the position gain with the slave in free motion, and Y_{s0} is the slave admittance and G_f is the force gain with the master constrained.

Alternative MCS block descriptions that are useful can be written in terms of the admittance operator Y mapping $f = [f_h^T\ f_e^T]^T$ to $v = [v_m^T\ v_s^T]^T$ and the scattering operator $S = (I - Y)(I + Y)^{-1}$ mapping the input wave into the output wave $f - v = S(f + v)$. The admittance operator is proper so it fits in the linear controller design formalism better than the hybrid operator

[4, 22], while the norm or structured singular value of the scattering operator provides important passivity/absolute stability conditions [3, 11].

2.2 Robust Stability Conditions

The goal of the teleoperation controller is to maintain stability against various environment and operator impedances, while achieving performance or transparency as will be defined below.

For passive operator and environment blocks, a sufficient condition for stability is passivity of the teleoperator (e.g. [3]).

It was shown in [11] that the bilateral teleoperator shown in Fig. 2.2 is stable for all passive operator and environment blocks if and only if the scattering operator S of the $2n$-port MCS is bounded-input-bounded-output stable and satisfies $\sup_\omega \mu_\Delta(j\omega) \leq 1$. The structured singular value μ_Δ is taken with respect to the 2-block structure $\text{diag}\{S_h, S_e\}$, where S_h and S_e are the scattering matrices of strictly passive Z_h and Z_e, respectively. This result is an extension of Llewellyn's criterion for absolute stability of 2-ports terminated by passive impedances [32]. Various extensions of the result to nonlinear systems are discussed in [11].

Following common practice, robust stability conditions can be obtained from bounds on nominal operator and environment dynamic models using structured singular values [10].

2.3 Performance Specifications

For the operator to be able to control the slave, a kinematic correspondence law must be defined. In *position control mode*, this means that the unconstrained motion of the slave must follow that of the master modulo some pre-defined or programmable scaling. Position tracking does not need to perform well at frequencies above 5-10 Hz (the human hand cannot generate trajectories with significant frequency content above this range). In terms of the hybrid parameters defined in Eq. 2.1), G_p should approximate a position gain $n_p I$ at low frequencies.

Forces encountered by the constrained slave should be transmitted to the hand by the master in a frequency range of up to a few hundred Hz. In terms of the hybrid parameters defined in Eq. 2.1, G_f should approximate a force gain $n_f I$.

Teleoperation system transparency can be quantified in terms of the match between the mechanical impedance of the environment encountered by the slave and the mechanical impedance transmitted to or felt by the operator at the master [17, 28], or by the requirement that the position/force responses of the teleoperator master and slave be identical [48]. If $f_e = Z_e v_s$, the impedance transmitted to the operator's hand $f_h = Z_{th} v_m$ is given by

$$Z_{th} = Z_{m0} + G_f Z_e (I - Y_{s0} Z_e)^{-1} G_p^{-1}. \tag{2.2}$$

The teleoperation system is said to be *transparent* if the slave follows the master, i.e. $G_p = n_p I$, and Z_{th} is equal to $(n_f/n_p)Z_e$ for any environment impedance Z_e. For transparency, $Y_{s0} = 0$, $Z_{m0} = 0$ and $G_f = n_f I$. Note that the above definition of transparency is symmetric with respect to the MCS teleoperation block. If the teleoperation system is transparent, then the transmitted impedance from the operator's hand to the environment is the scaled operator impedance $Z_{te} = (n_p/n_f)Z_h$.

Alternatively, transparency can be similarly defined by imposing transmission of the environment impedance added to a "virtual tool" [22, 26] or "intervenient" or "centering" impedance [41, 48] Z_{t0}, i.e. $Z_{th} = Z_{t0} + (n_f/n_p)Z_e$, for all Z_e. Z_{t0} can be taken to be the master impedance [50].

A transparent "model reference" for scaled teleoperation can be defined as shown in Fig. 2.3 [22]. It can be shown that for any constant, real, posi-

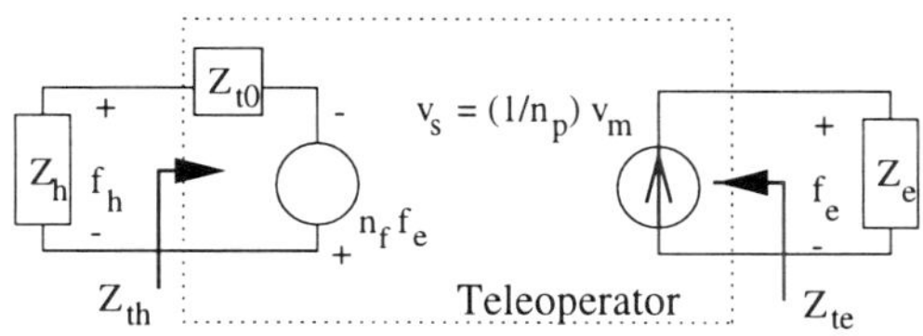

Fig. 2.3. Model reference for ideal scaled teleoperation

tive scalings n_p and n_f, and any passive tool impedance Z_{t0}, the structured singular value of the scattering matrix of this system is less than one at all frequencies, so the model reference is stable when in contact with any passive operator and environment.

It has been argued that since various impedance elements (mass, damper, stiffness) do not scale with dimension in the same way, scaling effects should be added for an ideal response instead of the "kinematic" scaling used above [11].

2.4 Four-Channel Controller Architecture

It has been shown in [28] that in order to achieve transparency as defined by impedance matching, a *four-channel* architecture using the sensed master and slave forces and positions is required as illustrated in Fig. 2.4, where Z_m and Z_s are the master and slave manipulator impedances, C_m and C_s are local master and slave manipulator controllers, and C_1 through C_4 are bilateral teleoperation control blocks. The forces f_h^a and f_e^a are exogenous operator and environment forces. Tradeoffs between teleoperator transparency and stability robustness have also been examined empirically in [28]. The observation that a "four channel" architecture is required for transparency is important as various teleoperation controller architectures have been presented in the past based on what flow or effort is sensed or actuated by the MCS block shown in Fig. 2.2 [8]. In Fig. 2.4, setting C_2 and C_3 to zero yields the "position-position" architecture, setting C_1 and C_2 to zero yields the "position-force" architecture, etc. For certain systems having very accurate models of the master and the slave [41], sensed forces can be replaced by estimated forces using observers [15]. However, observer-derived force estimates are band-limited (typically less than 50 Hz) by noise and model errors so are more suitable in estimating hand forces to be fed forward to the slave than environment forces to be returned to the master. In terms of the parameters

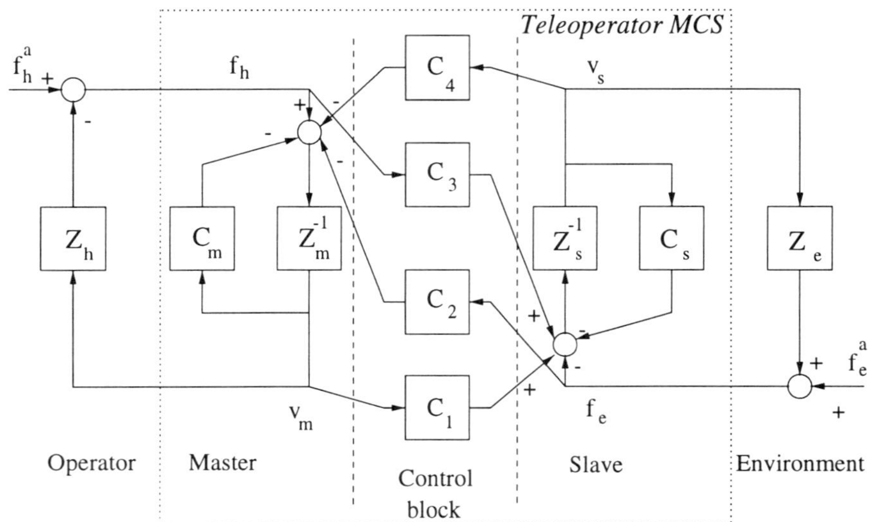

Fig. 2.4. Four-channel teleoperation system

from Fig. 2.4 the environment impedance transmitted to the operator is given by

$$\begin{aligned} Z_{th} &= [I-(C_2Z_e+C_4)(Z_s+C_s+Z_e)^{-1}C_3]^{-1} \times \\ &\quad [(C_2Z_e+C_4)(Z_s+C_s+Z_e)^{-1}C_1+(Z_m+C_m)]. \end{aligned} \tag{2.3}$$

2.5 Controller Design via Standard Loop Shaping Tools

A number of controller synthesis approaches using standard "loop-shaping" tools have been proposed to design teleoperation controllers assuming that the operator and environment impedances are known and fixed (with uncertainties described as magnitude frequency bounds or by additive noise signals). The system in Fig. 2.4 can be transformed into the standard augmented plant form (e.g. [7]) shown in Fig. 2.5.

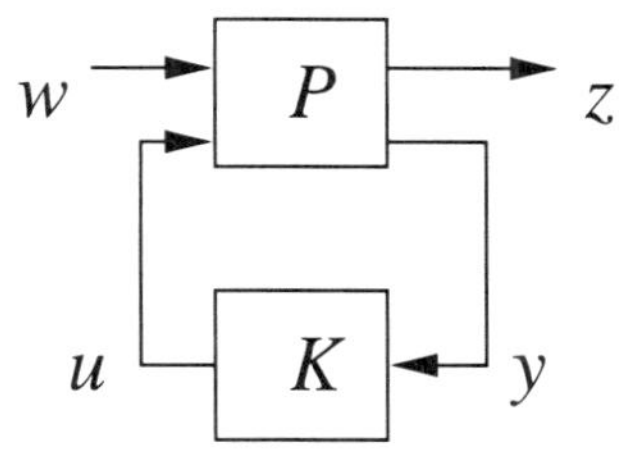

Fig. 2.5. Standard system for controller synthesis

In [24], a $2n$-by-$2n$ H^∞-optimal controller K based on contact forces was designed to minimize a weighted error between actual desired transfer functions for positions and forces using H^∞ control theory.

In [31], the μ-synthesis framework was used to design a teleoperator which is stable for a pre-specified time delay while optimizing performance characteristics.

A general framework for the design of teleoperation controllers using H^∞ optimization and a $2n$-by-$4n$ controller block K is presented in [46]. All the controller blocks C_m, C_s, and C_1 through C_4 are included as required for transparency [28]. Additional local force-feedback blocks from f_e to the input of Z_s^{-1}, and from f_h to the input of Z_m^{-1}, are also included in this controller. In addition to the exogenous hand and environment forces f_e^a and f_h^a, noise signals are considered in the input vector. Typical error outputs in the vector z of Fig. 2.5 include $z_1 = W_1(f_h^a - n_f f_e)$, designed to maximize "force transparency at the master", $z_2 = W_2(x_s - Z_m^{-1}(f_h + n_f f_e)/n_p)$, designed to "maximize the position transparency at the slave", and $z_3 = W_3(x_m - n_p x_s)$ designed to maximize kinematic correspondence. The weight functions W_1, W_2, W_3 are low-pass. Delays have also been included in the model as Padé all-pass approximations with output errors modified accordingly. Various performance and performance vs. stability tradeoffs have been examined.

2.6 Parametric Optimization-based Controller Design

Ideally, one would like to find a $2n$-by-$4n$ teleoperation controller K that solves the optimization problem

$$\min_{\text{stabilizing } K} \|W(Y(K) - Y_d)\|_\infty \text{ such that } \sup_\omega \mu_\Delta S(K)(j\omega) \leq 1, \tag{2.4}$$

where Y_d is a proper, stable, reference admittance model derived from the system in Fig. 2.3, $Y(K)$ is the teleoperation MCS system admittance matrix, $S(K)$ is the teleoperation MCS scattering matrix and W is a weighting function. If such a problem had a solution, the resulting system would perform within a known bound from the reference model and would be stable against any passive operator and environment dynamics. Even though this problem does not account for other plant uncertainties, it cannot be solved by current techniques.

A controller synthesis approach that optimizes a measure of transparency subject to a "distance to passivity" as defined in [45] is presented in [4]. The design is accomplished by using semi-infinite optimization (see, for example, [37]) to solve an optimization problem that is not necessarily convex.

Another approach has been developed using the Youla parameterization of stabilizing controllers and convex optimization [22]. Since the variation in human impedance is relatively small by comparison to the change in environment impedance, it was assumed that the hand impedance is known and fixed. High order controllers were designed by solving a *convex* optimization problem of the form

$$\min_{\text{stabilizing } K} \|W_H(Y_H(K) - Y_{Hd})\|_\infty \quad \text{such that} \quad \inf_{\omega}\{\mathrm{Re} Y_{te}(K)(j\omega) \geq 0, \tag{2.5}$$

where Y_H and Y_{Hd} are admittance transfer functions (designed and desired, respectively) and Y_{te} is the MCS block admittance seen from the environment, *with a known operator impedance* Z_h.

If the hand impedance is equal to that for which the system was designed, the constraint on Y_{te} ensures that the environment faces a passive system and hence it is stable for any strictly passive environment. Design examples showing performance tradeoffs or transparency/robustness tradeoffs and experimental results have been presented.

2.7 Nonlinear Transparent Control

A nonlinear teleoperation scheme that is transparent at high gain was presented in [44]. The approach uses the nonlinear rigid body dynamics of the master and slave manipulators but neglects the operator dynamics. Measured master and slave forces are used in the master controller. A stability proof and bounded position and force tracking errors have been obtained.

2.8 Passivation for Delays and Interconnectivity

In outer-space or sub-sea applications, significant delays appear in the control/communication block implemented by C_1, C_2, C_3 and C_4 in Fig. 2.4 and lead to instability by causing the scattering matrix of the MCS system to have infinite norm [3]. Instead of transmitting forces and velocities as in Fig. 2.4, the active control can be modified to mimic a lossless transmission line [3].

Stability of the system can be ensured if each of the manipulator/controller blocks is made passive. Reflections in the lossless transmission line between the master and the slave manipulator can lead to poor performance that can be alleviated somewhat by matched terminations [34].

The idea of building modular robot systems by making each of the building blocks passive lead to a sophisticated system that allows teleoperated and shared control of multiple robots for programming and teleoperation [1]. In [2], it is shown that passivity of the modules can be preserved after discretization by using wave variables instead of forces and velocities and applying a discretization that preserves the norm of the scattering matrix (Tustin's method).

The performance loss derived from preserving modularity via passivity is not yet clear. An experimental study of a teleoperator using a passive interconnection of passive systems showed rather poor performance [27].

Other methods have been presented in order to deal with the communication delay problem. For delays of a couple of seconds or less, the dual hybrid teleoperation approach [38] described below provides some kinesthetic feedback while maintaining stability. For larger delays, the use of *predictive displays* has been proposed and demonstrated [6, 20]. The user is presented with a graphical display of a robot and world model, possibly superimposed over current camera images. Force feedback information is conveyed by the dynamic simulation of the environment, which is updated based on sensory information.

The concept of *teleprogramming* was also introduced to deal with the problem of delays [13]. In this approach, the master and slave have local high-level supervisory controllers and the bilateral controllers (blocks C_1 through C_4 in Fig. 2.4) are replaced with communication modules that transmit only high level programs. Based on the completion report of remotely executed programs, the operator can make manipulation decisions. All force feedback information is generated by the master controller based on the environment model.

2.9 Adaptive Teleoperation Control

The controllers designed for fixed operator and environment impedance are too complex and require too many adjustments of design weights for them to be computed on-line easily. It is possible that complex gain-scheduling schemes could be developed to cover the broad range of operating conditions encountered for different operator and slave environments, but these would be quite complicated (up to six-dimensional frequency-dependent matrices Z_h and Z_e must be accommodated). As an alternative, techniques using environment identification have been proposed [17, 18].

A bilateral adaptive impedance control architecture has been proposed in [17]. The idea is to use operator and environment impedance estimators at the master and slave and local master and slave controllers (C_m and C_s

in Fig. 2.4) to duplicate the environment impedance at the master and the operator impedance at the slave. If the impedance estimators do converge, the scheme would provide transparency the way a four-channel architecture does. In addition, the estimated impedances could be processed in order to avoid stability problems caused by delays or modeling errors. This scheme is very attractive but relies on accurate impedance estimators that are difficult to obtain.

In [18], a transparent bilateral control method is presented using the above "impedance reflection" idea. Environment position, velocity and acceleration are used to estimate environment impedance. The estimated impedance is used in the slave controller for good tracking performance and by the master controller to achieve transparency. With the conventional identification approach employed, it was found that environment identification converges slowly, has fairly high sensitivity to delays, and therefore is unsuitable when the environment changes fast, as is the case when manipulating objects in the presence of hard constraints [18].

An adaptive slave motion controller has been proposed in [34], where the adaptive control method of [43] is used for the slave unconstrained motion, with the constrained slave direction being controlled in stiffness mode.

2.10 Dual Hybrid Teleoperation

For directions in which Z_e is known, the environment impedance does not need to be identified. In particular, in directions in which Z_e is known to be small (e.g. free-motion), the master should act as a force source/position sensor and have low impedance, while the slave should behave as a position source/force sensor and have high impedance. Thus, in directions in which Z_e is small, positions are sent to the slave and forces are returned to the master, with C_1 and C_2 having unity transmission, and C_3, C_4 having zero transmission. The dual situation applies in directions in which Z_e is known to be large, (e.g. stiff contact or constraints). In those directions, the master should act as a force sensor/position source and have high impedance, with forces being sent to the slave and positions being returned to the master. Thus, in directions in which Z_e is large, C_1 and C_2 should have zero transmission, while C_3 and C_4 should be close to unity. From Eq. 2.3, it can be seen that the above insures that along very small or very large values of Z_e, the transmitted impedance equals that of the master with local controller $Z_m + C_m$, which can be set to the minimum or maximum achievable along required directions.

This concept of "dual hybrid teleoperation" has been introduced, studied and demonstrated experimentally in [38]. It has been shown that when the geometric constraints for a teleoperation task are known, the master and slave workspaces can be split into dual position-controlled and force-controlled subspaces, and information can be transmitted unilaterally in these orthogonal subspaces, while still providing useful kinesthetic feedback to the operator.

2.11 Velocity Control with Force Feedback

For some teleoperation systems, such as remotely-controlled excavators [36], position control is not a realistic option due to issues of safety and vastly different master and slave manipulator workspaces that would imply very poor motion resolution if scaling were to be used [50]. Instead, *velocity control mode* is used, in which the slave velocity follows the master position, so ideally $G_p = n_p sI$ in Eq. 2.1. Transparency based on transmitted impedance can be defined in a similar manner, and requires that the derivative of the environment force be returned to the master, so ideally $G_f = n_f sI$ in Eq. 2.1 [50]. To avoid returning the derivative of environment force that could be very noisy, velocity mode control can be modified to include a low-pass filter making G_p and G_f proper.

Experiments with velocity-mode teleoperation systems have indeed shown that direct force feedback leads to poor transparency and poor stability margins, especially when stiff environments are encountered. As an alternative, a new approach called "stiffness feedback" has been proposed. Instead of returning direct force information, the master stiffness is modulated by the environment force, from a minimum positive stiffness corresponding to the minimum expected force to a maximum positive stiffness corresponding to the maximum expected force. In order to avoid blocking the slave against a stiff environment, the stiffness law applies only when the environment force opposes slave motion. It can be shown that this control scheme is locally transparent when the environment force opposes slave motion and experimental results have been very positive [30, 36].

3. Teleoperation Control Design Challenges

In spite of the significant amount of research in the area of teleoperation, there are still very few applications in which the benefits of transparent bilateral teleoperation have been clearly demonstrated, in spite of areas of great potential, such as teleoperated endoscopic surgery, microsurgery, or the remote control of construction, mining or forestry equipment. Whether this is due to fundamental physical limitations of particular teleoperator systems or due to poorly performing controllers is still not clear. From this perspective, probably the single most important challenge ahead is a better understanding of the limits of performance of teleoperation systems. Towards this goal, it would be useful to have a benchmark experimental system and task to be completed for which various controllers could be tested. Unfortunately, it would be very difficult to do this entirely through simulation, as the dynamic algorithms necessary to develop a reasonable array of tasks would be just as much under test as the teleoperation control schemes themselves. Furthermore, the minimum number of degrees of freedom for reasonably representative tasks would have to be at least three, e.g. planar master/slave systems.

Specific improvements could be made to the fixed teleoperation controllers designed via conventional loop shaping or parametric optimization. In particular, a class of operator impedances that is broader than a single fixed impedance but narrower than all passive impedances should be developed with associated robust stability conditions. Since the control design problem was formulated as a constrained "semi-infinite" optimization problem, different algorithms could be tested or new ones developed. Like many other multi-objective optimal control problems, robust teleoperator controller design problems are likely to be hard to solve.

There seems to be much promise in the design of adaptive bilateral teleoperation controllers with relatively simple and physically motivated structures. In particular, indirect adaptive schemes based on Hannaford's architecture [17] are likely to succeed. Whereas fast or nonlinear environment identification techniques are necessary to accommodate contact tasks and these seem quite difficult to develop, operator dynamics identification seems to be quite feasible [16]. Some of the difficulties encountered in developing identification algorithms may be circumvented by the use of dual hybrid teleoperation or newly developed variants that are not based on orthogonal decomposition of the task space into position and force controlled spaces. Another interesting research area is the automatic selection of the position and force controlled subspaces.

4. Teleoperation in Virtual Environments

Manipulation in virtual environments has potential applications in training systems, computer-aided mechanical design and ergonomic design. For virtual environments, the master (more often called haptic interface in this context) control algorithms differ from bilateral teleoperation control algorithms in that the slave manipulator and its environment become a dynamic simulation. The simulation of systems dynamics for graphical or haptic rendering is a topic of substantial research. See, for example, [14] and other articles in the same proceedings.

Two approaches have been proposed for interfacing haptic devices to dynamic simulations. The *impedance display*, used by most researchers, taking sensed motions as inputs, passing them through a "virtual coupler" [9] to the dynamic simulator, and returning forces to the device, and the *admittance display*, taking sensed forces as input and returning positions to the haptic device. The relative advantages of these display modes have barely been touched upon, with the ability to build modular systems ("summing forces and distributing motion") [49] with non penetration constraints [47] presented in favor of the admittance approach.

Looking back at the debate on teleoperation "architectures", it seems that a four-channel coupling of haptic interface and dynamic simulation via a virtual coupler should be used. This would allow the haptic interface to behave as a force sensor or position sensor depending on the impedance of the task. The implication on dynamic simulators remains to be determined, but there is no reason why forces from the virtual coupler could not be added to sensed forces.

From a control point of view, the existence of a full dynamic model of the slave has both advantages and disadvantages. On the one hand, the design becomes easier because no environment identification is necessary. On the other, the design becomes more difficult because dynamic simulations require significant computing power which is often distributed, so one can expect to deal with multiple rate asynchronous systems. The argument for building complex systems using passive building blocks [1] is quite compelling, especially since techniques for passive implementations of multi body simulations are being developed [9].

Better understanding of hybrid systems is needed for the control of haptic interfaces, as manipulation of objects in the presence of non penetration constraints often require switching of controller/simulation states [40, 49].

5. Conclusion

A survey of teleoperation control for scaled manipulation and manipulation in virtual environments has been presented in this chapter. It seems that contributions from the areas of systems identification, adaptive control, multi objective optimal control and hybrid systems could be integrated in novel ways to provide solutions to problems of transparent bilateral control. The scope of the survey was quite limited. Interesting work in the design of haptic interfaces, novel ways of achieving passivity using nonholonomic systems, and issues of dynamic systems simulation for virtual reality have not been addressed.

References

[1] Anderson R J 1995 SMART: A modular control architecture for telerobotics. *IEEE Robot Automat Soc Mag.* 2(3):10–18

[2] Anderson R J 1996 Building a modular robot control system using passivity and scattering theory. In: *Proc 1996 IEEE Int Conf Robot Automat.* Minneapolis, MN, pp 1626–1629

[3] Anderson R J, Spong M W 1989 Bilateral control of operators with time delay. *IEEE Trans Automat Contr.* 34:494–501

[4] Andriot C A, Fournier R 1992 Bilateral control of teleoperators with flexible joints by the H^∞ approach. In: *Proc 1993 SPIE Conf Telemanip Tech.* pp 80–91

[5] Batter J J, Brooks F P Jr 1972 GROPE-1: A computer display to the sense of feel. In: *Proc IFIP.* pp 759–763

[6] Bejczy A, Kim W S 1990 Predictive displays and shared compliance control for time-delayed manipulation. *Proc IEEE/RSJ Int Work Intel Robot Syst.* Tsuchiura, Japan, pp 407–412
[7] Boyd S P, Barratt C H 1991 *Linear Controller Design: Limits of Performance.* Prentice-Hall, Englewood Cliffs, NJ
[8] Brooks T, 1990 Telerobotic Response Requirements. Tech Rep STX/ROB/90-03, STX Robotics
[9] Brown J M, Colgate J E 1997 Passive implementation of multibody simulations for haptic display. *1997 ASME Int Mech Eng Congr Exp.* Dallas, TX
[10] Chiang R Y, Safonov M G 1992 Robust Control Toolbox for Use with Matlab. The MathWorks, Inc.
[11] Colgate J E 1993 Robust impedance shaping telemanipulation. *IEEE Trans Robot Automat.* 9:374–384
[12] Desoer C A, Vidyasagar M 1975 *Feedback Systems.* Academic Press, New York
[13] Funda J and Paul R P 1990 Teleprogramming: Overcoming communication delays in remote manipulation. In: *Proc 1990 IEEE Int Conf Syst Man Cyber.* Los Angeles, CA, pp 873–875
[14] Gillespie R B, Colgate J E 1997 A survey of multibody dynamics for virtual environments. *1997 ASME Int Mech Eng Congr Exp.* Dallas, TX
[15] Hacksel P, Salcudean S E 1994 Estimation of environment forces and rigid-body velocities using observers. In: *Proc 1994 IEEE Int Conf Robot Automat.* San Diego, CA, pp 931–936
[16] Hajian A Z, Howe R D 1994 Identification of the mechanical impedance at the human finger tip. In: *Proc 1994 ASME Int Mech Eng Congr Exp.* Chicago, IL, DSC-vol 55-1, pp 319–327
[17] Hannaford B 1989 A design framework for teleoperators with kinesthetic feedback. *IEEE Trans Robot Automat.* 5:426–434
[18] Hashtrudi-Zaad K, Salcudean S E 1996 Adaptive transparent impedance reflecting teleoperation. In: *Proc 1996 IEEE Int Conf Robot Automat.* Minneapolis, MN, pp 1369–1374
[19] Hayward V, Astley O R 1995 Performance measures for haptic interfaces. In: Giralt G, Hirzinger G (eds) *Robotics Research: The Seventh International Symposium.* Springer-Verlag, London, UK
[20] Hirzinger G, Brunner B, Dietrich J, Heindl J 1993 Sensor-based space robotics – ROTEX and its telerobotic features. *IEEE Trans Robot Automat.* 9:649–663
[21] Hogan N 1989 Controlling impedance at the man/machine interface. In: *Proc 1989 IEEE Int Conf Robot Automat.* Scottsdale, AZ, pp 1626–1631
[22] Hu Z, Salcudean S E, Loewen P D, 1996 Optimization-based teleoperation controller design. In: *Proc 13th IFAC World Congr.* San Francisco, CA, vol D, pp 405-410
[23] Hunter I W, Lafontaine S, Nielsen P M F, Hunter P J, Hollerbach J M 1989 A microrobot for manipulation and dynamical testing of single living cells. In: *Proc IEEE Micro Electr Mech Syst.* Salt Lake City, UT, pp 102–106
[24] Kazerooni H, Tsay T-I, Hollerbach K 1993 A controller design framework for telerobotic systems. *IEEE Trans Contr Syst Tech.* 11:105–116
[25] Kelley A J, Salcudean S E 1994 The development of a force feedback mouse and its integration into a graphical user interface. In: *Proc 1994 ASME Int Mech Eng Congr Exp.* Chicago, IL, DSC-vol 55-1, pp 287–294
[26] Kosuge K, Itoh T, Fukuda T, Otsuka M 1995 Scaled telemanipulation system using semi-autonomous task-oriented virtual tool. *Proc 1995 IEEE/RSJ Int Conf Intel Robot Syst.* Pittsburgh, PA, pp 124–129

[27] Lawn C A, Hannaford B 1993 Performance testing of passive communications and control in teleoperation with time delay. In: *Proc 1993 IEEE Int Conf Robot Automat.* Atlanta, GA, vol 3, pp 776–783

[28] Lawrence D A 1993 Stability and transparency in bilateral teleoperation. *IEEE Trans Robot Automat.* 9:624–637

[29] Lawrence D A, Chapel J D 1994 Performance trade-offs for hand controller design. In: *Proc 1994 IEEE Int Conf Robot Automat.* San Diego, CA, pp 3211–3216

[30] Lawrence P D, Salcudean S E, Sepehri N, Chan D, Bachmann S, Parker N, Zhu M, Frenette R 1995 Coordinated and force-feedback control of hydraulic excavators. In: Khatib O, Salisbury J K (eds) *Experimental Robotics IV.* Springer-Verlag, London, UK, pp 181–194

[31] Leung G M H, Francis B A, Apkarian A 1995 Bilateral controller for teleoperators with time delay via μ-synthesis. *IEEE Trans Robot Automat.* 10:105–116

[32] Llewellyn F B 1952 Some fundamental properties of transmission systems. In: *Proc IRE.* 40:271

[33] Mitsuishi M, Watanabe H, Nakanishi H, Kubota H, Iizuka Y 1997 Dexterity enhancement for a tele-micro-surgery system with multiple macro-micro co-located operation point manipulators and understanding of the operator's intention. In: *Proc 1st Joint Conf CVRMED II & MRCAS.* Grenoble, France, pp 821–830

[34] Niemeyer G, Slotine J-J E 1991 Stable adaptive teleoperation. *IEEE J Ocean Eng.* 16:152–162

[35] Ouh-Young M, Pique M, Hughes J, Srinivasan N, Brooks F P Jr 1988 Using a manipulator for force display in molecular docking. In: *Proc 1988 IEEE Int Conf Robot Automat.* Philadelphia, PA, pp 1824-1829

[36] Parker N R, Salcudean S E, Lawrence P D 1993 Application of force feedback to heavy duty hydraulic machines. In *Proc 1993 IEEE Int Conf Robot Automat.* Atlanta, GA, vol 1, pp 375–381

[37] Polak E 1977 *Optimization: Algorithms and Consistent Approximations.* Springer-Verlag, New York

[38] Reboulet C, Plihon Y, Briere Y 1995 Interest of the dual hybrid control scheme for teleoperation with time delays. In: Khatib O, Salisbury J K (eds) *Experimental Robotics IV.* Springer-Verlag, London, UK, pp 498–506

[39] Salcudean S E, Ku S, Bell G 1997 Performance measurement in scaled teleoperation for microsurgery. In: *Proc 1st Joint Conf CVRMED II & MRCAS.* Grenoble, France, pp 789–798

[40] Salcudean S E, Vlaar T 1994 On the emulation of stiff walls and static friction with a magnetically levitated input–output device. In *Proc 1994 ASME Int Mech Eng Congr Exp.* Chicago, IL, DSC-vol 55-1, pp 303–309

[41] Salcudean S E, Wong N M, Hollis R L 1995 Design and control of a force-reflecting teleoperation system with magnetically levitated master and wrist. *IEEE Trans Robot Automat.* 11:844–858

[42] Satava R M, Jones S B 1997 Virtual environments for medical training and education. *Presence.* 6:139–146

[43] Slotine J-J E, Li W 1989 Composite adaptive control of robot manipulators. *Automatica.* 25:509–519

[44] Strassberg Y, Goldenberg A A, Mills J K 1993 A new control scheme for bilateral teleoperating systems: stability and robustness analysis. *ASME J Dyn Syst Meas Contr.* 115:345–351

[45] Wen J T 1988 Robustness analysis based on passivity. In: *Proc 1988 Amer Contr Conf.* Atlanta, GA, pp 1207–1212

[46] Yan J, Salcudean S E, 1996 Teleoperation controller design using H^∞ optimization. *IEEE Trans Contr Syst Tech.* 4:244–258

[47] Yokokohji Y, Hollis R L, Kanade T 1996 What you see is what you can feel – Development of a visual/haptic interface to virtual environment. In: *Proc IEEE Virtual Reality Annual Int Symp.* Santa Clara, CA pp 46–60

[48] Yokokohji Y, Yoshikawa T 1994 Bilateral control of master-slave manipulators for ideal kinesthetic coupling. *IEEE Trans Robot Automat.* 10:605–620

[49] Yoshikawa T, Hitoshi U 1997 Module-based architecture of world model for haptic virtual reality. In: *Prepr 5th Int Symp Experim Robot.* Barcelona, Spain, pp 111–122

[50] Zhu M, Salcudean S E 1995 Achieving transparency for teleoperator systems under position and rate control. *Proc 1995 IEEE/RSJ Int Conf Intel Robot Syst.* Pittsburgh, PA, pp 7–12

Recent Progress in Fuzzy Control

Feng-Yih Hsu and Li-Chen Fu

Department of Electrical Engineering, National Taiwan University, ROC

Fuzzy control has become a pervasively popular approach to the task of controller design because of its conceptual simplicity and easy realization but also because of its appealing performance demonstrated in a variety of practical applications. Through extensive and intensive research on the field, remarkable progress has been made in the recent literature. This chapter is aimed at reviewing such research progress and introducing some up-to-date results.

1. Introduction

In this chapter, we will review the most recent progress in the literature of fuzzy control. Up to now, fuzzy control has become a pervasively popular approach to the task of controller design. This is so not only because its theories are conceptually so straightforward that it is easily acceptable to the vast control literature, but also because it has demonstrated remarkable performance in a variety of practical applications. Theoretically speaking, the approach arises from an origin, where fuzzy control is usually referred to as an interpolated rule-based control. To be more persuasive, the inverted pendulum and the robot arm are usually taken as the testbed. However, for the testing purpose, one is more concerned with how much the so-designed controller and the human expert can be alike, rather than with the stability and the robustness of the controlled system. Of course, one can also incorporate some artificial intelligence techniques, such as a genetic algorithm or learning to achieve enhanced control [13, 22]. The genetic algorithm can provide a faster solution in searching for the best fuzzy rules via extensive simulations or experiments over the controlled system which can be regarded as a black-box system. On the other hand, a learning algorithm is constructed to extract some knowledge from the behavioral law of the controlled system learning, or from the neural nets. However, when the underlying system is too complex to be described, it is difficult to find a suitable learning algorithm to improve the fuzzy rules. Recently, a linguistic learning-based fuzzy control (LLBFC) with a sequential learning mechanism has been proposed to solve the above problems by imitating the procedure of controller design generally adopted by human beings [10]. The key spirit is that a sequential learning mechanism can first decompose the system into several subsystems, each of which can be easily described using some linguistic rules, and then establish the control by sequentially learning the control strategies of the individual subsystems.

With advances of the relevant theories [21, 16, 19], one has gradually realized fuzzy mechanisms can play the role as the so-called universal approximator which facilitates one to parameterize the system vagueness or the system uncertainties naturally. This explains the reason why the current trend of the theoretical developments in this regard is to combine the above-mentioned techniques with some conventional control theories into the hybrid fuzzy control approach such as fuzzy model analysis, adaptive fuzzy control, fuzzy variable structure control, fuzzy H^∞ control [21, 9]. The control using a fuzzy model approach is to represent the system dynamics in terms of a collection of linear systems with embedding of fuzzy if-then rules. Then, the stability of the overall system can be analyzed by LMI theorem [20]. Adaptive fuzzy control is to parameterize the fuzzy rules as products of some unknown rule parameters and some known regressor function so that adaptive technique can be applied to on-line update those rules [21]. Fuzzy variable structure control is to design the fuzzy rules which can behave as a variable structure control after setting some rule parameters [6, 9]. Fuzzy control can solve the problem of H^∞ performance with a prescribed disturbance attenuation level by adaptively updating some rule parameter [3].

In order to make the developed fuzzy controllers more convincing, it becomes a trend in demonstrating the controller performance in practice. Particularly, adaptive variable structure control is applied to robot manipulators to solve the problem in position tracking control, hybrid force/position control, contour-following control, and the deburring robot control, [6, 9]. It is worth noting that the structure of the fuzzy controller can sometimes be realized as a neural network one [16, 5], and both controllers can nowadays be implemented as some computer chip with fast parallel computing [23, 24].

2. Mathematical Foundations

Consider a fuzzy rule base, for instance, with input $x = [x_1, \cdots, x_n]^T$ and output $y = [y_1, \cdots, y_m]^T$, and then the j-th fuzzy rule is represented and inferred as follows:

$$\begin{array}{ll} \text{rule}[j]: & \text{if } x_1 \text{ is } A_1^{(j)} \text{ and} \cdots x_n \text{ is } A_n^{(j)}, \text{then } y_1 \text{ is } B_1^{(j)} \text{ and} \cdots y_m \text{ is } B_m^{(j)} \\ \text{fact:} & x_1 \text{ is } A_1' \text{ and} \cdots x_n \text{ is } A_n' \\ \hline \text{conclusion:} & y_1 \text{ is } B_1' \text{ and} \cdots y_m \text{ is } B_m' \end{array} \tag{2.1}$$

where A_i,A_i',B_i and B_i' are called fuzzy sets. Generally, a fuzzy control law consists of a fuzzy rule base with crisp inputs and crisp outputs, so that the fuzzy inference can be derived as some approximator f with constant parameters θ and α as [21]:

$$y = f(x, \theta, \alpha) = \frac{\sum_{j=1}^{p} \theta_j w_j(x, \alpha)}{\sum_{j=1}^{p} w_j(x, \alpha)} = \sum_{j=1}^{p} \theta_j \nu_j(x, \alpha) = \theta^T \nu, \tag{2.2}$$

where $\theta_j \in \Re^m$ is a parameter vector representing the numerical values associated with the fuzzy sets $B_1^{(j)}, \cdots, B_m^{(j)}$, and $w_j(x, \alpha)$ is a weighting function, expressed as follows:

$$w_j(x,\alpha) = \begin{cases} \mu_{A_1^{(j)}}(x_1,\alpha) \cdot \cdots \cdot \mu_{A_n^{(j)}}(x_n,\alpha) & \text{if sup-product operator,} \\ \min\{\mu_{A_1^{(j)}}(x_1,\alpha), \cdots \cdot \mu_{A_n^{(j)}}(x_n,\alpha)\} & \text{if sup-min operator;} \end{cases} \tag{2.3}$$

$\mu_A(x, \alpha)$ is the membership function characterized by constant parameter α, and ν_j is called fuzzy basis function (fuzzy regressor) defined as follows:

$$\nu_j(x,\alpha) = \frac{w_j(x,\alpha)}{\sum_{j=1}^{p} w_j(x,\alpha)} \tag{2.4}$$

with p being the total number of fuzzy rules. Note that, from expression (2.3), the combining operator 'and' can be implemented in two alternatives, either sup-product operator or sup-min operator.

3. Enhanced Fuzzy Control

Apparently, applying the fuzzy rule base (2.1) or the approximator (2.2) as a means to representation in the fuzzy control are equivalent. However, the fuzzy controllers designed based on (2.1) and on (2.2) mean different design approaches. In the former, one first constructs a reasonable fuzzy rule base with fuzzy sets determined by experts using some linguist variables (e.g. slow, very slow). These fuzzy sets are then realized after being assigned suitable membership functions which symbolize mappings from linguist variables to specific numerical values (as θ, α, in (2.2)). The latter is regarded as some interpolation scheme to approximate the involved nonlinear functions by seeking suitable parameters θ and α. However, lacking the systematic searching approach and the specification of the domain of interest, such fuzzy control approach is usually required to be combined with the other powerful methodologies (theories) to facilitate one to locate appropriate parameters or, equivalently, to determine appropriate rules.

3.1 Learning-based Fuzzy Control

Some fuzzy controllers can automatically update their fuzzy rules by incorporating some artificial intelligence techniques, such as genetic algorithm or learning algorithm. The genetic algorithm can provide a faster solution in searching for the best fuzzy rules via extensive simulations or experiments over the controlled system which can be regarded as a black-box system. On the other hand, the learning algorithm is constructed to extract some knowledge from the behavioral law of controlled system, or from the neural nets

[12, 1]. However, when the underlying system is too complex to be clearly described, it is difficult to construct the suitable learning algorithm to improve the fuzzy rules. Recently, a linguistic learning-based fuzzy control (LLBFC) with a sequential learning mechanism is proposed to solve the above problems by imitating the procedure of controller design generally adopted by human being [10]. The key spirit is that a sequential learning mechanism can decompose the system into several subsystems, each of which can be easily described using some linguistic rules, and then establish the control by sequentially learning the control strategies of the individual subsystems.

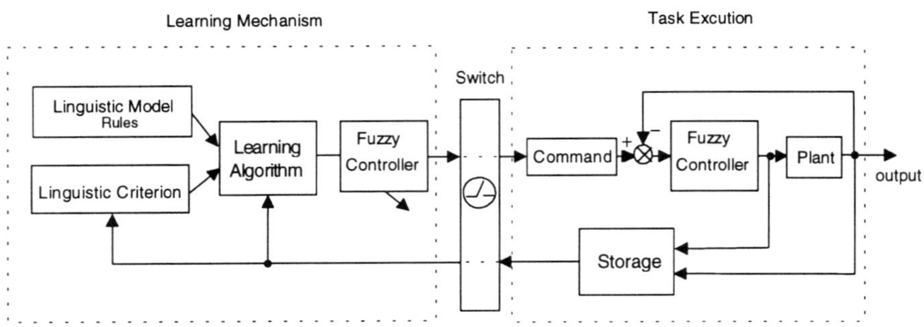

Fig. 3.1. Linguistic learning-based fuzzy control

The architecture of LLBFC mainly consists of the following five parts (see Fig. 3.1):

- fuzzy controller: consists of *If-then* rules and drives the system to meet the specified goal,
- linguistic criterion: declares the specification of the system performance,
- linguistic model rules: consists of *If-then* rules that describes the behavior of the controlled system,
- storage: saves some measurable system states and the control input during task running,
- learning algorithm: updates the fuzzy control rules.

To demonstrate the above-mentioned control schemes [10, 1], an inverted pendulum system depicted in Fig. 3.2 is taken as a testbed, which is often viewed as the level of ability of the control skills. The set-up of an inverted pendulum system consists of a DC motor and a cart carrying a pole, where the motor is to drive the cart so that the pole will not fall down. In Fig. 3.2, θ denotes the angle displacement of the pole, x denotes the position of the cart, and u is the control input.

A sequential learning procedure for the inverted pendulum is given as in Fig. 3.3. Then, the overall system is decomposed into two subsystems, one for

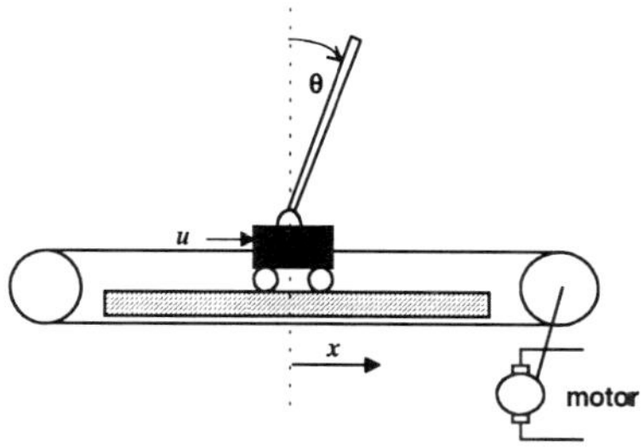

Fig. 3.2. The diagram of an inverted pendulum system

the angular displacement of the pole and the other for the position of the cart. The sequence is arranged such that the subcontroller (with output u_θ) for the angular position of the pole starts the learning process first till the response of the pole subsystem meets the linguistic criterion, and then the learning process of the other subcontroller (with output u_x) for the position of the cart is conducted via the help of some learning condition function $h_x(\theta, \dot{\theta}, x, \dot{x})$. Finally, the overall controller after the learning can be expressed as follows:

$$u = u_\theta + h_x(\theta, \dot{\theta}, x, \dot{x}) u_x \tag{3.1}$$

The control objective here is to regulate both the angular displacement of the pole and the position of the cart at the origin value subject to linguistic performance criteria: *Given the admissible overshoot, undershoot and system constraints, the controlled system should have the shortest risetime.*

3.1.1 Learning procedure for LLBFC. The learning procedure is initiated to set the fuzzy controller u_θ as a bang-bang controller, and after that, the learning mechanism starts to update those fuzzy rules. A large oscillation occurs for the angular displacement of the pole in Fig. 3.4a. Fig. 3.4b shows the simulation results for the fuzzy controller u_θ after 16 times of running with a satisfactory result and the result of the control input is similar to an optimum strategy by a well trained human operator. Figure 3.4d shows the overall learning process in the phase plane θ-$\dot{\theta}$ for two extreme initial conditions. At the beginning, the system response has a large oscillation and gradually converges to an optimum response when the number of iteration of learning increases.

After u_θ is trained completely, the hitting condition function, h_x, is implemented as follows:

$$h_x(\theta, s_\theta, s_x) = \begin{cases} -1, & \text{if } |\theta| \le \epsilon,\ |s_\theta| \le \epsilon_\theta \text{ and } |s_x| \ge \epsilon_x \\ 0, & \text{otherwise,} \end{cases} \tag{3.2}$$

where $s_\theta = \dot{\theta} + \lambda_\theta \theta$ and $s_x = \dot{x} + \lambda_x x$ are two augmented variables with constants $\lambda_\theta > 0$ and $\lambda_x > 0$; ϵ, ϵ_θ and ϵ_x are some small positive constants. Note

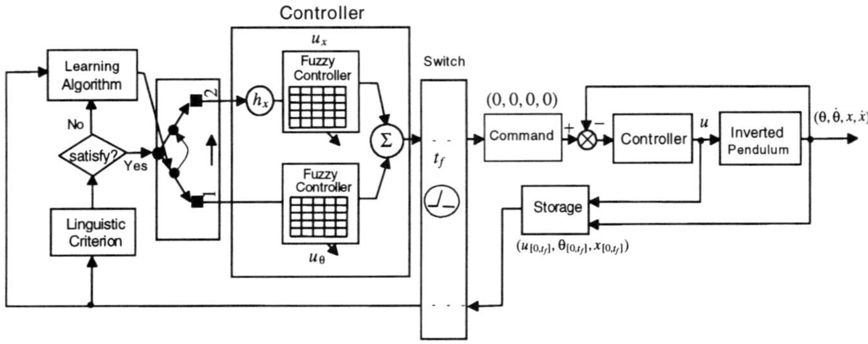

Fig. 3.3. The sequential learning mechanism for an inverted pendulum

that $s_\theta \leq \epsilon_\theta$ is utilized to characterize the desired pole dynamics, whereas $s_x \geq \epsilon_x$ is to determine whether the hitting force is necessary (because the cart can naturally move back to the neighborhood of the origin, if $|s_x| < \epsilon_x$). Figure 3.5 shows the simulation results for the overall fuzzy controller after the task runs 18 times and Figure 3.5b shows the response of the position of the cart with satisfactory performance. In fact, from Fig. 3.5c, which shows the response of the angular displacement for the controlled system, we find it similar to Fig. 3.4c, the response of the control input u_θ. This fact illustrates that u_x is similar to an optimal strategy for a well trained human operator.

Remark 3.1. Simulation results showed that the proposed controller not only has the fast convergence in learning algorithm but also has satisfactory performance after sound training of the controller. When being compared with other researches in literature [12, 1], the proposed learning algorithm only needs to take a few tens of times ($16 + 18 = 34$) to complete the process of learning a designated controller and to achieve appealing system performance.

3.2 Approximation-based Fuzzy Control

While the fuzzy control is regarded as some approximator (2.2), many conventional control schemes can in fact be combined into the hybrid fuzzy control approach to enhance the approximating capability, such as approximate some a *priori* unknown function or even to approximate a designated conventional robust controllers. These hybrid control approaches can be listed as follows:

- Fuzzy model analysis [20]: a model which represents system dynamics in terms of a collection of linear systems with embedding of fuzzy if-then rules. From (2.2), the stability of the overall system can be analyzed by an LMI theorem.

- Adaptive fuzzy control [21]: a fuzzy control for which the rule parameter θ is regarded as unknown constant and ν is regarded as the regressor function so that adaptive control can be applied to update θ on-line. Based on Lyapunov theory, the stability of the controlled system can be proven.
- Fuzzy variable structure control [6, 9]: a fuzzy control which can behave as a variable structure control by providing that the parameter θ and the membership functions.
- Fuzzy H^{∞} control [3]: the fuzzy control which can achieve H^{∞} performance with a prescribed disturbance attenuation level via adaptive update of θ.

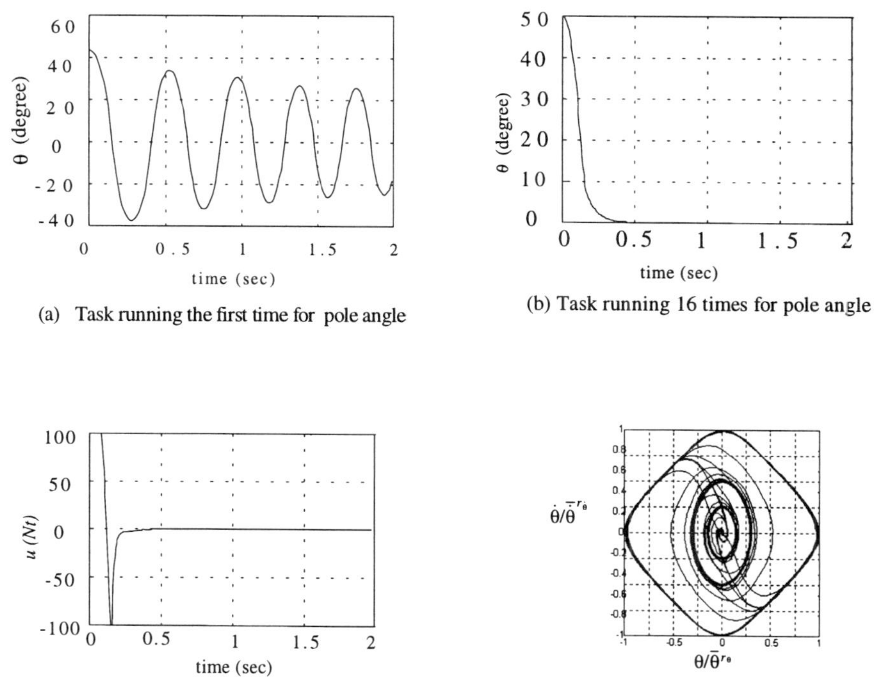

(a) Task running the first time for pole angle

(b) Task running 16 times for pole angle

(c) Task running 16 times for control input

(d) Learning process in the phase plane

Fig. 3.4. Learning procedure of LLBFC for pole controller

A trend in demonstrating the above fuzzy controllers to be persuasive is to apply them in practice. A very popular demonstration example is a robot manipulator, which will be used to investigate the fuzzy control below.

3.2.1 Control problems of robot manipulators. Consider an n degree-of-freedom articulated robot manipulator equipped with a cutting tool performing contour-following motion in order to remove burrs from a part, as depicted in Fig. 3.6. Its dynamic model in joint coordinates can be derived as follows:

$$M(q)\ddot{q} + C(q,\dot{q})\dot{q} + G(q) + D(\dot{q}) = \tau + \tau_f, \tag{3.3}$$

where $q \in \Re^n$ is the joint vector, $M(q) \in \Re^{n \times n}$ is the inertia matrix, $C(q,\dot{q})\dot{q}$ is the vector representing the centrifugal and Coriolis forces satisfying $\dot{M} - 2C$ is a skew-symmetric matrix, $G(q)$ is the vector of gravitational forces, $D(\dot{q})$ is the vector of friction forces, τ_f is the vector of external contact forces and moments, and τ is the vector of control input forces and moments.

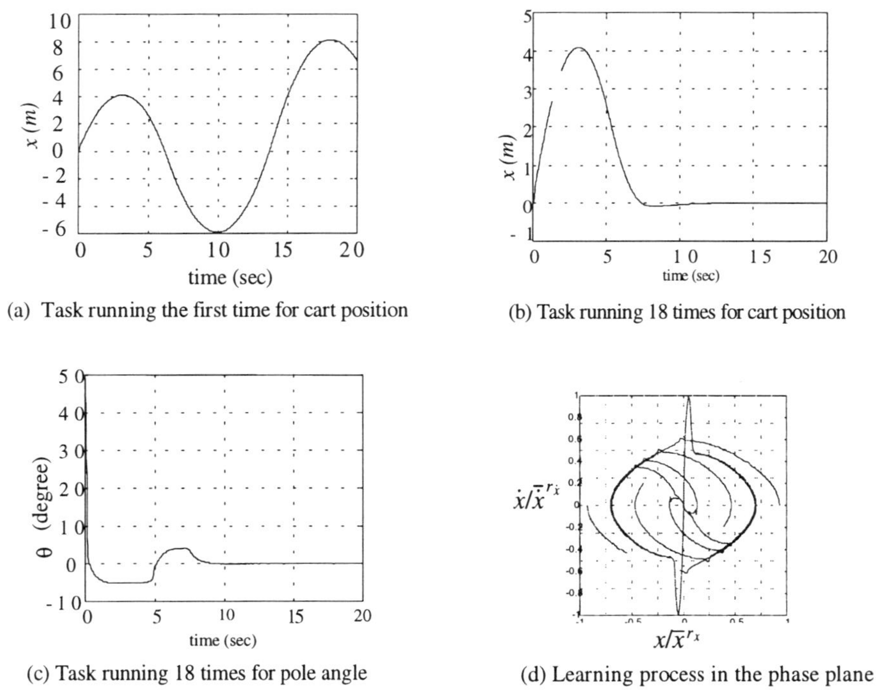

(a) Task running the first time for cart position

(b) Task running 18 times for cart position

(c) Task running 18 times for pole angle

(d) Learning process in the phase plane

Fig. 3.5. Learning procedure of LLBFC for cart controller

Fig. 3.6. Robot manipulator performing control task

To ease the controller design for the task, we re-express the dynamic model in the Cartesian coordinates. First, we assume that the Cartesian coordinate of the cutting tool, namely, x, is with respect to the world frame $\{W\}$, so that x can be represented as a function of its joint coordinates, q in the reference frame, i.e.

$$x = H(q), \tag{3.4}$$

where $x = [x_1, \cdots, x_6]^T = [x_p^T, x_o^T]^T$, with $x_p \in \Re^3$ being its position vector of cutting tool and $x_o \in \Re^3$ being the orientation vector, and $q = [q_1, \cdots, q_n]^T$. Differentiating Eq. (3.4), we then get

$$\dot{x} = \frac{\partial H(q)}{\partial q}\dot{q} = J(q)\dot{q}, \tag{3.5}$$

where $J(q) \in \Re^{6\times n}$ is a Jacobian transform matrix and is assumed to be of full rank for q lying in a compact set in the joint space, so that there exists a one-to-one mapping between x and q in a properly defined compact set. Thus, J has a pseudo-inverse matrix J^+, satisfying $JJ^+ = I$. Then, letting $M_x = J^{+T}MJ^+$, $C_x = J^{+T}CJ^+ - J^{+T}MJ^+\dot{J}J^+$, $G_x = J^{+T}G$, $D_x = J^{+T}D$, $f_\tau = J^{+T}\tau$ and $f = J^{+T}\tau_f$, we can derive the dynamics of the robot manipulator in the world frame as follows:

$$M_x(x)\ddot{x} + C_x(x,\dot{x})\dot{x} + G_x(x) + D_x(\dot{x}) = f_\tau + f. \tag{3.6}$$

Here, the torque vector τ in joint coordinates can be derived as $\tau = J^T f_\tau$.

Apparently, the robot manipulator has to achieve some suitable dynamics to perform the task with desired contour motion and force, while contacting the parts. The control problems are the uncertainties in robot dynamics and unknown contact environment. The proposed adaptive fuzzy variable structure control can efficiently solve the above problems, as positioning tracking control [6], hybrid force/position control [7], contour following with unknown objects [8], and deburring robot [9].

3.2.2 Adaptive fuzzy variable structure control. Assume the system (3.6) with the uncertainties which can be bounded by the function vector $g(s) = [g_1(s), \cdots, g_n(s)]^T$, where $s = [s_1, \cdots, s_n]^T$ is the sliding mode vector. Here, our goal is to design a fuzzy controller $u_f = [u_{f_1}, \cdots, u_{f_n}]^T$ which can compensate for the uncertainties. Then, consider a fuzzy controller with control input u_f, consisting of n ($n = 6$) multi-input single-output (MISO) fuzzy controllers, which are respectively characterized by

$$u_{f_i} \triangleq u_{f_i}(s) : \Omega_1 \times \cdots \times \Omega_n \to \Re$$

where u_{f_i} is the i-th fuzzy controller, $s = [s_1, \cdots, s_n]^T$ is the input fuzzy vector, $\Omega_1 \equiv [-\Upsilon\Delta_1, \Upsilon\Delta_1]$, $\cdots$, $\Omega_n \equiv [-\Upsilon\Delta_n, \Upsilon\Delta_n]$ with Υ being a positive integer and can be set arbitrarily large to constitute enlarge enough compact sets, and Δ_i being some positive real numbers. Here, each of the membership functions is given as an m-th order multiple dimension central B-spline function (as depicted in Fig. 3.7), of which the j-th dimension is defined as follows:

$$N_{m_j}(s_j) = \sum_{k=0}^{m+1} \frac{(-1)^k}{m!} \binom{m+1}{k} \left[(s_j + (\frac{m+1}{2} - k)\Delta_j)_+\right]^m \tag{3.7}$$

where we use the notation

$$x_+ := \max(0, x) \tag{3.8}$$

The m-th order B-spline type of membership function has the following properties:

- an $(m-1)$-th order continuously differentiable function, i.e. $N_{m_j}(s_j) \in C^{m-1}$;
- local compact support, i.e. $N_{m_j}(s_j) \neq 0$ only for $s_j \in [-\frac{m+1}{2}\Delta_j, \frac{m+1}{2}\Delta_j]$
- $N_{m_j}(s_j) > 0$ for $s_j \in (-\frac{m+1}{2}\Delta_j, \frac{m+1}{2}\Delta_j)$
- symmetric with respect to the center point (zero point)
- $\sum_{i_1=-\infty}^{\infty} \cdots \sum_{i_j=-\infty}^{\infty} N_{m1}(s_1 - i_1\Delta_1) \cdots N_{m_j}(s_j - i_j\Delta_j) = 1$, for $j \in \mathcal{Z}^+$

Based on the definition of the local compact support, the above property can be rewritten as

$$\sum_{i_1 \in I_{c1}(s_1)} \cdots \sum_{i_j \in I_{cj}(s_j)} N_{m1}(s_1 - i_1\Delta_1) \cdots N_{m_j}(s_j - i_j\Delta_j) = 1, \text{ for } j \in \mathcal{Z}^+ \tag{3.9}$$

where $I_{cj}(s_j)$ is an integer set, defined as follows:

$$I_{cj}(s_j) \equiv \{i : \frac{s_j}{\Delta_j} - \frac{m+1}{2} < i < \frac{s_j}{\Delta_j} + \frac{m+1}{2}, i \in \mathcal{Z}, j \in \mathcal{Z}^+\} \tag{3.10}$$

Then the membership functions for the j-th fuzzy variable s_j are defined as follows:

$$\mu_{j_i}(s_j) = N_{mj}(s_j - i\Delta_j) \tag{3.11}$$

whose compact support is given as:

$$\Omega_{j_i} = [(i - \frac{m+1}{2})\Delta_j, \ (i + \frac{m+1}{2})\Delta_j], \tag{3.12}$$

for $j = 1, \cdots, n$, and $i = -\Upsilon, \cdots, 0, \cdots, \Upsilon$, which means that $s_j \in \text{int}(\Omega_{j_i})$ implies that $\mu_{j_i}(s_j) > 0$ and $\Omega_j \equiv \cup_{i \in \{-\Upsilon, \cdots, \Upsilon\}} \Omega_{s_{j_i}}$.

Apparently, it is possible that $\Omega_{j_i} \cap \Omega_{j_k} \neq \emptyset$, for some $i \neq k$, i.e. s_j can simultaneously fall into several compact supports. The indices labeling those supports are equivalent to those in the definition of (3.10) and hence can be rewritten as:

$$\begin{aligned} I_{cj}(s_j) &\equiv \{i : s_j \in \text{int}(\Omega_{j_i}), i \in \mathcal{Z}, -\Upsilon \leq i \leq \Upsilon\} \\ &\equiv \{i : \Omega_{j_i} \subset \Omega_{c_j}(s_j)\} \end{aligned} \tag{3.13}$$

where $\Omega_{c_j}(s_j)$ is the union set of those compact supports, defined as follows:

$$\Omega_{c_j}(s_j) \equiv \cup_{i \in I_{cj}(s_j)} \Omega_{j_i} \tag{3.14}$$

which means that $i \in I_{cj}(s_j)$ is equivalent to $s_j \in \Omega_{c_j}(s_j)$.

From (2.2), we can represent the above fuzzy controllers as follows:

$$\begin{aligned} u_f &= \frac{\sum_{i_1=-\Upsilon}^{\Upsilon} \cdots \sum_{i_n=-\Upsilon}^{\Upsilon} \mu_{1i_1}(s_1) \cdots \mu_{ni_n}(s_n) \theta_{i_1 i_2 \cdots i_n}}{\sum_{i_1=-\Upsilon}^{\Upsilon} \cdots \sum_{i_n=-\Upsilon}^{\Upsilon} \mu_{1i_1}(s_1) \cdots \mu_{ni_n}(s_n)} \\ &= \sum_{i_1=-\Upsilon}^{\Upsilon} \cdots \sum_{i_n=-\Upsilon}^{\Upsilon} \nu_{i_1 \cdots i_n}(s_1, \cdots, s_n) \theta_{i_1 \cdots i_n} \\ &= \sum_{i_1 \in I_{c1}(s_1)} \cdots \sum_{i_n \in I_{cn}(s_n)} \nu_{i_1 \cdots i_n}(s_1, \cdots, s_n) \theta_{i_1 \cdots i_n} = \theta^T \nu, \end{aligned} \tag{3.15}$$

where $i_1, \cdots, i_n$ are integer indices, $\nu_{i_1 \cdots i_n}(s_1, \cdots, s_n)$ is the fuzzy basis function, and $\theta_{i_1 \cdots i_n} \in \Re^n$ is the parameter vector.

Define a new vector s_Δ as follows:

$$s_{\Delta i} = \begin{cases} s_i, & \text{as } s_i < -\Delta_i \text{ or } s_i > \Delta_i; \\ 0, & \text{otherwise (i.e. } s_i \in [-\Delta_i, \Delta_i]); \end{cases} \tag{3.16}$$

so that

$$\dot{s}_\Delta = \dot{s} \text{ for } s_\Delta \neq 0.$$

Here, our goal is to design u_f to satisfy the following

$$u_{f_i}(s) = \begin{cases} k_{f_i}(s)\text{sgn}(s_i), & \text{if } s_i \notin [-\Delta_i, \Delta_i]; \\ k_{si}(s), & \text{otherwise}; \end{cases} \tag{3.17}$$

where $k_f(s) \geq g_i(s)$, $k_s(s) = [k_{s1}(s), \cdots, k_{sn}]^T$ is a smooth function vector to make u_f smooth, and$[-\Delta_i, \Delta_i]$ is regarded as a designated dead-zone range which can be arbitrarily set.

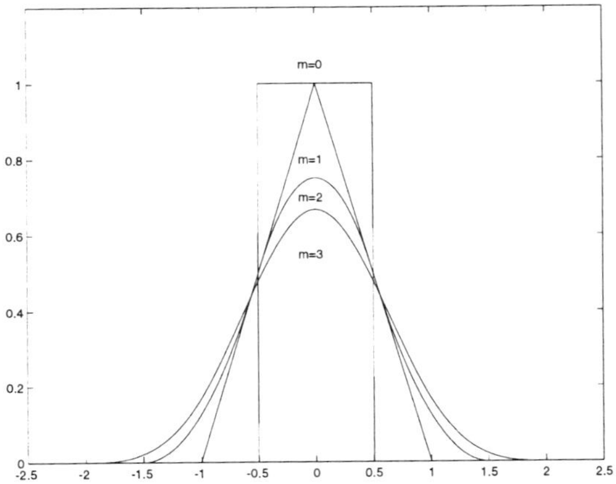

Fig. 3.7. The m-th order B-spline basis for m=0, 1, 2, and 3

Proposition 3.1. *If the fuzzy control law u_f is given as in (3.15), then there exist a class of the fuzzy controller which can satisfy the expression (3.17).*

Proof. To prove this, we have to assure that u_f satisfies two properties, namely, (a) $\mathrm{sgn}(u_{f_j}) = \mathrm{sgn}(s_j)$, and (b) $|u_{f_j}| \geq |g_j|$, when $s_j \in \Omega_j \backslash \Omega_{j0}$, for $j = 1, \cdots, n$, respectively, from variable structure control theory.
Proof of (a): From the definition I_{cj} in (3.10), when $s_j \in \Omega_j \backslash \Omega_{j0}$, it follows that

$$\begin{cases} I_{cj}(s_j) \subset \{-\Upsilon, \cdots, -1\}, & s_{\Delta j} < 0 \text{ (i.e. } s_j < -\frac{m+1}{2}\Delta_j) \\ I_{cj}(s_j) \subset \{1, \cdots, \Upsilon\}, & s_{\Delta j} > 0 \text{ (i.e. } s_j > \frac{m+1}{2}\Delta_j); \end{cases} \tag{3.18}$$

for $j = 1, \cdots, \rho$, respectively, where $\{-\Upsilon, \cdots, -1\}$ and $\{1, \cdots, \Upsilon\}$ are both integer sets. Then, the representation of the j-th fuzzy controller can be rewritten as follows:

$$u_{f_j} = \begin{cases} \sum_{i_1=-\Upsilon}^{\Upsilon} \cdots \sum_{i_{n-1}=-\Upsilon}^{\Upsilon} \sum_{i_n=-\Upsilon}^{-1} \nu_{i_1 \cdots i_n} \theta_{i_1 \cdots i_n} & \text{as } s_{\Delta j} < 0; \\ \sum_{i_1=-\Upsilon}^{\Upsilon} \cdots \sum_{i_{n-1}=-\Upsilon}^{\Upsilon} \sum_{i_n=1}^{\Upsilon} \nu_{i_1 \cdots i_n} \theta_{i_1 \cdots i_n j} & \text{as } s_{\Delta j} > 0; \end{cases}$$

Since $\nu_{i_1 \cdots i_n}$ is always positive, the sign of u_{f_j} can be determined by $\theta_{i_1 \cdots i_n j}$'s. Hence, we set $\theta_{i_1 \cdots i_n j} < 0$, for $-\Upsilon \leq i_k \leq \Upsilon$, $k \neq j$, $-\Upsilon \leq i_j \leq -1$ and $\theta_{i_1 \cdots i_j} > 0$, for $-\Upsilon \leq i_k \leq \Upsilon$, $k \neq j$, $1 \leq i_j \leq \Upsilon$

As a result, we can conclude that $\mathrm{sgn}(u_{f_j}) = \mathrm{sgn}(s_j)$, for $j = 1, \cdots, n$.

Proof of (b): Give the following definitions for $j = 1, \cdots, n$:

$$\begin{aligned} k_{j\max}(s) &= \max\{sup_{x \in \Omega_{c1}(s_1) \times \cdots \times \Omega_{cj}(s_j)} g_j(x), x \in \Re^n\}, \\ \theta_{j\min}(s) &= \min\{|\theta_{i_1 \cdots i_n j}|, i_1 \in I_{c1}(s_1), \cdots, i_n \in I_{cn}(s_n)\}, \end{aligned}$$

By setting $\theta_{j_{\min}} \geq k_{j_{\max}}$, we will obtain the following inequality:

$$|\theta_{i_1 \cdots i_n j}| \geq \theta_{j_{\min}} \geq k_{j_{\max}} \text{ for } i_1 \in I_{c1}(s_1), \cdots, i_n \in I_{cn}(s_n)$$

By virtue of the fact

$$\sum_{i_1=-\Upsilon}^{\Upsilon} \cdots \sum_{i_n=-\Upsilon}^{\Upsilon} \nu_{i_1 \cdots i_n} = \sum_{i_1 \in I_{c1}(s_1)} \cdots \sum_{i_n \in I_{cn}(s_n)} \nu_{i_1 \cdots i_n} = 1 \tag{3.19}$$

and in the cases where $s_{j_\Delta} \neq 0$, for $j = 1, \cdots, n$, we can derive the following result:

$$\begin{aligned} |u_{f_j}| &= \sum_{i_1 \in I_{c1}(s_1)} \cdots \sum_{i_n \in I_{cn}(s_n)} \nu_{i_1 \cdots i_n}(s) |\theta_{i_1 \cdots i_n j}| \\ &\geq \sum_{i_1 \in I_{c1}(s_1)} \cdots \sum_{i_n \in I_{cn}(s_n)} \nu_{i_1 \cdots i_n}(s) \theta_{j_{\min}} \\ &= \theta_{j_{\min}}(s) \geq k_{j_{\max}}(s) \\ &\geq |g_j|, \text{ as } s_{j_\Delta} \neq 0, \end{aligned}$$

for $j = 1, \cdots, n$. This completes our proof.

The following adaptive law to seek the proper θ to meet the condition of Proposition 3.1 will be necessary.

$$\dot{\theta} = r\nu(s) s_\Delta^T, \text{ for } s \in \Omega_1 \times \cdots \times \Omega_n \tag{3.20}$$

Remark 3.2. The fuzzy controller (3.15) with the adaptive law (3.20) takes the following advantages:

- Locally weighted fuzzy controller: Only rules supported by compact set Ω_{cj} required to be updated so that those rules will be locally weighted.
- Smooth fuzzy controller: Apparently, the fuzzy controller (3.15) can behave as a smoother controller if we can choose the membership functions to be smoother high order B-spline functions.

3.2.3 Example. A five degree-of-freedom (DOF) articulated robot arm is applied to perform a contour-following for an unknown elliptical cylinder object is located on the table, expressed as $\frac{{x_{c1}}^2}{50^2} + \frac{(x_{c2}-300)^2}{25^2} = 1$ and $0 \leq x_{c3} \leq 50$. The desired position trajectories are given as $x_{c1}(t) = 50\cos(0.5\pi t)$, $x_{c2}(t) = 300 + 50\sin(0.5\pi t)$ and $x_{c3}(t) = 25(1 - \cos(0.5\pi t))$. The desired force magnitude trajectory f_{dm} is given as $f_{dm} = 10 - 5\exp(-t)$ N with initial contact force 5 N being given [8].

At the beginning, since initial matrix parameters are set to zero in the first period of tracking motion, after the first period, adaptive variable structure control is initiated. The position error and force error are given in Fig. 3.8; we can find that the error is converging to zero.

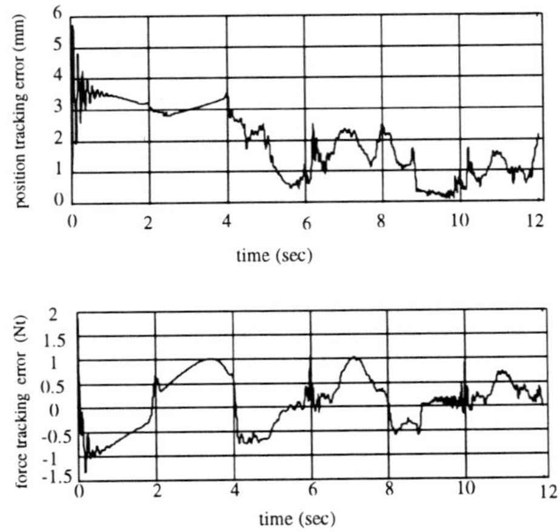

Fig. 3.8. Robot manipulator performing contour following for unknown object

4. Conclusion

In this chapter, we reviewed the most recent progress in the literature of fuzzy control and present some up-to-date research results. With all the advances of the theories, fuzzy control not only is a practically powerful control scheme but also becomes a theoretically complete control field. It is highly expected that such control will some day evolve as a generic control tool which can be so tangible to human mind.

References

[1] Barto A, Sutton R, Anderson C W 1983 Neuronlike adaptive elements that can solve difficult learning control problems. *IEEE Trans Syst Man Cyber.* 13:834–846

[2] Berenji H R, Khedkar P 1992 Learning and tuning fuzzy logic controllers through reinforcements. *IEEE Trans Neural Net.* 3:724–739

[3] Chen B-S, Lee C-H, Chang Y-C 1996 H^∞ tracking design of uncertain nonlinear SISO systems: Adaptive fuzzy approach. *IEEE Trans Fuzzy Syst.* 4:32-43

[4] Clouse J A, Utgoff P E 1992 A teaching method for reinforcement learning. In: *Proc Mach Learn Conf.*

[5] Han M-W, Kopacek P 1996 Neuro-fuzzy approach in service robotics. In: *Prepr 13th IFAC World Congr.* San Francisco, CA

[6] Hsu F-Y, Fu L-C 1994 Adaptive robust fuzzy control for robot manipulators. In: *Proc 1994 IEEE Int Conf Robot Automat.* San Diego, CA, pp 649–654

[7] Hsu F-Y, Fu L-C 1995 A new design of adaptive fuzzy hybrid force/position controller for robot manipulators. In: *Proc 1995 IEEE Int Conf Robot Automat.* Nagoya, Japan, pp 863–868

[8] Hsu F-Y, Fu L-C 1996 An adaptive fuzzy hybrid control for robot manipulators following contours of an uncertain object. In *Proc 1996 IEEE Int Conf Robot Automat.* Minneapolis, MN, pp 2232–2237

[9] Hsu F-Y, Fu L-C 1996 Intelligent robot deburring using adaptive fuzzy hybrid control. In: *Proc 27th Int Symp Industr Robot.* Milan, Italy, pp 847–852

[10] Hsu F-Y, Fu L-C 1997 Intelligent fuzzy controller with a sequential learning mechanism. *36th IEEE Conf Decision Contr.* San Diego, CA

[11] Juditsky A, Hjalmarsson H, Benveniste A, Delyon B, Ljung L, Sjoberg J, Zhang Q 1995 Nonlinear black-box models in system identification: Mathematical foundations. *Automatica.* 31:1725–1750

[12] Kim J, Zeigler B P 1996 Design fuzzy logic controllers using a multiresolution search paradigm. *IEEE Trans Fuzzy Syst.* 4:213–216

[13] Kwong W A, Passino K M 1996 Dynamic focused fuzzy learning control. *IEEE Trans Syst Man Cyber – Part B: Cyber.* 26:53–74

[14] Lee C C 1991 A self-learning rule-based controller employing approximate reasoning and neural net concepts. *Int J Intel Syst.* 71–93

[15] Lin C-J, Teng C 1996 Reinforcement learning for an ART-based fuzzy adaptive learning control network. *IEEE Trans Neural Net.* 7:1–23

[16] Lin C T, Lee C S G 1991 Neural-network-based fuzzy logic control and decision systems. *IEEE Trans Comp.* 40:1320–1326

[17] Ordonez R, Spooner J T, Passino K M 1996 Stable multi-input multi-output adaptive fuzzy control. In: *Proc 35th IEEE Conf Decision Contr.* Kobe, Japan, pp 610–615

[18] Patrikar A, Provence J 1993 Control of dynamic systems using fuzzy logic and neural networks. *Int J Intel Syst.* 727–748

[19] Sjoberg J, Zhang Q, Ljung L, Benveniste A, Delyon B, Glorennec P, Hjalmarsson H, Juditsky A 1995 Nonlinear black-box modeling in system identification: a unified overview. *Automatica.* 31:1691–1724

[20] Wang H O, Tanaka K, Griffin M F 1996 An approach to fuzzy control of nonlinear systems: Stability and design issues. *IEEE Trans Fuzzy Syst.* 4:14–23

[21] Wang L X 1994 *Adaptive Fuzzy Systems and Control: Design and Stability Analysis.* Prentice-Hall, Englewood Cliffs, NJ

[22] Whitley D, Dominic S, Das R, Anderson C W 1993 Genetic reinforcement learning for neurocontrol problems. *Mach Learn.* 259–284

[23] 1995 Fuzzy hardware. *IEEE Micro*

[24] 1994 Analog VLSI and neural network. *IEEE Micro*

Trajectory Control of Flexible Manipulators

Alessandro De Luca

Dipartimento di Informatica e Sistemistica, Università degli Studi di Roma "La Sapienza", Italy

We present some feedback control techniques recently developed for the exact solution of trajectory tracking problems for manipulators with flexible elements. Two classes are considered: *i)* robots with rigid links but with elastic transmissions, in which flexibility is concentrated at the joints, and *ii)* robots with lightweight and/or long arms, where flexibility is distributed along the links. For robots with elastic joints, we introduce a generalized inversion algorithm for the synthesis of a dynamic feedback control law that gives input-output decoupling and full state linearization. For robots with flexible links, the end-effector trajectory tracking problem is solved based on the iterative computation of the link deformations associated with the desired output motion, combined with a state trajectory regulator. For both robot models, the control design is performed directly on the second-order dynamic equations.

1. Introduction

Modeling robot manipulators as rigid mechanical systems is an idealization that becomes unrealistic when higher performance is requested. Tasks involving fast motion and/or hard contact with the environment are expected to induce deflections in the robot components, eventually exciting an oscillatory behavior. There are two sources of vibration in robot manipulators: *joint flexibility*, due to the elasticity of motion transmission elements such as harmonic drives, belts, or long shafts [26], and *link flexibility*, introduced by a long reach and slender/lightweight construction of the arm [6, 17]. In order to be able to counteract the negative effects of flexibility, advanced robot control systems should be designed on the basis of a more complete dynamic model of the robot (see, e.g. [27] and [5])

In robotic systems with flexible elements, output trajectories are typically defined beyond the structural flexibility, i.e. in terms of link motion for robots with elastic joints or at the level of manipulator tip for robots with link flexibility. We address in this chapter the stable and accurate reproduction of such trajectories using model-based state feedback control. Standard tools for solving trajectory tracking problems in nonlinear systems, such as feedback linearization, input-output decoupling, or inversion control (see, e.g. [19]), are not sufficient in these cases, so that the application of more advanced control techniques should be investigated.

In particular, the complete dynamic model of robots with elastic joints fails to satisfy the necessary conditions for input-output decoupling and/or full linearization by static state feedback [15], as opposed to the case of rigid robots for which these methods are equivalent to the well-known computed torque technique. Use of the larger class of *dynamic state feedback* controllers is helpful, because elastic joint robots are in fact input-output invertible systems without zero dynamics. However, the linearizing dynamic compensator has been derived so far only for very simple manipulators, while a general synthesis method is still missing. In Sect. 2. we will provide a constructive answer to this problem. Furthermore, we will be able to characterize in a precise way a tight upper bound for the dimension of the needed dynamic compensator.

On the other hand, the mapping between the joint torque input and the end-effector position output in robots with flexible links is associated with an unstable zero dynamics [14], the nonlinear equivalent of non-minimum phase zeros in a linear setting. The straightforward application of inversion-based control leads in this case to an unbounded increase of the internal arm deformation and, eventually, to control explosion. Different approaches have been presented in order to overcome this problem while exactly tracking the desired tip motion of multi-link manipulators: inversion in the frequency domain, iterative learning control, nonlinear regulation, or a combination of these. In all cases, the key feature is the computation of the *bounded link deformations* (and of the joint motions) producing the desired trajectory of the manipulator tip. Based on this idea, Sect. 3. presents in a unified framework three different but rather equivalent solutions to the end-effector trajectory tracking problem. We report also some illustrative experimental results obtained on a prototype two-link planar manipulator with flexible forearm, available in our Robotics Laboratory [10].

The chapter is organized so that its two main parts, devoted respectively to elastic joint and flexible link manipulators, are self-contained and independent.

2. Robots with Elastic Joints

We study the dynamic feedback linearization problem for robot arms with elastic joints. In particular, we consider a specific class of dynamic models which is, however, general enough so as to include many interesting instances, like robots moving on a plane. All robots within this class cannot be linearized nor input-output decoupled using only static state feedback. The design of the dynamic control law is presented in a constructive way, without resorting to state-space equations. The obtained result enables to solve the trajectory tracking problem in a global sense and with a prescribed linear error dynamics.

2.1 Dynamic Modeling

Consider an open kinematic chain of $N+1$ rigid bodies, interconnected by N joints undergoing elastic deformation. The robot is actuated by electrical drives which are assumed to be located at the joints. Let $q \in \mathbb{R}^N$ be the link positions, and $\theta \in \mathbb{R}^N$ be the motor (i.e. rotor) positions, as reflected through the gear ratios. With this choice, the difference $q_i - \theta_i$ is the ith joint deformation and the direct kinematics of the whole arm will be a function of the link variables q only. The following quite general assumptions are made:

Assumption 2.1. Joint deformations are small, so that elasticity in the joint is modeled as a linear spring.

Assumption 2.2. The rotors of the motors are modeled as uniform bodies having their center of mass on the rotation axis.

Assumption 2.2 implies that both the inertia matrix and the gravity term in the dynamic model will be independent from the position θ of the motors.

Following the Lagrangian approach, we compute the *kinetic energy* of the robot structure (including links and motors as rigid bodies) as

$$T = \frac{1}{2} \begin{bmatrix} \dot{q}^T & \dot{\theta}^T \end{bmatrix} \begin{bmatrix} B(q) & S(q) \\ S^T(q) & J \end{bmatrix} \begin{bmatrix} \dot{q} \\ \dot{\theta} \end{bmatrix}, \tag{2.1}$$

where all blocks of the inertia matrix are $N \times N$ matrices: $B(q)$ contains the inertial properties of the rigid links, $S(q)$ accounts for the inertial couplings between motors and links, while $J = \text{diag}\{J_1, \dots, J_n\}$, $J_i > 0$, is the matrix of the effective rotor inertias of the motors.

Consider the standard case in which the ith motor is mounted on link $i-1$ and moves link i. Since the kinetic energy of the ith motor does not depend on the motion of the ith link and of the subsequent ones, we have the following strong model property:

Property 2.1. Matrix $S(q)$ has the upper triangular structure

$$\begin{bmatrix} 0 & S_{12}(q_1) & S_{13}(q_1, q_2) & \cdots & S_{1N}(q_1, \dots, q_{N-1}) \\ 0 & 0 & S_{23}(q_2) & \cdots & S_{2N}(q_2, \dots, q_{N-1}) \\ \vdots & \vdots & \vdots & \ddots & \vdots \\ 0 & 0 & 0 & \cdots & S_{N-1,N}(q_{N-1}) \\ 0 & 0 & 0 & \cdots & 0 \end{bmatrix}, \tag{2.2}$$

where the most general cascade dependence is shown for each single term.

For the ease of presentation, we will focus on a particular situation:

Assumption 2.3. Matrix S in Eq. (2.1) is *constant*.

For instance, Assumption 2.3 is valid for planar robots with any number of rotational joints or for a spatial $3R$ elbow manipulator. In the former case, it can be shown that the expression of the elements of S is $S_{ij} = J_j$, apart from those entries that are structurally zero.

The *potential energy* is given by the sum of the gravitational energy, for both motors and links, and of the elastic energy stored at the joints. By virtue of Assumptions 2.1 and 2.2, we have

$$U = U_g(q) + \frac{1}{2}(q-\theta)^T K(q-\theta), \tag{2.3}$$

in which $K = \mathrm{diag}\{K_1, \ldots, K_n\}$, $K_i > 0$ being the elastic constant of joint i.

The robot dynamic model is obtained from the Euler-Lagrange equations for the Lagrangian $L = T - U$. Under the above assumptions, the $2N$ second-order differential equations have the form (see, e.g. [15] for a detailed derivation)

$$B(q)\ddot{q} + S\ddot{\theta} + c(q,\dot{q}) + g(q) + K(q-\theta) = 0 \tag{2.4}$$

$$S^T\ddot{q} + J\ddot{\theta} + K(\theta - q) = \tau, \tag{2.5}$$

where $c(q,\dot{q})$ are Coriolis and centrifugal terms, $g(q) = (\partial U_g/\partial q)^T$ are gravity terms, and $\tau \in I\!R^N$ are the torques supplied by the motors.

We note explicitly that in the case of a single link and for some other special kinematic structures with elastic joints (e.g. a $2R$ polar robot) it is found that $S = 0$, implying no inertial couplings between the link and the motor dynamics. The same situation is forced by a modeling assumption introduced in [25], namely by considering in the angular part of the kinetic energy of each rotor only the part due to its relative rotation. When $S = 0$, the model is always feedback linearizable by *static* state feedback. In the following, we will assume that at least one element in matrix S is different from zero. In this case, the control property of linearization by static feedback is always destroyed and we should look for a more general *dynamic* feedback controller in order to achieve full state linearization and input-output decoupling.

2.2 Generalized Inversion Algorithm

Let $q_d(t)$ be a desired smooth reference trajectory for the link variables q. A dynamic state feedback control for the input torques τ in Eq. (2.5) is a law of the form

$$\tau = \alpha(x,\xi) + \beta(x,\xi)v \tag{2.6}$$

$$\dot{\xi} = \gamma(x,\xi) + \delta(x,\xi)v \tag{2.7}$$

where $x = (q, \theta, \dot{q}, \dot{\theta}) \in I\!R^{4N}$ is the state of the robot, $\xi \in I\!R^M$ is the state of the dynamic compensator (of order M to be defined), and $v \in I\!R^N$ is the new control input. Our objective is to design such a control law so that the closed-loop system made by Eqs. (2.4)–(2.5) and (2.6)–(2.7) is fully represented by decoupled chains of input-output integrators, i.e.

$$\frac{d^{r_i} q_i}{dt^{r_i}} = v_i, \qquad i = 1, \ldots, N, \tag{2.8}$$

with the additional requirement that

$$\sum_{i=1}^{N} r_i = 4N + M, \tag{2.9}$$

where r_i the closed-loop relative degree of the output variable q_i.

This problem formulation asks for both input-output decoupling and full state linearization (in the proper coordinates) of the closed-loop system. Sufficient conditions for the existence of a solution to this problem exist [20] and general algorithms for constructing the required control law can be found in [19].

It has been shown in [9] that the general model of robots with elastic joints (and thus, in particular, Eqs. (2.4)–(2.5)) can be linearized and input-output decoupled via dynamic state feedback. However, the actual construction of the dynamic controller has been a difficult task until now and was performed only on a case-by-case basis. An example of such a controller for a $2R$ planar robot can be found in [7]. One difficulty in deriving a systematic method for the synthesis of the controller (2.6)–(2.7) for robots with elastic joints is due to the fact that all available algorithms are defined in terms of a state-space representation of the system. The transformation of Eqs. (2.4)–(2.5) into first-order state equations, though simple, hides the physical role of the following algorithmic steps and makes them computationally more complex. In addition, it has been found [8] that the dimension M of the dynamic controller and the degrees r_i of the obtained linear input-output relations (2.8) depend on the number of joints as well as on the kinematic structure of the robot.

With the above limitations in mind, we propose a new general algorithm that proceeds in an incremental way by solving a series of partial linearization and input-output decoupling problems, directly defined on the robot dynamic model (2.4)–(2.5).

2.2.1 Step 1: Input-output decoupling with respect to $\boldsymbol{\theta}$. From the structure of Eq. (2.5), we define the following control law for τ

$$\tau = Ju + S^T \ddot{q} + K(\theta - q) \tag{2.10}$$

where $u \in I\!\!R^N$ is the new control input. This control law imposes the dynamics

$$\begin{aligned} B(q)\ddot{q} + Su + c(q, \dot{q}) + g(q) + K(q - \theta) &= 0 & (2.11)\\ \ddot{\theta} &= u. & (2.12) \end{aligned}$$

The implementation of the control law (2.10) by state feedback requires the elimination of the link acceleration $\ddot{q}$. Solving for $\ddot{q}$ in Eq. (2.11) and substituting in Eq. (2.10) gives

$$\tau = \left[J - S^T B^{-1}(q) S\right] u - S^T B^{-1}(q) \left[c(q,\dot{q}) + g(q) + K(q-\theta)\right] + K(\theta - q). \tag{2.13}$$

Equation (2.12) shows that a linear and decoupled relation has been obtained between each input component u_i and each output θ_i $(i = 1, \ldots, N)$, by using a static state feedback law $\tau = \tau(q, \theta, \dot{q}, u)$. In the closed-loop system, we have $2N$ states (namely, q and $\dot{q}$) that are unobservable from the output θ.

2.2.2 Step 2: Input-output decoupling with respect to f. By defining a new output f as

$$f = B(q)\ddot{q} + c(q,\dot{q}) + g(q) + Kq, \tag{2.14}$$

Equation (2.4) can be rewritten as

$$f(q,\dot{q},\ddot{q}) + Su - K\theta = 0, \tag{2.15}$$

where Eq. (2.12) has been used. We note that output f has the dimension of a generalized force.

Differentiating twice Eq. (2.15), we obtain

$$\ddot{f}(q,\dot{q},\ddot{q}) + S\ddot{u} - Ku = 0. \tag{2.16}$$

By defining the following control law for u

$$u = K^{-1}\left[S\ddot{u} + w'\right], \tag{2.17}$$

where $w' \in I\!R^N$ is the new control input, we would simply get

$$\ddot{f}(q,\dot{q},\ddot{q}) = w', \tag{2.18}$$

i.e. a linear and decoupled relation between each input w'_i and each output f_i $(i = 1, \ldots, N)$.

Owing to the Property 2.1 of matrix S, the control law (2.17) inherits a hierarchical structure and is thus well defined, even if its implementation requires input differentiation. To avoid input differentiation, we proceed in a different way by adding on each input channel u_i a string of integrators. In particular, $2(i-1)$ integrators, with states ϕ_{ij}, are put on the ith channel $(i = 2, \ldots, N;\ j = 1, \ldots, 2(i-1))$:

$$
\begin{array}{llllll}
u_1 = \overline{w}_1 & & & & & \\
u_2 = \phi_{21}, & \dot{\phi}_{21} = \phi_{22}, & \dot{\phi}_{22} = \overline{w}_2 & & & \\
\quad\vdots & \quad\vdots & & \ddots & & \\
u_i = \phi_{i1}, & \dot{\phi}_{i1} = \phi_{i2}, & \ldots & & \dot{\phi}_{i,2(i-1)} = \overline{w}_i & \\
\quad\vdots & \quad\vdots & & & \ddots & \\
u_N = \phi_{N1}, & \dot{\phi}_{N1} = \phi_{N2}, & \ldots & & \ldots & \dot{\phi}_{N,2(N-1)} = \overline{w}_N,
\end{array}
\tag{2.19}
$$

where $\overline{w} \in I\!R^N$ is a temporary control input. The total number of added integrators is $N(N-1)$. Denote by ϕ the vector collecting the states of all these integrators.

Differentiating $2(i-1)$ times the ith scalar equation in (2.16) (or, equivalently, $2i$ times the ith equation in (2.15)), and keeping into account the dynamic extension (2.19), we obtain

$$
\begin{aligned}
\frac{d^{2i} f_i}{dt^{2i}} &= -\sum_{j=i+1}^{N} S_{ij} \frac{d^{2i} u_j}{dt^{2i}} + K_i \frac{d^{2(i-1)} u_i}{dt^{2(i-1)}} \\
&= -S_{i,i+1} \overline{w}_{i+1} - \sum_{j=i+2}^{N} S_{ij} \phi_{j,2i+1} + K_i \overline{w}_i,
\end{aligned}
\tag{2.20}
$$

for $i = 1, \ldots, N$. By defining recursively the following control law for $\overline{w}$

$$
\begin{aligned}
K_N \overline{w}_N &= w_N \\
K_{N-1} \overline{w}_{N-1} &= S_{N-1,N} \overline{w}_N + w_{N-1} \\
K_i \overline{w}_i &= S_{i,i+1} \overline{w}_{i+1} + \textstyle\sum_{j=i+2}^{N} S_{ij} \phi_{j,2i+1} + w_i \\
& \quad (i = N-2, N-3, \ldots, 1),
\end{aligned}
\tag{2.21}
$$

we obtain

$$
\frac{d^{2i} f_i}{dt^{2i}} = w_i, \qquad i = 1, \ldots, N. \tag{2.22}
$$

Equation (2.22) shows again a linear and decoupled relation between each input w_i and each output f_i ($i = 1, \ldots, N$), resulting now from the application of the linear dynamic compensator $u = u(\phi, w)$ obtained through Eqs. (2.19) and (2.21). Indeed, when combining this compensator with Eq. (2.13), a nonlinear dynamic state feedback $\tau = \tau(q, \theta, \dot{q}, \phi, w)$ is defined for the original robot torque input. Note also that the total number of states of the robot and of the compensator is $4N + N(N-1) = N(N+3)$ whereas, from Eqs. (2.22), the number of states on the input-output channels is $N(N+1)$. Therefore, in the closed-loop system we have still $2N$ states that are unobservable from the output f.

2.2.3 Step 3: Input-output decoupling with respect to q. As the last algorithmic step, we tackle the input-output decoupling and linearization problem for the original output q. The mapping from f to q, represented by Eq. (2.14), contains the main nonlinearities of the robot link dynamics. In order to cancel them in a well-defined way, we need to dynamically balance the input-output relations in Eqs. (2.22). In fact, differentiating $2(N-i)$ times the ith equation in (2.22) we get

$$\frac{d^{2(N-i)}}{dt^{2(N-i)}}\frac{d^{2i}f_i}{dt^{2i}} = \frac{d^{2N}}{dt^{2N}}\left(b_i^T(q)\ddot{q}+c_i(q,\dot{q})+g_i(q)+K_iq_i\right) = \frac{d^{2(N-i)}w_i}{dt^{2(N-i)}}, \quad (2.23)$$

for $i=1,\ldots,N$, where $b_i(q)$ the ith column of the link inertia matrix $B(q)$. To avoid differentiation of the input w, we add $2(N-i)$ integrators, with states ψ_{ij}, on the ith channel ($i=1,\ldots,N-1$; $j=1,\ldots,2(N-i)$):

$$\begin{array}{lllllll}
w_N &=& \overline{v}_N & & & & \\
w_{N-1} &=& \psi_{N-1,1}, & \dot{\psi}_{N-1,1} &=& \psi_{N-1,2}, & \dot{\psi}_{N-1,2} = \overline{v}_{N-1} \\
\vdots & & & \vdots & & & \ddots \\
w_i &=& \psi_{i1}, & \dot{\psi}_{i1} &=& \psi_{i2}, & \ldots \quad \dot{\psi}_{i,2(i-1)} = \overline{v}_i \\
\vdots & & & \vdots & & & \ddots \\
w_1 &=& \psi_{11}, & \dot{\psi}_{11} &=& \psi_{12}, & \ldots \quad \psi_{1,2(N-1)} = \overline{v}_1
\end{array} \quad (2.24)$$

where $\overline{v} \in I\!R^N$ is a temporary control input. The total number of integrators is again $N(N-1)$. Denote by ψ the vector collecting the states of all these integrators.

Resume the vector notation and rewrite Eqs. (2.23), using Eqs. (2.24), as

$$\frac{d^{2N}}{dt^{2N}}\left(B(q)\ddot{q}+c(q,\dot{q})+g(q)+Kq\right) = \overline{v}. \quad (2.25)$$

Performing differentiation term by term gives

$$B(q)q^{\{2(N+1)\}} + n(q,\dot{q},\ldots,q^{\{2N+1\}}) = \overline{v}, \quad (2.26)$$

where

$$n = \sum_{k=1}^{2N}\binom{2N}{k}B^{\{k\}}(q)q^{\{2(N+1)-k\}}+c^{\{2N\}}(q,\dot{q})+g^{\{2N\}}(q)+Kq^{\{2N\}}, \quad (2.27)$$

and we have used the compact notation $x^{\{i\}} = d^ix/dt^i$. Therefore, by defining the linearizing control law

$$\overline{v} = B(q)v + n(q,\dot{q},\ldots,q^{\{2N+1\}}), \quad (2.28)$$

we finally obtain

$$\frac{d^{2(N+1)}q_i}{dt^{2(N+1)}} = v_i, \qquad i = 1, \ldots, N. \tag{2.29}$$

Note that Eq. (2.28) can be seen a generalization of the computed torque method for rigid robots and is globally defined thanks to the positive definiteness of the link inertia matrix $B(q)$.

Input-output decoupling and linearization has been achieved by means of the nonlinear dynamic feedback $w = w(q, \dot{q}, \ldots, q^{\{2N+1\}}, \psi, v)$, obtained from Eqs. (2.24) and (2.28). Note that the dependence of this control law on $\ddot{q}$ and on higher derivatives can be eliminated recursively, in terms of the robot states $(q, \theta, \dot{q}, \dot{\theta})$ and of the compensator states (ϕ, ψ).

Define the total state of the dynamic compensator as $\xi = (\phi, \psi)$, which is of dimension $M = 2N(N-1)$. By combining Eqs. (2.13), (2.19), (2.21), (2.24), and (2.28) we obtain a nonlinear dynamic state feedback control law $\tau = \tau(x, \xi, v)$ with the structure (2.6)–(2.7). Furthermore, Eqs. (2.29) are in the form (2.8) with uniform relative degrees $r_i = 2(N+1)$, for all $i = 1, \ldots, N$. Condition (2.9) on the sum of the relative degrees, which guarantees full state linearization beside input-output decoupling, is fulfilled. In fact, the number of states on the input-output channels $(2N(N+1))$ equals the sum of the number of states of the robot $(4N)$ and of the compensator $(2N(N-1))$. Thus, we have no more unobservable states left in the closed-loop system, which is in turn completely described by the linear dynamics (2.29).

A number of final remarks are in order.

Remark 2.1. The stable tracking of output reference trajectories $q_{di}(t)$ ($i = 1, \ldots, N$) is realized by any standard control technique for linear single input-single output systems. Using, e.g. pole assignment, we design

$$v_i = q_{di}^{\{2(N+1)\}} + \sum_{j=0}^{2N+1} a_{ij} \left(q_{di}^{\{i\}} - q_i^{\{i\}} \right), \qquad i = 1, \ldots, N, \tag{2.30}$$

where the a_{ij}'s are coefficients of Hurwitz polynomials

$$s^{2(N+1)} + a_{i,2N+1}s^{2N+1} + \cdots + a_{i2}s^2 + a_{i1}s + a_{i0}, \qquad i = 1, \ldots, N, \tag{2.31}$$

having prescribed roots in the complex right-half plane. From Eqs. (2.30) it also follows that perfect tracking requires $2(N+1)$-times differentiable trajectories (degree of smoothness).

Remark 2.2. In the implementation of the above tracking controller based on linearization and input-output decoupling via dynamic feedback, the main computational effort is concentrated in the evaluation of the term (2.27), which in turn requires the explicit expressions of the linearizing coordinates $q^{\{i\}}$ ($i = 2, \ldots, 2N+1$) (see also Eqs. (2.30)). These computations are easily customized for a specific robot arm since all components of the control law are defined in terms of the available dynamic model elements. Moreover, we require to invert the link inertia matrix $B(q)$ only once. This inverse can be stored and then used repeatedly in computing the expressions of the linearizing coordinates.

Remark 2.3. We have implicitly assumed that all the strictly upper triangular elements of matrix S in Eq. (2.4) are different from zero. If some of these elements vanish, the dimension of the required dynamic compensator will decrease together with the lengths of the input-output integrators chains (2.8). These output relative degrees may also not be equal to each other. Therefore, the value $M = 2N(N-1)$ is in general only an upper bound to the dimension of the linearizing dynamic controller, in agreement with the results obtained in [8]. For the planar $2R$ robot with elastic joints considered in [7], we have $N = 2$ and a constant non-zero value ($S_{12} = J_2$) for the single nontrivial element in matrix S. The upper bound is then attained in this case: a dynamic compensator of order 4 leads to two chains of 6 input-output integrators.

Remark 2.4. It is a simple exercise to verify that, when $S = 0$, the three steps of the above algorithm build up the static feedback linearizing controller of [25]. In particular, the dynamic extensions in Eqs. (2.19) and (2.24) vanish.

3. Robots with Flexible Links

We consider the inverse dynamics problem for robot arms with flexible links, i.e. the computation of the input torque that allows exact tracking of a trajectory defined for the manipulator end-effector. We restrict ourselves to finite-dimensional dynamic models of flexible manipulators. A stable inversion controller is derived numerically, based on the computation of bounded link deformations and, from these, of the required feedforward torque associated with the desired tip motion. Three different algorithms are presented for this computation, all defined on the second-order robot dynamic equations. Stable trajectory tracking is then obtained by adding a (partial) state feedback, within a nonlinear regulation approach.

3.1 Dynamic Modeling

Consider an open kinematic chain structure, with a fixed base and N moving flexible links, interconnected by N (rigid) rotational joints. Each link deformation is distributed in nature and would be best described by an infinite-dimensional model, typically that of an Euler beam with proper boundary conditions at the two ends [23, 2]. However, for all but the most simple structures, it is impossible to solve for the exact time evolution of the arm deflection. Therefore, the use of approximated finite-dimensional models is preferred for multi-link flexible manipulators.

In the following, some simplifying assumptions are made:

Assumption 3.1. Link deformations are small, so that only linear elastic effects are present.

Assumption 3.2. For each link, flexibility is limited to the plane of nominal rigid motion, i.e. the plane normal to the preceding joint axis.

Assumption 3.2 implies that each link can only bend in one lateral direction, being stiff with respect to axial forces and to torsion. In view of this, the bending deformation $w_i(x_i, t)$ at a generic point $x_i \in [0, \ell_i]$ along the ith link of length ℓ_i is modeled, using separation in time and space, as

$$w_i(x_i, t) = \sum_{j=1}^{N_{ei}} \phi_{ij}(x_i)\delta_{ij}(t), \qquad i = 1, \ldots, N, \tag{3.1}$$

where the N_{ei} spatial components $\phi_{ij}(x_i)$ are *assumed modes* of deformation satisfying geometric and/or dynamic boundary conditions, while $\delta_{ij}(t)$ are the associated generalized coordinates.

Let $\theta \in I\!\!R^N$ be the vector of joint angular positions, and $\delta \in I\!\!R^{N_e}$ the vector of link deformations, where $N_e = \sum_{i=1}^N N_{ei}$. The arm kinematics and its kinetic and potential energy can be described in terms of θ, δ, and their first derivatives. The Euler-Lagrange equations provide the dynamic model of an N-link flexible manipulator in the form of $N + N_e$ second-order differential equations (see, e.g. [5] for details):

$$\begin{bmatrix} B_{\theta\theta}(\theta,\delta) & B_{\theta\delta}(\theta,\delta) \\ B_{\theta\delta}^T(\theta,\delta) & B_{\delta\delta}(\theta,\delta) \end{bmatrix} \begin{bmatrix} \ddot{\theta} \\ \ddot{\delta} \end{bmatrix} + \begin{bmatrix} c_\theta(\theta,\delta,\dot{\theta},\dot{\delta}) \\ c_\delta(\theta,\delta,\dot{\theta},\dot{\delta}) \end{bmatrix} + \begin{bmatrix} g_\theta(\theta,\delta) \\ g_\delta(\theta,\delta) \end{bmatrix} + \begin{bmatrix} 0 \\ D\dot{\delta}+K\delta \end{bmatrix} = \begin{bmatrix} \tau \\ 0 \end{bmatrix}. \tag{3.2}$$

The positive-definite symmetric inertia matrix B is partitioned in blocks according to the rigid and flexible components, c is the vector of Coriolis and centrifugal forces, g is the vector of gravitational forces, $K > 0$ and $D \geq 0$ are diagonal matrices, of dimensions $N_e \times N_e$, representing the arm modal stiffness and damping, while τ is the torque at the joints. Note that no input torque appears in the right-hand side of the last N_e equations (3.2), because link deformations in Eqs. (3.1) are described in the reference frames clamped at each link base.

Remark 3.1. The model structure (3.2) holds for any finite-dimensional approximation of distributed flexibility. However, specific choices for the assumed modes ϕ_{ij} may imply convenient simplifications in the block $B_{\delta\delta}$ of the inertia matrix. In particular, orthonormality of the modes of each link induces a decoupled structure for the diagonal inertia subblocks of $B_{\delta\delta}$, which in turn may collapse into a constant diagonal matrix.

Remark 3.2. A rather common approximation is to evaluate the total kinetic energy of the system in the undeformed configuration $\delta = 0$. This implies that the inertia matrix B, and thus also the Coriolis and centrifugal terms

c, are independent of δ, and that the velocity terms c_δ lose the quadratic dependence on $\dot{\delta}$. If, in addition, $B_{\delta\delta}$ is constant, also c_θ loses the quadratic dependence on $\dot{\delta}$, while each component of c_δ becomes a quadratic function of $\dot{\theta}$ only.

For the sake of simplicity, we will consider in the next section the following dynamic model for control design

$$\begin{bmatrix} B_{\theta\theta}(\theta) & B_{\theta\delta}(\theta) \\ B_{\theta\delta}^T(\theta) & B_{\delta\delta} \end{bmatrix} \begin{bmatrix} \ddot{\theta} \\ \ddot{\delta} \end{bmatrix} + \begin{bmatrix} c_\theta(\theta, \dot{\theta}, \dot{\delta}) \\ c_\delta(\theta, \dot{\theta}) \end{bmatrix} + \begin{bmatrix} 0 \\ D\dot{\delta} + K\delta \end{bmatrix} = \begin{bmatrix} \tau \\ 0 \end{bmatrix}, \tag{3.3}$$

where gravity effects are not included and the previous remarks have been taken into account. In particular, the dependence of c_θ on $\dot{\delta}$ is linear.

Since our objective is the tracking of an end-effector trajectory, we conveniently define as system output

$$y = \theta + \Phi_e \delta, \tag{3.4}$$

where the constant $N \times N_e$ matrix Φ_e is defined as

$$\Phi_e = \text{block diag}\left\{\phi_{i1}(\ell_i)/\ell_i, \ldots, \phi_{i,N_{ei}}(\ell_i)/\ell_i\right\}. \tag{3.5}$$

The output y_i is a linear approximation of the angle pointing from the ith link base to its end. According to Assumption 3.2, the direct kinematics of the flexible manipulator, i.e. the position and orientation of the arm tip, can be written only in terms of the components of y.

We finally point out that, in the presence of uniform mass distribution for each link, any dynamic model of the form (3.2) (or (3.3)) retains the same relevant control feature: the zero dynamics associated with output (3.4) is always unstable (see, however, [12]).

3.2 Stable Inversion Control

Inversion control is effective for tracking joint trajectories of a flexible link manipulator (i.e. with $y = \theta$ in place of Eq. (3.4)), because the zero dynamics is stable in this case [13]. The direct extension of an inversion control law to the tip output (3.4) leads to closed-loop instabilities, due to non-admissible feedback cancellation effects.

Frequency domain inversion has been proposed in [3, 4] as one of the first solutions to this instability problem. By working in the Fourier domain, this method defines the required open-loop control torque in one step (for linear models of one-link flexible arms) or in few iterations (in multi-link manipulators). *Learning control* has been applied in [24, 22] for iteratively building the input torque over repeated trials on the same desired output trajectory. In both approaches, the generated torque is noncausal: a non-zero input is applied in time *before* the actual start of the output trajectory. This preloading effect brings the flexible manipulator in the proper initial state that enables reproduction of the desired trajectory, while preserving an overall bounded link deformation. *Nonlinear regulation* has been used in [11, 21]; asymptotic output tracking is obtained by closing a stabilizing state feedback around the reference state trajectory. Finally, separation of stable and unstable zero dynamics and noncausal operation are the main features of the *stable inversion* approach proposed in [28, 16].

All the above methods share a common idea: in order to exactly reproduce an end-effector trajectory, the links of a flexible manipulator should experience a specific output-related bounded deformation history. Any attempt to control the arm deformation in a different way, e.g. trying to reduce as much as possible link deformation like in vibration damping control [14], destroys exact tracking and/or induces closed-loop instability.

Let $y_d(t)$ be a desired smooth reference trajectory for the tip, defined in a closed finite interval $[0, T]$. From Eq. (3.4), we can eliminate θ and $\dot{\theta}$ in the last N_e equations of (3.3), obtaining

$$B_{\delta\delta}\ddot{\delta} + D\dot{\delta} + K\delta + B_{\theta\delta}^T(y - \Phi_e\delta)\left(\ddot{y} - \Phi_e\ddot{\delta}\right) + c_\delta(y - \Phi_e\delta, \dot{y} - \Phi_e\dot{\delta}) = 0, \quad (3.6)$$

which is a dynamic constraint to be always satisfied by tip motion and link deformations. Plugging the desired evolution $y_d(t)$ in Eq. (3.6) gives a set of nonlinear differential equations for the only unknown function $\delta(t)$. Suppose that a bounded solution $\delta_d(t)$ can be found (together with its first and second time derivatives). We can use then the first N equations in (3.3) for defining the nominal input torque

$$\tau_d = B_{\theta\theta}(\theta_d)\ddot{\theta}_d + B_{\theta\delta}(\theta_d)\ddot{\delta}_d + c_\theta(\theta_d, \dot{\theta}_d, \dot{\delta}_d), \quad (3.7)$$

where

$$\theta_d(t) = y_d(t) - \Phi_e\delta_d(t) \quad (3.8)$$

is the required joint motion. As a result, the main bottleneck is the computation of a bounded solution $\delta_d(t)$ to Eq. (3.6) evaluated at $y = y_d(t)$. In the following, we present three alternative numerical methods.

3.2.1 Method 1: Approximate nonlinear regulation. In nonlinear output regulation [19], the control law is formed by two contributions: a feedforward term driving the system output along its desired evolution, and a state feedback term necessary to stabilize the closed-loop dynamics around the reference state trajectory. For a flexible link manipulator, the feedforward term is given by Eq. (3.7) while the desired link deformation $\delta_d(t)$ is part of the reference state trajectory to be computed.

The output reference trajectory should be generated by an exosystem with state denoted by Y_d. Each component of the reference state trajectory will be specified as a nonlinear function of the exosystem state Y_d. For a flexible manipulator it is then sufficient to determine $\delta_d = \pi(Y_d)$ and $\dot{\delta}_d = (\partial\pi/\partial Y_d)\dot{Y}_d$.

In particular, the vector function $\pi(Y_d)$ should satisfy Eq. (3.6), evaluated along the reference output evolution:

$$B_{\delta\delta}\ddot{\pi}(Y_d) + D\dot{\pi}(Y_d) + K\pi(Y_d) + B_{\theta\delta}^T(y_d - \Phi_e\pi(Y_d))\left(\ddot{y}_d - \Phi_e\ddot{\pi}(Y_d)\right)$$
$$+ c_\delta(y_d - \Phi_e\pi(Y_d), \dot{y}_d - \Phi_e\dot{\pi}(Y_d)) = 0. \qquad (3.9)$$

An approximate solution $\widehat{\pi}(Y_d)$ to Eq. (3.9) can be obtained by a numerical approach, as in many practical nonlinear regulation problems [18]. If in this approximation we use a class of basis elements that are bounded functions of their arguments, $\widehat{\pi}(Y_d)$ will necessarily be a bounded function over time as long as the trajectory $y_d(t)$ and its derivatives are bounded.

In [11], a polynomial function of Y_d was used as $\widehat{\pi}(Y_d)$. The nonlinear terms in Eq. (3.9) are expanded in Taylor series and the constant coefficients in $\widehat{\pi}(Y_d)$ are determined through the polynomial identity principle. Computational savings are obtained via a recursive procedure, solving Eq. (3.9) for increasing expansion orders until the final desired precision is obtained.

As noted in [16], this approach does not allow the use of noncausal inputs. Therefore, although the computed link deformation $\delta_d(t) = \widehat{\pi}(Y_d(t))$ may be different from zero at time $t = 0$, there is no way to preload the manipulator to such a value.

3.2.2 Method 2: Iterative inversion in the frequency domain. For linear non-minimum phase mechanical systems, a stable inversion algorithm in the frequency domain has been introduced in [3] by regarding both the input $\tau(t)$ and the output $y_d(t)$ as periodic functions. Provided that the involved signals are Fourier-transformable, all quantities will automatically be bounded over time.

An extended interval of definition is considered for the output reference trajectory, namely with $t \in [-\Delta, T + \Delta]$, where Δ gives enough time to preload and discharge the internal deformation in the flexible manipulator, without motion of its end-effector. Indeed, we have: $y_d(t) = y_d(0)$, for $t \in [-\Delta, 0]$, and $y_d(t) = y_d(T)$, for $t \in [T, T + \Delta]$.

When the system is nonlinear the inversion algorithm is applied repeatedly, using successive linear approximations of the whole flexible manipulator equations around the nominal trajectory [4].

The same idea can be used in a simpler fashion, namely iterating the linearization process on the flexible dynamics only. For, rewrite Eq. (3.6) as

$$B_{\delta\delta}\ddot{\delta} + D\dot{\delta} + K\delta + f(y, \dot{y}, \ddot{y}, \delta, \dot{\delta}, \ddot{\delta}) = 0, \qquad (3.10)$$

with

$$f = B_{\theta\delta}^T(y - \Phi_e\delta)\left(\ddot{y} - \Phi_e\ddot{\delta}\right) + c_\delta(y - \Phi_e\delta, \dot{y} - \Phi_e\dot{\delta}). \qquad (3.11)$$

We compute a bounded link deformation $\delta_d(t)$ associated with the end-effector motion $y_d(t)$ by the following algorithm:

1. Choose an initial $\delta^{(0)}(t)$, with first and second time derivatives, over the time interval $[-\Delta, T+\Delta]$. Typically, $\delta^{(0)}(t) \equiv 0$. Set $k=0$.
2. Using Eq. (3.11), define the forcing term

$$f^{(k)}(t) = f(y_d(t), \dot{y}_d(t), \ddot{y}_d(t), \delta^{(k)}(t), \dot{\delta}^{(k)}(t), \ddot{\delta}^{(k)}(t)), \tag{3.12}$$

   and solve

$$B_{\delta\delta}\ddot{\delta} + D\dot{\delta} + K\delta + f^{(k)}(t) = 0 \tag{3.13}$$

   using the FFT method as in [3]. Denote the solution as $\delta^{(k+1)}(t)$, defined for $t \in [-\Delta, T+\Delta]$.
3. If $\|\delta^{(k+1)}(t) - \delta^{(k)}(t)\| \leq \epsilon_\delta$ for all $t \in [-\Delta, T+\Delta]$, set $\delta_d(t) = \delta^{(k+1)}(t)$ and stop. Else, set $k = k+1$ and go to step 2.

3.2.3 Method 3: Iterative learning in the time domain. Robot learning control allows to acquire from experiments (or from simulations on an accurate dynamic model) the input torque needed for reproducing a desired output trajectory [1]. The trajectory is repeated several times and, at the end of each trial, the tracking error is used for updating the command to be applied at the next iteration. This method is well established for rigid robots, with simple PD-like updates of the input command.

In the presence of link flexibility, additional filtering of high-frequency signal components is needed to guarantee convergence. Since the tracking error processing is performed off-line, noncausal filtering is allowed (i.e. we can update the command at a given instant using also error samples at later instants of the previous trial), as well as anticipated shifting of signals in time. In this way, we can learn the input torque to be applied for $t \in [-\Delta, T+\Delta]$, even outside the interval of actual definition of the output trajectory [24].

A similar approach is proposed here for the numerical solution of Eq. (3.6). Again, we limit the learning process to the flexible dynamics. Instead of using the tracking error, define a *deformation torque error* as

$$e = B_{\delta\delta}\ddot{\delta} + D\dot{\delta} + K\delta + B_{\theta\delta}^T(y_d - \Phi_e\delta)\left(\ddot{y}_d - \Phi_e\ddot{\delta}\right) + c_\delta(y_d - \Phi_e\delta, \dot{y}_d - \Phi_e\dot{\delta}), \tag{3.14}$$

namely the left-hand side of Eq. (3.6), evaluated on the desired output trajectory y_d. Indeed, an admissible link deformation history $\delta(t)$ satisfies

$$e(\delta, \dot{\delta}, \ddot{\delta}, t) = 0, \qquad \forall t \in [-\Delta, T+\Delta]. \tag{3.15}$$

According to the iterative learning paradigm, we compute the link deformation $\delta_d(t)$ associated with the end-effector motion $y_d(t)$ by the following algorithm:

1. Choose an initial $\delta^{(0)}(t)$, with first and second time derivatives, over the time interval $[-\Delta, T+\Delta]$. Typically, $\delta^{(0)}(t) \equiv 0$. Set $k=0$.

2. Using Eq. (3.14), define

$$e^{(k)}(t) = e(\delta^{(k)}(t), \dot{\delta}^{(k)}(t), \ddot{\delta}^{(k)}(t), t). \tag{3.16}$$

If

$$\|e^{(k)}(t)\| \leq \epsilon_e, \qquad \forall t \in [-\Delta, T + \Delta], \tag{3.17}$$

set $\delta_d(t) = \delta^{(k)}(t)$ and stop. Else, process the error $e^{(k)}(t)$ by finite-impulse response (FIR) filters as in [24], obtaining a filtered version $e_f^{(k)}(t)$ and its derivative $\dot{e}_f^{(k)}(t)$.

3. Update by the following PD-like learning rule

$$\delta^{(k+1)}(t) = \delta^{(k)}(t) - K_{LP}\, e_f^{(k)}(t) - K_{LD}\, \dot{e}_f^{(k)}(t), \tag{3.18}$$

with sufficiently small learning gains $K_{LP} > 0$ and $K_{LD} > 0$. Set $k = k+1$ and go to step 2.

A number of final remarks are in order.

Remark 3.3. Although a complete convergence analysis is lacking for Methods 2 and 3, their success is supposed to depend on the strong underlying linear structure of equation (3.13) and definition (3.16), respectively. In particular, the dependence on $\delta(t)$ in the forcing term $f^{(k)}$ of Eq. (3.13), in view of Assumption 3.1, is a small perturbation affecting the reference output trajectory $y_d(t)$. Similarly, the nonlinear time-varying part in $e^{(k)}$ is mainly due to $y_d(t)$ and thus, being repetitive in nature, is well handled by the learning process. Moreover, the update (3.18) can be seen as a step of the gradient method for solving Eq. (3.15).

Remark 3.4. In all methods, the evaluation of $\delta_d(t)$ and $\dot{\delta}_d(t)$ at time $t = 0$, together with the use of Eq. (3.8) for the robot joint angles, provides the correct initial state producing a bounded evolution for the link deformation. If the flexible manipulator starts in this deformed state, use of Eq. (3.7) yields *exact* tracking of the end-effector trajectory.

Remark 3.5. If the initial state is not on its computed reference trajectory, a stabilizing term should be added in order to drive the state towards this solution, and only *asymptotic* output tracking can be guaranteed. This can be accomplished —at least locally— using a linear state feedback regulator, characterized by a matrix F,

$$\tau = \tau_d + F \begin{bmatrix} \theta_d - \theta \\ \dot{\theta}_d - \dot{\theta} \\ \delta_d - \delta \\ \dot{\delta}_d - \dot{\delta} \end{bmatrix}, \tag{3.19}$$

with τ_d given by Eq. (3.7). One can also use a simpler stabilizing matrix F in Eq. (3.19), as in the *partial state* feedback controller

$$\tau = \tau_d + F_P(\theta_d - \theta) + F_D(\dot{\theta}_d - \dot{\theta}), \tag{3.20}$$

with positive definite (diagonal) matrices F_P and F_D [14]. Note that control (3.19) (as well as (3.20)) can be also applied over the entire interval $[-\Delta, T + \Delta]$, yielding a more robust version of a noncausal trajectory regulator.

3.3 Experimental Results

We report here some experimental results obtained for the end-effector trajectory tracking of FLEXARM, a two-link planar manipulator with a flexible forearm and direct-drive DC motors available at the Robotics Laboratory of DIS. The dynamic model of the arm can be found in [10], where two modes are used for describing the forearm bending in the horizontal plane of motion. The essential data are as follows: the length of the first rigid link and of the flexible forearm are, respectively, $\ell_1 = 0.3$ m and $\ell_2 = 0.7$ m; the forearm weight is 1.8 kg and its first two eigenfrequencies are at 4.7 and 14.4 Hz.

Since the first link is rigid, the output is defined as (see Eq. (3.4))

$$y = \begin{bmatrix} \theta_1 \\ \theta_2 + \phi_{21}(\ell_2)\delta_{21}/\ell_2 + \phi_{22}(\ell_2)\delta_{22}/\ell_2 \end{bmatrix}. \tag{3.21}$$

The reference trajectory is a 7th-order polynomial with zero initial and final velocity, acceleration, and jerk for both scalar outputs. The first output (joint 1) moves 45°, while the second output (tip of flexible forearm) moves 90° in 2 s.

The control law is given by Eq. (3.20) with $F_{P1} = 100$, $F_{P2} = 130$, $F_{D1} = 6$, and $F_{D2} = 8$. Method 3 was used for the computation of link deformation, extending learning over 3 s ($\Delta = 0.5$ s). Convergence of the deformation torque error with tolerance $\epsilon_e = 10^{-6}$ Nm in Eq. (3.17) is reached within 30 iterations on the nominal model.

Figures 3.1 and 3.2 show the desired and actual trajectory for the two outputs, with maximum errors of about 0.8° and 1° respectively. The applied joint torques are given in Fig. 3.3 (off-set values are due to noise and to some residual gravity effects caused by imperfect balancing of the structure).

In Fig. 3.4, the computed velocity profiles of the two joints ($\dot{\theta}_{d1}$ and $\dot{\theta}_{d2}$ from Eq. (3.8)) and of the angular deformation at the tip ($\dot{w}_2(\ell_2)/\ell_2 = (\phi_{21}(\ell_2)\dot{\delta}_{d,21} + \phi_{22}(\ell_2)\dot{\delta}_{d,22})/\ell_2$) are compared with the actual ones. The differences come from the inaccuracy of the model used for control computations. Before $t = 0.5$ s and after $t = 2.5$ s, it is possible to appreciate the preloading and discharging effects.

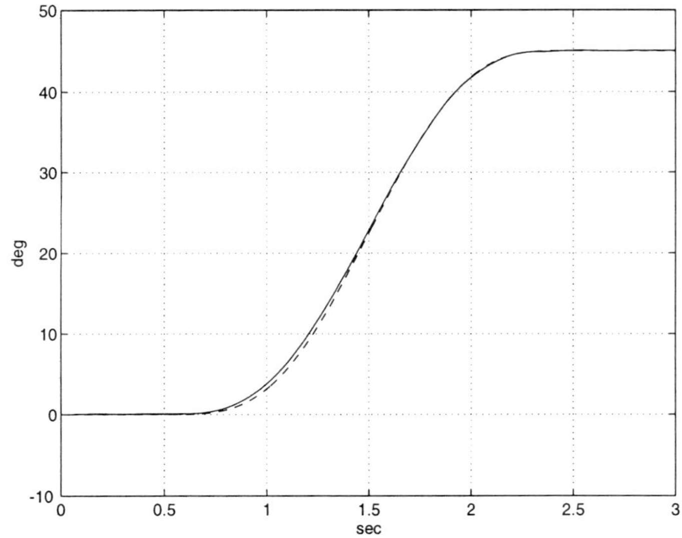

Fig. 3.1. Desired and actual trajectory for joint 1

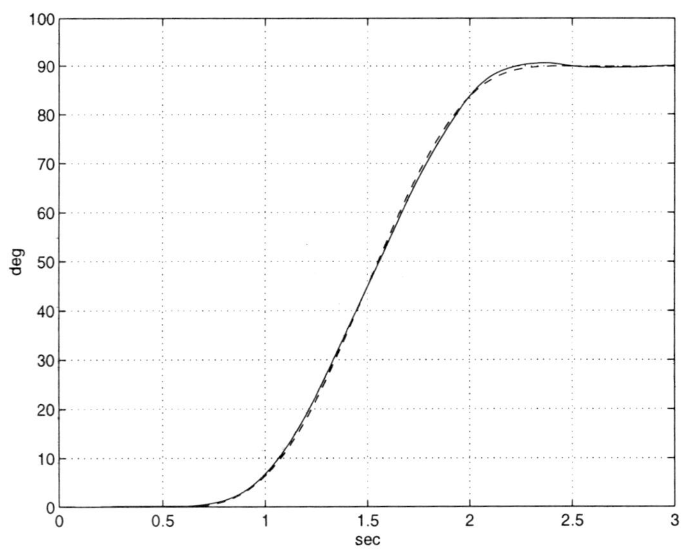

Fig. 3.2. Desired and actual trajectory for tip of link 2

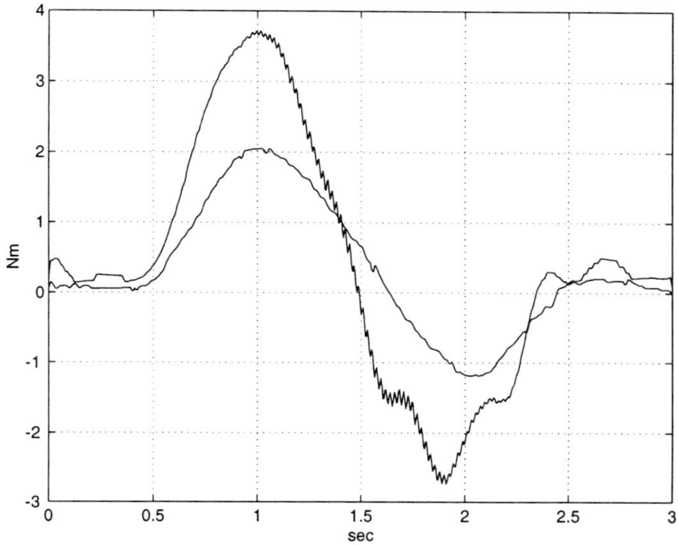

Fig. 3.3. Joint torques

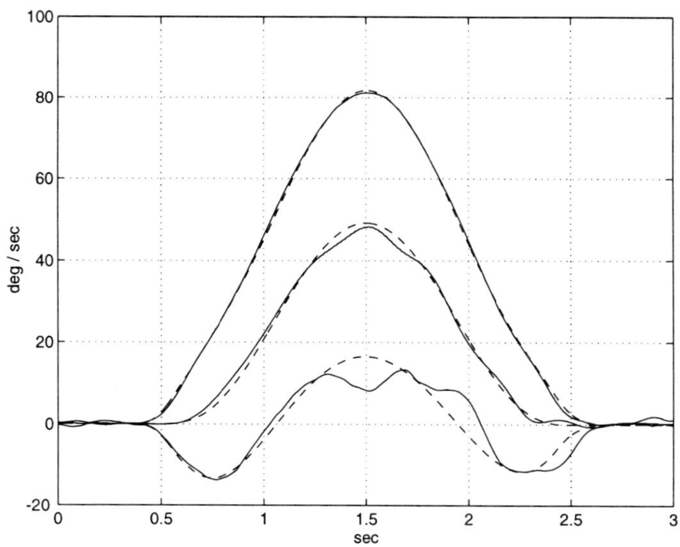

Fig. 3.4. Computed and actual velocity of joint 2 (top), joint 1 (center), and tip deformation (bottom)

4. Conclusions

The consideration of joint and link flexibility in robot manipulators gives rise to interesting theoretical issues when designing controllers for accurately tracking end-effector trajectories.

For a class of robots with elastic joints in which static state feedback fails to achieve exact linearization and input-output decoupling, we have introduced a new general algorithm for the synthesis of a dynamic feedback law reaching the same control goal.

In the case of robots with flexible links, an example of nonlinear systems with unstable zero dynamics, we have shown that some recently developed control techniques address essentially the same problem, namely the characterization of the bounded link deformation associated with a specified end-effector trajectory. For this critical computation three algorithms have been proposed.

Although we have assumed throughout the chapter ideal conditions, with perfect knowledge of the robot dynamic models and availability of full state measures, the presented results may be a starting point for the definition of adaptive and robust controllers, possibly using only output feedback.

For both types of flexible manipulators, only the second-order robot equations have been used in the analysis and in the control design. More physical insight is gained by working directly with the dynamic model terms, while control computations can be quite reduced, especially in the case of elastic joints. We regard this as a step toward the algorithmic design of advanced nonlinear controllers for general mechanical systems, without resorting to the state-space format.

Acknowledgement. The algorithm in Sect. 2.2 is the result of a joint work with Pasquale Lucibello. The experimental results on the two-link flexible arm in Sect. 3.3 were obtained by Stefano Panzieri. This work is supported by *MURST 40%* and *CNR 95.00106.CT07* funds.

References

[1] Arimoto S, Kawamura S, Miyazaki F 1984 Bettering operation of robots by learning. *J Robot Syst.* 1:123–140

[2] Barbieri E, Özgüner Ü 1988 Unconstrained and constrained mode expansions for a flexible slewing link. *ASME J Dyn Syst Meas Contr.* 110:416–421

[3] Bayo E 1987 A finite-element approach to control the end-point motion of a single-link flexible robot. *J Robot Syst.* 4:63–75

[4] Bayo E, Serna M A, Papadopoulus P, Stubbe J 1989 Inverse dynamics and kinematics of multi-link elastic robots: An iterative frequency domain approach. *Int J Robot Res.* 8(6):49–62

[5] Book W J 1984 Recursive Lagrangian dynamics of flexible manipulator arms. *Int J Robot Res.* 3(3):87–101

[6] Book W J 1990 Modeling, design, and control of flexible manipulator arms: A tutorial review. In: *Proc 29th IEEE Conf Decision Contr.* Honolulu, HI, pp 500–506

[7] De Luca A 1988 Dynamic control of robots with joint elasticity. In: *Proc 1988 IEEE Int Conf Robot Automat.* Philadelphia, PA, pp 152–158

[8] De Luca A 1988 Control properties of robot arms with joint elasticity. In: Byrnes C I, Martin C F, Saeks R E (eds) *Analysis and Control of Nonlinear Systems.* North-Holland, Amsterdam, The Netherlands, pp 61–70

[9] De Luca A, Lanari L 1995 Robots with elastic joints are linearizable via dynamic feedback. In: *Proc 34th IEEE Conf Decision Contr.* New Orleans, LA, pp 3895–3897

[10] De Luca A, Lanari L, Lucibello P, Panzieri S, Ulivi G 1990 Control experiments on a two-link robot with a flexible forearm. In: *Proc 29th IEEE Conf Decision Contr.* Honolulu, HI, pp 520–527

[11] De Luca A, Lanari L, Ulivi G 1991 End-effector trajectory tracking in flexible arms: Comparison of approaches based on regulation theory. In: Canudas de Wit C (ed) *Advanced Robot Control.* Springer-Verlag, Berlin, Germany, pp 190–206

[12] De Luca A, Lucibello P, Ulivi G 1989 Inversion techniques for trajectory control of flexible robot arms. *J Robot Syst.* 6:325–344

[13] De Luca A, Siciliano B 1993 Inversion-based nonlinear control of robot arms with flexible links. *AIAA J Guid Contr Dyn.* 16:1169–1176

[14] De Luca A, Siciliano B 1996 Flexible Links. In: Canudas de Wit C, Siciliano B, Bastin G (eds) *Theory of Robot Control.* Springer-Verlag, London, UK, pp 219–261

[15] De Luca A, Tomei P 1996 Elastic Joints. In: Canudas de Wit C, Siciliano B, Bastin G (eds) *Theory of Robot Control.* Springer-Verlag, London, UK, pp 179–217

[16] Devasia S, Chen D, Paden B 1996 Nonlinear inversion-based output tracking. *IEEE Trans Automat Contr.* 41:930–942

[17] Fraser A R, Daniel R W 1991 *Perturbation Techniques for Flexible Manipulators.* Kluwer, Boston, MA

[18] Huang J, Rugh W J 1992 An approximation method for the nonlinear servomechanism problem. *IEEE Trans Automat Contr.* 37:1395–1398

[19] Isidori A 1995 *Nonlinear Control Systems.* 3nd Edition, Springer-Verlag, London, UK

[20] Isidori A, Moog C H, De Luca A 1986 A sufficient condition for full linearization via dynamic state feedback. In: *Proc 25th IEEE Conf Decision Contr.* Athens, Greece, pp 203–208

[21] Lucibello P, Di Benedetto M D 1993 Output tracking for a nonlinear flexible arm. *ASME J Dyn Syst Meas Contr.* 115:78–85

[22] Lucibello P, Panzieri S 1996 End point trajectory control with internal stability of a flexible link by learning. In: *Proc 1996 IEEE Int Conf Robot Automat.* Minneapolis, MN, pp 2117–2123

[23] Meirovitch L 1967 *Analytical Methods in Vibrations.* Macmillan, New York

[24] Panzieri S, Ulivi G 1995 Disturbance rejection of iterative learning control applied to trajectory tracking for a flexible manipulator. In: *Proc 3rd Euro Contr Conf.* Roma, Italy, pp 2374–2379

[25] Spong M W 1987 Modeling and control of elastic joint robots. *ASME J Dyn Syst Meas Contr.* 109:310–319

[26] Sweet L M, Good M C 1985 Redefinition of the robot motion control problem. *IEEE Contr Syst Mag.* 5(3):18–24

[27] Tomei P 1991 A simple PD controller for robots with elastic joints. *IEEE Trans Automat Contr.* 36:1208–1213

[28] Zhao H, Chen D 1993 Exact and stable tip trajectory tracking for multi-link flexible manipulator. In: *Proc 32nd IEEE Conf Decision Contr.* San Antonio, TX, pp 1371–1376

Dynamics and Control of Bipedal Robots

Yildirim Hurmuzlu

Mechanical Engineering Department, Southern Methodist University, USA

A sound understanding of the dynamic principles governing legged locomotion is an essential requirement in developing high performance all terrain vehicles, designing highly mobile legged robots, and in the diagnosis and treatment of gait problems. Synthesis and analysis of bipedal locomotion is a complex task which requires knowledge of the dynamics of multi-link mechanisms, collision theory, control theory, and nonlinear dynamical systems theory. There is a rich body of literature that concerns the design and development of bipedal robots. Investigators in the field have conducted several numerical as well as experimental analyses to design robotic mechanisms capable of generating stable gait patterns. Desired gait patterns have been commonly generated by specially tailored objective functions or simply by imposing kinematic profiles of human gait patterns on the mechanisms under investigation. Subsequently, various control strategies have been developed to realize the desired patterns. Several investigators have approached the problem from a totally different perspective. It has been shown that unactuated, very simple bipedal mechanisms were able to walk on descending surfaces. This type of gait is called "passive walking" and has been receiving growing interest in the area. This line of research offers the potential of leading to simple machines that do not require the high power actuators that are requested by the active control schemes. Such an achievement will open new possibilities in the field and eliminate the two main obstacles: complexity and requirement of high power; that has impeded the development of autonomous walking robots. In this chapter we will review the main results in the development and analysis of bipedal robots. We will identify some of the main problems and seek to provide directions for future research.

1. How Does a Multi-link System Achieve Locomotion?

One may describe any bipedal robot as multi-link system that can ambulate on a given walking surface. We first must first understand the mechanisms that lead to progression of the system from one point to another before we design and develop bipedal locomotion robots. In this section we outline the main features of a locomotion system. We describe the main dynamic characteristic of such systems and correlate the cyclic motions of kinematic chains with the walking action.

1.1 Inverted Pendulum Models

Inverted pendulum models of various complexities have been extensively used in the modeling of gait of humans and bipedal walking machines. Although, the dynamics of bipedal locomotion is intuitively similar to that of an inverted pendulum, there is a fundamental question that has to be addressed in assessing the validity of the model. The question arises from the inherent instability of inverted pendulums in upright positions. It is well known that one can change the structural stability of inverted pendulum systems by applying torques at various joints. Consider a single degree of freedom model that consists of one mass and a torsional spring at the pivot point. Typical modes of motion include oscillations about two static equilibria and a third mode where the pendulum undergoes cyclic motions. These three modes are depicted in Fig. 1.1 and labeled as A, B and C respectively. If the ground surface is included, the third mode of behavior is ruled out since this leads to the collapse of the pendulum. When we add a second link to the simple system that we have considered and coordinate the motion of various members by applying appropriate joint moments the dynamic behavior remains similar to we have observed in the single member case [16]. In the presence of the walking surface, the system may still operate in either mode A or B, as long as all the parts of the system remain above the ground, Fig. 1.2a and Fig. 1.2b. When the system is in mode C then the swing limb will contact the ground and prompt a chain of events that may lead to stable progression. The importance of this contact event can be better understood if the motion is depicted in the phase space of the state variables.

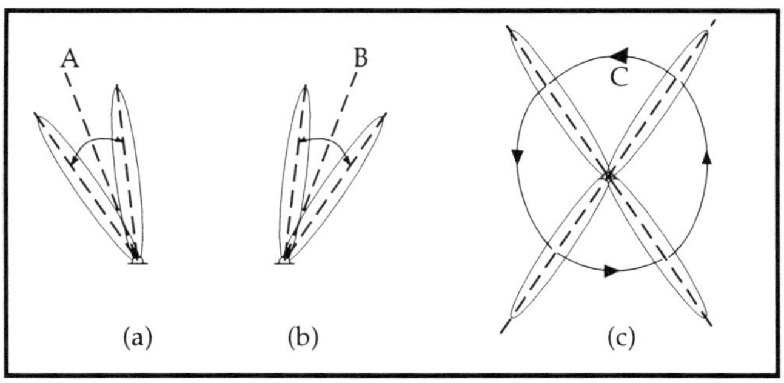

Fig. 1.1. Modes of oscillation of a simple pendulum

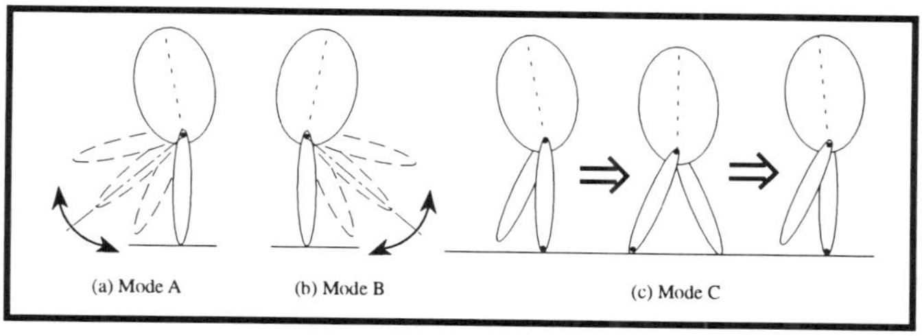

Fig. 1.2. Modes of oscillation of a three element biped

1.2 Impact and Switching

We simplify the present discussion by describing the events that lead to stable progression of a biped for a single degree of freedom system, however this approach can be generalized to higher order models. The phase plane portrait corresponding to the dynamic behavior that is described in the previous section is depicted Fig. 1.3a. The sample trajectories corresponding to each mode of behavior are labeled accordingly. The vertical dashed lines represent the values of the coordinate depicted in the phase plane for which the contact occurs. For the motions depicted in this figure, the only trajectory that leads to contact is C. The contact event for this simple model produces two simultaneous events:

1. Impact, which is represented by a sudden change in generalized velocities.
2. Switching due to the transfer of pivot to the point of contact.

At this instant, the role of the limbs are exchanged, the old stance limb becomes the new swing limb and the old swing limb becomes the new stance limb. This exchange is reflected by sudden changes in the values of generalized positions and velocities.

The combined effect of impact and switching on the phase plane portrait is depicted in Fig. 1.3b. As shown in the figure, the effect of the contact event will be a sudden transfer in the phase from point 1 to point 2, which is generally located on a different dynamic trajectory than the original one. If the destination of this transfer is on the original trajectory, then the resulting motion becomes periodic, Fig. 1.3c. This type of periodicity has unique advantage when the inverted pendulum system represents a biped. Actually, this is the only mode of behavior that this biped can achieve progression. The most striking aspect of this particular mode of behavior is that the biped achieves periodicity by utilizing only a portion of a dynamic trajectory. The impact and switching modes provide the connection between the cyclic motions of the kinematic chain and the walking action.

We can clearly observe from the preceding discussion that the motion of a biped involves continuous phases separated by abrupt changes resulting from

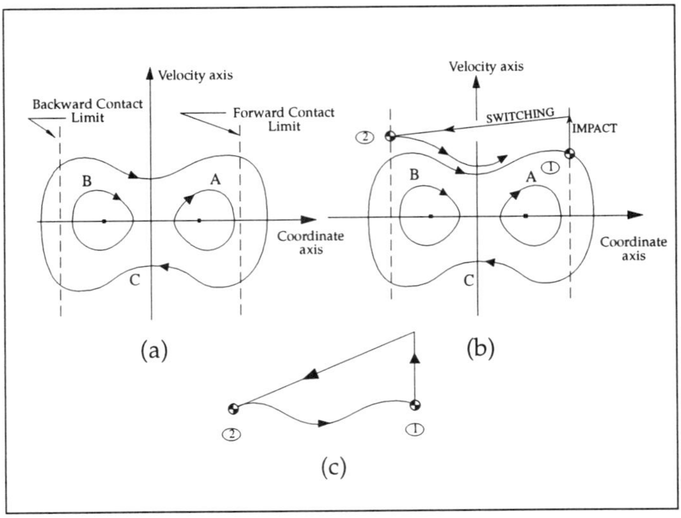

Fig. 1.3. Impact and switching on phase plane portrait

impact of the feet with the walking surface. During the continuous phase, we may have none, one, or two feet in simultaneous contact with the ground. In the case of one or more feet contacts, the biped is a dynamical system that is subject to unilateral constraints. When a foot impacts the ground surface, we face the impact problem of a multi-link chain with unilateral constraints. In fact, the overall motion of the biped may include a very complex sequence of continuous and discontinuous phases. This poses a very challenging control problem, with an added complication of continuously changing motion constrains and large velocity perturbations resulting from ground impacts.

2. Equations of Motion and Stability

We now outline the equations of motion and stability conditions for bipedal robots.

2.1 Equations of Motion During the Continuous Phase of Motion

Here, we consider a planar bipedal model that has n members connected to one another by purely rotational joints. There is an actuator located at each joint. We assume massless feet to simplify our presentation. Although we neglect the dynamics of the feet, we assume that the biped can apply torques at the ankles. The locomotion of the biped includes three modes of motion: single support phase, double support phase, and instances where both lower limbs are above the ground surface. Accordingly, the resulting motion is classified under two categories. If only the two former modes are present, the motion will be classified as walking. Otherwise, we have running or another form of non-locomotive action such as jumping or hopping.

Equations of motion during the continuous phase can be written in the following general form

$$\dot{\mathbf{x}} = \mathbf{f}(\mathbf{x}) + \mathbf{b}(\mathbf{x})\mathbf{u} \tag{2.1}$$

where $\mathbf{x}$ is the $n+2$ dimensional state vector, $\mathbf{f}$ is an $n+2$ dimensional vector field, $\mathbf{b}(\mathbf{x})$ is an $n+2$ dimensional vector function, and $\mathbf{u}$ is the n dimensional control vector. Equation (2.1) is subject to m constraints of the form:

$$\phi(\mathbf{x}) = \mathbf{0} \tag{2.2}$$

depending on the number of feet contacting the walking surface.

2.2 Impact and Switching Equations

During locomotion, when the swing limb (i.e. the limb that is not on the ground) contacts the ground surface (heel strike), the generalized velocities will be subject to jump discontinuities resulting from the impact event. Also, the roles of the swing and the stance limbs will be exchanged, resulting in additional discontinuities in the generalized coordinates and velocities [15]. The individual joint rotations and velocities do not actually change as the result of switching. Yet, from biped's point of view, there is a sudden exchange in the role of the swing and stance side members. This leads to a discontinuity in the mathematical model. The overall effect of the switching can be written as the follows:

$$\mathbf{x}^* = \mathbf{S_w}\, \mathbf{x} \tag{2.3}$$

where the superscripts $\mathbf{x}^*$ is the state immediately after switching and the matrix $\mathbf{S_w}$ is the switch matrix with entries equal to 0 or 1.

Using the principles of linear and angular impulse and momentum, we derive the impact equations containing the impulsive forces experienced by the system. However, applying these principles require some prior assumptions about the impulsive forces acting on the system during the instant of impact. Contact of the tip of the swing limb with the ground surface initiates the impact event. Therefore, the impulse in the y direction at the point of contact should be directed upward. Our solution is subject to the condition that the impact at the contact point is perfectly plastic (i.e. the tip of the swing limb does not leave the ground surface after impact). A second underlying assumption is that the impulsive moments at the joints are negligible. When contact takes place during the walking mode, the tip of the trailing limb is contacting the ground and has no initial velocity. This is always true when the motion is no-slip locomotion. The impact can lead to two possible outcomes in terms of the velocity of the tip of the trailing limb immediately after contact. If the subsequent velocity of the tip in the y direction is positive (zero), the tip will (will not) detach from the ground, and the case is called "single impact" ("double impact"). We identify the proper solution by checking a set of conditions that must be satisfied by the outcome of each case (see Fig. 2.1).

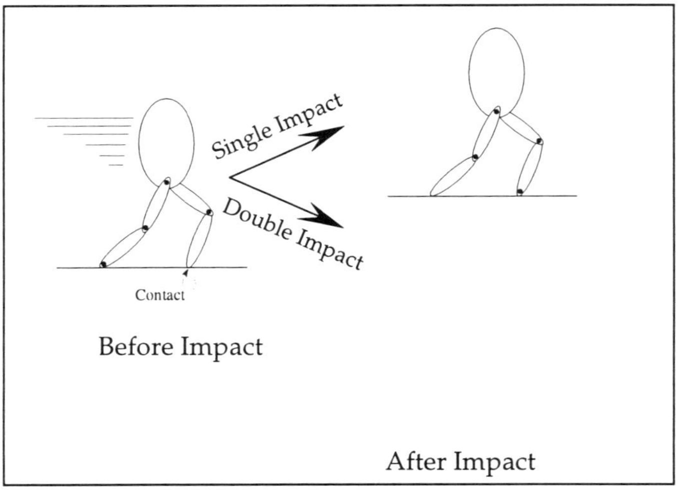

Fig. 2.1. Outcomes of the impact event

Solution of the impact equations (see [20] for details) yields:

$$\mathbf{x}^{+} = \mathbf{I_m}(\mathbf{x}^{-}) \tag{2.4}$$

where $\mathbf{x}^{-}$ and $\mathbf{x}^{+}$ are the state vector before and after impact respectively, and the matrix $\mathbf{I_m}(\mathbf{x}^{-})$ is the impact map.

2.3 Stability of the Locomotion

In this chapter, the approach to the stability analysis takes into account two generally excepted facts about bipedal locomotion. The motion is discontinuous because of the impact of the limbs with the walking surface [15, 18, 28]. The dynamics is highly nonlinear and linearization about vertical stance should be avoided [17, 27].

Given the two facts that have been cited above we propose to apply discrete mapping techniques to study the stability of bipedal locomotion. This approach has been applied previously to study of the dynamics of bouncing ball [8], to the study of vibration dampers [24, 25], and to bipedal systems [16]. The approach eliminates the discontinuity problems, allows the application of the analytical tools developed to study nonlinear dynamical systems, and brings a formal definition to the stability of bipedal locomotion.

The method is based on the construction of a first return map by considering the intersection of periodic orbits with an $k-1$ dimensional cross section in the k dimensional state space. There is one complication that will arise in the application of this method to bipedal locomotion. Namely, different set of kinematic constraints govern the dynamics of various modes of motion. Removal and addition of constraints in locomotion systems has been studied before [11]. They describe the problem as a two-point boundary value problem where such changes may lead to changes in the dimensions of the state space required to describe the dynamics. Due to the basic nature of discrete maps, the events that occur outside the cross section are ignored. The situation can be resolved by taking two alternative actions. In the first case a mapping can be constructed in the highest dimensional state space that represents all possible motions of the biped. When the biped exhibits a mode of motion which occurs in a lower dimensional subspace, extra dimensions will be automatically included in the invariant subspace. Yet, this approach will complicate the analysis and it may not be always possible to characterize the exact nature of the motion. An alternate approach will be to construct several maps that represent different types motion, and attach various conditions that reflect the particular type of motion. We will adopt the second approach in this chapter. For example, for no slip walking, without the double support phase, a mapping $\mathbf{P}_{nsls}$ is obtained as a relation between the state $\mathbf{x}$ immediately after the contact event of a locomotion step and a similar state ensuing the next contact. This map describes the behavior of the intersections of the phase trajectories with a Poincaré section Σ_{nsls} defined as

$$\Sigma_{nsls} = \{(\mathbf{x}, \dot{\mathbf{x}}) \in \Re^n \mid x_T(\mathbf{x}) > 0, y_T(\mathbf{x}) = 0 \mid \dot{y}_T^+(\mathbf{x}) > 0, \left|\frac{\hat{F}_{rx}}{\hat{F}_{ry}}\right| < \mu, \left|\frac{F_{rx}}{F_{ry}}\right| < \mu, F_{ry} > 0\}, \quad (2.5)$$

where x_T and y_T are the x and y coordinates of the tip of the swing limb respectively, μ is the coefficient of friction, and $\mathbf{F}$ and $\hat{\mathbf{F}}$ are ground reaction force and impulse respectively. The first two conditions in Eq. (2.5) establish the Poincaré section (the cross section is taken immediately after foot contact during forward walking), whereas the attached four conditions denote no double support phase, no slip impact, no slippage of pivot during the single support phase and no detachment of pivot during the single support phase respectively. For example, to construct a map representing no slip running, the last condition will be removed to allow pivot detachments as they normally occur during running. We will not elaborate on all possible maps that may exist for bipedal locomotion, but we note that the approach can address a variety of possible motions by construction of maps with the appropriate set of attached conditions.

The discrete map obtained by following the procedure described above can be written in the following general form

$$\xi_i = \mathbf{P}(\xi_{i-1}) \tag{2.6}$$

where ξ is the n-1 dimensional state vector, and the subscripts denote the ith and $(i-1)$th return values respectively.

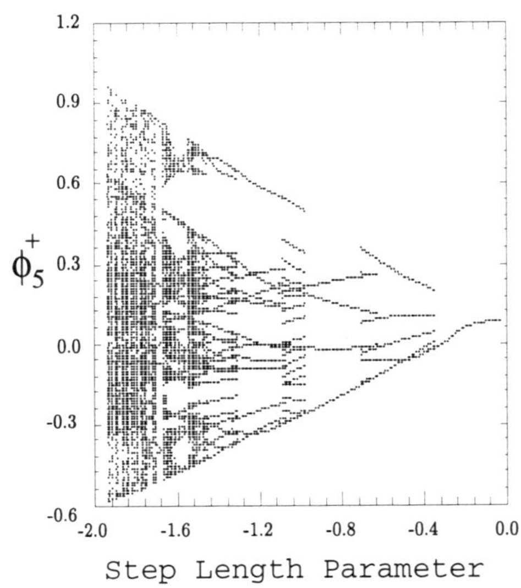

Fig. 2.2. Bifurcations of the single impact map of a five-element bipedal model

Periodic motions of the biped correspond to the fixed points of $\mathbf{P}$ where

$$\xi^* = \mathbf{P}^k(\xi^*). \tag{2.7}$$

where $\mathbf{P}^k$ is the kth iterate. The stability of $\mathbf{P}^k$ reflects the stability of the corresponding flow. The fixed point ξ^* is said to be stable when the eigenvalues ν_i, of the linearized map,

$$\delta\xi_i = \mathbf{DP}^k(\xi^*)\,\delta\xi_{i-1} \tag{2.8}$$

have moduli less that one.

This method has several advantages. First, the stability of gait now conforms with the formal stability definition accepted in nonlinear mechanics. The eigenvalues of the linearized map (Floquet multipliers) provide quantitative measures of the stability of bipedal gait. Finally, to apply the analysis to locomotion one only requires the kinematic data that represent all the relevant degrees of freedom. No specific knowledge of the internal structure of the system is needed.

The exact form of $\mathbf{P}$ cannot be obtained in closed form except for very special cases. For example, if the system under investigation is a numerical model of a man made machine, the equations of motion will be solved numerically to compute the fixed points of the map from kinematic data. Then stability of each fixed point will be investigated by computing the Jacobian using numerical techniques. This procedure was followed in [4]. We also note that this mapping may exhibit a complex set of bifurcations that may lead to periodic gaits with arbitrarily large number of cycles. For example, the planar, five-element biped considered in [12] leads to the bifurcation diagram depicted in Fig. 2.2 when the desired step length parameter is changed.

3. Control of Bipedal Robots

3.1 Active Control

Several key issues related to the control of bipedal robots remains unresolved. There is a rich body of work that addresses the control of bipedal locomotion systems. Furusho and Masubuchi [5] developed a reduced order model of a five element bipedal locomotion system. They linearized the equations of motion about vertical stance. Further reduction of the equations were performed by identifying the dominant poles of the linearized equations. A hierarchical control scheme based on local feedback loops that regulate the individual joint motions was developed. An experimental prototype was built to verify the proposed methods. Hemami et al. [11, 10] authored several addressing control strategies that stabilize various bipedal models about the vertical equilibrium. Lyapunov functions were used in the development of the control laws. The stability of the bipeds about operating points was guaranteed by constructing feedback strategies to regulate motions such as sway in the frontal plane. Lyapunov's method has been proved to be an effective tool in developing robust controllers to regulate such actions. Katoh and Mori [18] have considered a simplified five-element biped model. The model possesses three massive segments representing the upper body and the thighs. The lower segments are taken as telescopic elements without masses. The equations of motion were linearized about vertical equilibrium. Nonlinear feedback was used to assure asymptotic convergence to the stable limit cycle solutions of coupled van der Pol's equations. Vukobratovic et al. [27] developed a mathematical model to simulate bipedal locomotion. The model possesses massive lower limbs, foot structures, and upper-body segments such as head, hands etc.; the dynamics of the actuators were also included. A control scheme based on three stages of feedback is developed. The first stage of control guarantees the tracking in the absence of disturbances of a set of specified joint profiles, which are partially obtained from human gait data. A decentralized control scheme is used in the second stage to incorporate disturbances without considering the coupling effects among various joints. Finally, additional feedback loops are constructed to address the nonlinear coupling terms that are neglected in stage two. The approach preserves the nonlinear effects and the controller is robust to disturbances. Hurmuzlu [13] used five constraint relations that cast the motion of a planar, five-link biped in terms of four parameters. He analyzed the nonlinear dynamics and bifurcation patterns of a planar five-element model controlled by a computed torque algorithm. He demonstrated that tracking errors during the continuous phase of the motion may lead to extremely complex gait patterns. Chang and Hurmuzlu [4] developed a robust continuously sliding control scheme to regulate the locomotion of a planar, five element biped. Numerical simulation was performed to verify the ability of the controller to achieve steady gait by applying the proposed control scheme. Almost all the active control schemes often require very high torque actuation, severely limiting their practical utility in developing actual prototypes.

3.2 Passive Control

McGeer [21] introduced the so called passive approach. He demonstrated that simple, unactuated mechanisms can ambulate on downwardly inclined planes only with the action of gravity. His early results were used by recent investigators [6, 7] to analyze the nonlinear dynamics of simple models. They demonstrated that the very simple model can produce a rich set of gait patterns. These studies are particularly exciting, because they demonstrate that there is an inherent structural property in certain class of systems that naturally leads to locomotion. On the other hand, these types of systems cannot be expected to lead to actual robots, because they can only perform when the robot motion is assisted by gravitational action. These studies may, however, lead to the better design of active control schemes through effective coordination of the segments of the bipedal robots.

4. Open Problems and Challenges in the Control of Bipedal Robots

One way of looking at the control of bipedal robots is through the limit cycles that are formed by parts of dynamic trajectories and sudden phase transfers that result from impact and switching [15]. From this point of view, the biped may walk for a variety of schemes that are used to coordinate its segments. In essence, a dynamical trajectory that leads to the impact of the swing limb with the ground surface, will lead to a *locomotion step*. The question there remains is whether the coordination scheme can lead to a train of steps that can be characterized as gait. As a matter of fact, McGeer [21] has demonstrated that, for a biped that resembles the human body, only the action of gravity may lead to proper impacts and switches in order to produce steady locomotion. Active control schemes are generally based on trajectory tracking during the continuous phases of locomotion. For example, in [12, 4], the motion of biped during the continuous phase was specified in terms of five objective functions. These functions, however, were tailored only for the single support phase (i.e. only one limb contact with the ground). The controllers developed in these studies were guaranteed to track the prescribed trajectories during the continuous phases of motion. On the other hand, these controllers did not guarantee that the unilateral constraints that are valid for the single support phase would remain valid throughout the motion. If these constraint are violated, the control problem will be confounded by loss of controllability. While the biped is in the air, or it has two feet on the ground, the system is uncontrollable [2]. To overcome this difficulty, the investigators conducted numerical simulations to identify the parameter ranges that lead to single support gait patterns only. Stability of the resulting gait patterns were verified using the approach that was presented in Sect. 2.3. The open control problem is to develop a control strategy that *guarantees* gait stability throughout the locomotion. One of the main challenges in the field is to develop robust controllers that would also ensure the preservation of the unilateral constraints that were assumed to be valid during the system operation. Developing general feedback control laws and stability concept for hybrid mechanical systems, such as bipedal robots remains an open problem [2, 3].

A second challenge in developing controllers for bipeds is minimizing the required control effort in regulating the motion. Studying the passive (unactuated) systems is the first effort in this direction. This line of research is still in its infancy. There is still much room left for studies that will explore the development of active schemes that are based on lessons learned from the research of unactuated systems [6, 7].

Modeling of impacts of kinematic chains is yet another problem that is being actively pursued by many investigators [1, 20, 2]. Bipeds fall within a special class of kinematic chain problems where there are multiple contact points during the impact process [9, 14]. There has also been research efforts that challenge the very basic concepts that are used in solving impact problems with friction. Several definitions of the coefficient of restitution have been developed: kinematic [22], kinetic [23] and energetic [26]. In addition, algebraic [1] and differential [19] formulations are being used to obtain the equations to solve the impact problem. Various approaches may lead to significantly different results [20]. The final chapter on the solution of the impact problems of kinematic chains is yet to be written. Thus, modeling and control of bipedal machines would greatly benefit from future results obtained by the investigators in the field of collision research.

Finally, the challenges that face the researchers in the area of robotics are also present in the development of bipedal machines. Compact, high power actuators are essential in the development of bipedal machines. Electrical motors usually lack the power requirements dictated by bipeds of practical utility. Gear reduction solves this problem at an expense of loss of speed, agility, and the direct drive characteristic. Perhaps, pneumatic actuators should be tried as high power actuator alternatives. They may also provide the compliance that can be quite useful in absorbing the shock effect that are imposed on the system by repeated ground impacts. Yet, intelligent design schemes to power the pneumatic actuators in a mobile system seems to be quite a challenging task in itself. Future considerations should also include vision systems for terrain mapping and obstacle avoidance.

References

[1] Brach R M 1991 *Mechanical Impact Dynamics.* Wiley, New York

[2] Brogliato B 1996 *Nonsmooth Impact Mechanics; Models, Dynamics and Control.* Springer-Verlag, London, UK

[3] Brogliato B 1997 On the control of finite-dimensional mechanical systems with unilateral constraints. *IEEE Trans Automat Contr.* 42:200–215

[4] Chang T H, Hurmuzlu Y 1994 Sliding control without reaching phase and its application to bipedal locomotion. *ASME J Dyn Syst Meas Contr.* 105:447–455

[5] Furusho J, Masubichi M 1987 A theoretically reduced order model for the control of dynamic biped locomotion. *ASME J Dyn Syst Meas Contr.* 109:155–163

[6] Garcia M, Chatterjee A, Ruina A, Coleman M 1997 The simplest walking model: stability, and scaling. *ASME J Biomech Eng.* to appear

[7] Goswami A, Thuilot B, Espiau B 1996 Compass like bipedal robot part I: Stability and bifurcation of passive gaits. Tech Rep 2996, INRIA

[8] Guckenheimer J, Holmes P 1985 *Nonlinear Oscillations, Dynamical Systems, and Bifurcations of Vector Fields.* Springer-Verlag, New York

[9] Han I, Gilmore B J 1993 Multi-body impact motion with friction analysis, simulation, and experimental validation *ASME J Mech Des.* 115:412–422

[10] Hemami H, Chen B R 1984 Stability analysis and input design of a two-link planar biped. *Int J Robot Res.* 3(2)

[11] Hemami H, Wyman B F 1979 Modeling and control of constrained dynamic systems with application to biped locomotion in the frontal plane. *IEEE Trans Automat Contr.* 24

[12] Hurmuzlu Y 1993 Dynamics of bipedal gait; part I: Objective functions and the contact event of a planar five-link biped. *Int J Robot Res.* 13:82–92

[13] Hurmuzlu Y 1993 Dynamics of bipedal gait; part II: Stability analysis of a planar five-link biped. *ASME J Appl Mech.* 60:337–343

[14] Hurmuzlu Y, Marghitu D B 1994 Multi-contact collisions of kinematic chains with external surfaces. *ASME J Appl Mech.* 62:725–732
[15] Hurmuzlu Y, Moskowitz G D 1986 Role of impact in the stability of bipedal locomotion. *Int J Dyn Stab Syst.* 1:217–234
[16] Hurmuzlu Y, Moskowitz G D 1987 Bipedal locomotion stabilized by impact and switching: I. Two and three dimensional, three element models. *Int J Dyn Stab Syst.* 2:73–96
[17] Hurmuzlu Y, Moskowitz G D 1987 Bipedal locomotion stabilized by impact and switching: II. Structural stability analysis of a four-element model. *Int J Dyn Stab Syst.* 2:97–112
[18] Katoh R, Mori M 1984 Control method of biped locomotion giving asymptotic stability of trajectory. *Automatica.* 20:405–414
[19] Keller J B 1986 Impact with friction. *ASME J Appl Mech.* 53:1–4
[20] Marghitu D B, Hurmuzlu Y 1995 Three dimensional rigid body collisions with multiple contact points. *ASME J Appl Mech.* 62:725–732
[21] McGeer T 1990 Passive dynamic walking. *Int J Robot Res.* 9(2)
[22] Newton I 1686 *Philosophia Naturalis Principia Mathematica.* S Pepys, Reg Soc PRAESES
[23] Poisson S D 1817 *Mechanics.* Longmans, London, UK
[24] Shaw J, Holmes P 1983 A periodically forced piecewise linear oscillator *J Sound Vibr.* 90:129–155
[25] Shaw J, Shaw S 1989 The onset of chaos in a two-degree-of-freedom impacting system *ASME J Appl Mech.* 56:168–174
[26] Stronge W J 1990 Rigid body collisions with friction. In: *Proc Royal Soc.* 431:169–181
[27] Vukobratovic M, Borovac B, Surla D, Stokic D 1990 *Scientific Fundamentals of Robotics 7: Biped Locomotion.* Springer-Verlag, New York
[28] Zheng Y F 1989 Acceleration compensation for biped robots to reject external disturbances. *IEEE Trans Syst Man Cyber.* 19:74–84

Free-Floating Robotic Systems

Olav Egeland and Kristin Y. Pettersen

Department of Engineering Cybernetics, Norwegian University of Science and Technology, Norway

This chapter reviews selected topics related to kinematics, dynamics and control of free-floating robotic systems. Free-floating robots do not have a fixed base, and this fact must be accounted for when developing kinematic and dynamic models. Moreover, the configuration of the base is given by the Special Euclidean Group $SE(3)$, and hence there exist no minimum set of generalized coordinates that are globally defined. Jacobian based methods for kinematic solutions will be reviewed, and equations of motion will be presented and discussed. In terms of control, there are several interesting aspects that will be discussed. One problem is coordination of motion of vehicle and manipulator, another is in the case of underactuation where nonholonomic phenomena may occur, and possibly smooth stabilizability may be precluded due to Brockett's result.

1. Kinematics

A free-floating robot does not have a fixed base, and this has certain interesting consequences for the kinematics and for the equation of motion compared to the usual robot models. In addition, the configuration space of a free-floating robot cannot be described globally in terms of a set of generalized coordinates of minimum dimension, in contrast to a fixed base manipulator where this is achieved with the joint variables. In the following, the kinematics and the equation of motion for free-floating robots are discussed with emphasis on the distinct features of this class of robots compared to fixed-base robots.

A six-joint manipulator on a rigid vehicle is considered. The inertial frame is denoted by I, the vehicle frame by 0, and the manipulator link frames are denoted by $1, 2, \ldots, 6$.

The configuration of the vehicle is given by the 4×4 homogeneous transformation matrix

$$\boldsymbol{T}_0^I = \begin{pmatrix} \boldsymbol{R}_0^I & \boldsymbol{r}_0^I \\ \mathbf{0} & 1 \end{pmatrix} \in SE_3. \tag{1.1}$$

Here $\boldsymbol{R}_0^I \in SO(3)$ is the orthogonal rotation matrix from frame I to frame 0, and $\boldsymbol{r}_0^I$ is the position of the origin of frame 0 relative to frame I. The trailing superscript I denotes that the vector is given in I coordinates[1]. $SE(3)$ is the

[1] Throughout the chapter a trailing superscript on a vector denotes that the vector is decomposed in the frame specified by the superscript.

Special Euclidean Group of order 3 which is the set of all 4×4 homogeneous transformation matrices, while $SO(3)$ the Special Orthogonal Group of order 3 which is the set of all 3×3 orthogonal rotation matrices. It is well known that there is no three-parameter description of $SO(3)$ which is both global and without singularities (see e.g. [23, 34]).

The configuration of the manipulator is given by

$$\boldsymbol{\theta} = \begin{pmatrix} \theta_1 \\ \vdots \\ \theta_6 \end{pmatrix} \in \mathcal{R}^6 \tag{1.2}$$

which is the vector of joint variables. The configuration of the total system is given by $\boldsymbol{T}_0^I$ and $\boldsymbol{\theta}$, and the system has 12 degrees of freedom. Due to the appearance of the homogeneous transformation matrix $\boldsymbol{T}_0^I$ in the configuration space there is no set of 12 generalized coordinates that are globally defined. This means that an equation of motion of the form $\boldsymbol{M}_q(\boldsymbol{q})\ddot{\boldsymbol{q}}+\boldsymbol{C}_q(\boldsymbol{q},\dot{\boldsymbol{q}})\dot{\boldsymbol{q}} = \boldsymbol{\tau}_q$ will not be globally defined for this type of system. In the following it is shown that instead a globally defined equation of motion can be derived in terms of the generalized velocities of the system. Moreover, this model is shown to have the certain important properties in common with the fixed-base robot model; in particular, the inertia matrix is positive definite and the well-known skew-symmetric property is recovered.

A minimum set of generalized velocities for the system is given by the twelve-dimensional vector $\boldsymbol{u}$ defined by

$$\boldsymbol{u} = \begin{pmatrix} \boldsymbol{u}_0 \\ \dot{\boldsymbol{\theta}} \end{pmatrix} \tag{1.3}$$

where $\boldsymbol{u}_0$ is the six-dimensional vector of generalized velocities for the satellite given by

$$\boldsymbol{u}_0 = \begin{pmatrix} \boldsymbol{v}_0^0 \\ \boldsymbol{\omega}_0^0 \end{pmatrix} \tag{1.4}$$

where $\boldsymbol{v}_0^0$ is the three-dimensional velocity vector of the origin of the vehicle frame 0, and $\boldsymbol{\omega}_0^0$ is the angular velocity of the vehicle. Both vectors are given in vehicle coordinates. $\dot{\boldsymbol{\theta}}$ is the six-dimensional vector of generalized manipulator velocities.

The associated twelve-dimensional vector of generalized active forces from the actuators is given by

$$\boldsymbol{\tau} = \begin{pmatrix} \boldsymbol{\tau}_0 \\ \boldsymbol{\tau}_m \end{pmatrix}. \tag{1.5}$$

Here $\boldsymbol{\tau}_0$ is the six-dimensional vector of generalized active forces from reaction wheels and thrusters in the vehicle frame 0, while $\boldsymbol{\tau}_m$ is the six-dimensional vector of manipulator generalized forces.

2. Equation of Motion

The equation of motion presented in this section was derived using the Newton-Euler formalism in combination with the principle of virtual work in [8]. Here it is shown how to derive the result using energy functions as in Lagrange's equation of motion without introducing a set of generalized coordinates. The derivation relies heavily on [2] where Hamel-Boltzmann's equation is used for rigid body mechanisms, however, in the present derivation the virtual displacements are treated as vector fields on the relevant tangent planes. This allows for the use of well-established operations on vector fields, as opposed to the traditional formulation where the combination of the virtual displacement operator, quasi-coordinates, variations and time differentiation is quite difficult to handle [31].

The equation of motion is derived from d'Alembert's principle of virtual work, which is written as

$$\int_B (\ddot{\boldsymbol{r}} dm - d\boldsymbol{f})^{\mathrm{T}} \delta \boldsymbol{r} = 0 \tag{2.1}$$

where $\boldsymbol{r}$ is the position of the mass element dm in inertial coordinates, $d\boldsymbol{f}$ is the applied force, and $\delta\boldsymbol{r}$ is the virtual displacement. We introduce the generalized velocity vector $\boldsymbol{u}$ which is in the tangent plane of the configuration space, and a virtual displacement vector $\boldsymbol{\xi}$ in the same tangent plane as $\boldsymbol{u}$. The velocity $\dot{\boldsymbol{r}}$ and the virtual displacement $\delta\boldsymbol{r}$ satisfy

$$\dot{\boldsymbol{r}} = \frac{\partial \dot{\boldsymbol{r}}}{\partial \boldsymbol{u}} \boldsymbol{u}, \quad \text{and} \quad \delta \boldsymbol{r} := \frac{\partial \dot{\boldsymbol{r}}}{\partial \boldsymbol{u}} \boldsymbol{\xi} \tag{2.2}$$

In the case where there is a set of generalized coordinates $\boldsymbol{q}$ of minimum dimension, we will have $\boldsymbol{u} = \dot{\boldsymbol{q}}$ and $\boldsymbol{\xi} = \delta\boldsymbol{q}$. Next, we define the vector $\boldsymbol{\zeta}$ so that the time derivative of the virtual displacement $\delta\boldsymbol{r}$ is given by

$$\frac{d}{dt}(\delta \boldsymbol{r}) = \frac{\partial \dot{\boldsymbol{r}}}{\partial \boldsymbol{u}} \boldsymbol{\zeta} \tag{2.3}$$

The kinetic energy is

$$T := \frac{1}{2} \int_B \dot{\boldsymbol{r}}^{\mathrm{T}} \dot{\boldsymbol{r}} dm. \tag{2.4}$$

Equation (2.1) can be written

$$\frac{d}{dt}\left(\int_B \dot{\boldsymbol{r}}^{\mathrm{T}} dm \delta \boldsymbol{r}\right) - \int_B \dot{\boldsymbol{r}}^{\mathrm{T}} dm \frac{d}{dt}(\delta \boldsymbol{r}) - \int_B d\boldsymbol{f}^{\mathrm{T}} \delta \boldsymbol{r} = 0 \tag{2.5}$$

Consider the calculations

$$\frac{d}{dt}\left(\int_B \dot{\boldsymbol{r}}^{\mathrm{T}} dm \delta \boldsymbol{r}\right) = \frac{d}{dt}\left[\int_B \frac{\partial}{\partial \dot{\boldsymbol{r}}}\left(\frac{1}{2}\dot{\boldsymbol{r}}^{\mathrm{T}} dm \dot{\boldsymbol{r}}\right) \frac{\partial \dot{\boldsymbol{r}}}{\partial \boldsymbol{u}} \boldsymbol{\xi}\right] = \frac{d}{dt}\left(\frac{\partial T}{\partial \boldsymbol{u}} \boldsymbol{\xi}\right) \tag{2.6}$$

and

$$\int_B \dot{\boldsymbol{r}}^{\mathrm{T}} dm \frac{d}{dt}(\delta \boldsymbol{r}) = \int_B \frac{\partial}{\partial \dot{\boldsymbol{r}}} \left(\frac{1}{2} \dot{\boldsymbol{r}}^{\mathrm{T}} dm \dot{\boldsymbol{r}} \right) \frac{\partial \dot{\boldsymbol{r}}}{\partial \boldsymbol{u}} \boldsymbol{\zeta} = \frac{\partial T}{\partial \boldsymbol{u}} \boldsymbol{\zeta} \tag{2.7}$$

This gives

$$\frac{d}{dt} \left(\frac{\partial T}{\partial \boldsymbol{u}} \boldsymbol{\xi} \right) - \frac{\partial T}{\partial \boldsymbol{u}} \boldsymbol{\zeta} - \boldsymbol{\tau}^{\mathrm{T}} \boldsymbol{\xi} = 0 \quad \text{where} \quad \boldsymbol{\tau} = \int_B \frac{\partial \dot{\boldsymbol{r}}}{\partial u}^{\mathrm{T}} d\boldsymbol{f} \tag{2.8}$$

which is reminiscent of Hamel's central principle (*Zentralgleichung*) [31] except for the handling of the virtual displacements. This leads to the following equation of motion which is a modification of the Hamel-Boltzmann's equation of motion.

$$\left[\frac{d}{dt} \frac{\partial T}{\partial \boldsymbol{u}} - \boldsymbol{\tau}^{\mathrm{T}} \right] \boldsymbol{\xi} - \frac{\partial T}{\partial \boldsymbol{u}} (\boldsymbol{\zeta} - \dot{\boldsymbol{\xi}}) = 0 \tag{2.9}$$

Remark 2.1. This formulation is close to the Hamel-Boltzmann's equation of motion which is based on the use of quasi-coordinates. The advantage of the present formulation is that it relies on well-established computations in tangent planes, as opposed to the quite involved combination of variations, the virtual displacement operator δ and the differentiation operator d which is typical of the Hamel-Boltzmann's equation [31].

Remark 2.2. If the configuration is given by a Lie group G, then $\boldsymbol{u}$, $\boldsymbol{\xi}$ and $\boldsymbol{\zeta}$ are in the corresponding Lie algebra g, and the $\boldsymbol{\zeta} - \dot{\boldsymbol{\xi}}$ term is the rate of change of the vector field $\boldsymbol{\xi}$ due to the flow induced by $\boldsymbol{u}$. This is the Lie derivative of $\boldsymbol{\xi}$ with respect to $\boldsymbol{u}$ [25], which is found from the Lie bracket according to

$$\boldsymbol{\zeta} - \dot{\boldsymbol{\xi}} = [\boldsymbol{u}, \boldsymbol{\xi}] \tag{2.10}$$

When the Lie group is $SO(3)$ and $\boldsymbol{u} = \boldsymbol{\omega} \in so(3)$, the result is $\boldsymbol{\zeta} - \dot{\boldsymbol{\xi}} = [\boldsymbol{\omega}, \boldsymbol{\xi}] = -\boldsymbol{S}(\boldsymbol{\omega})\boldsymbol{\xi}$ which leads to the Euler equations for a rigid body. The case of $SE(3)$ and kinematic chains of rigid bodies is discussed below.

Remark 2.3. It is noted that when a vector $\boldsymbol{q}$ of generalized coordinates of minimum dimension is available, the equation is simplified by setting $\boldsymbol{u} = \dot{\boldsymbol{q}}$ and $\boldsymbol{\xi} = \delta \boldsymbol{q}$ in which case

$$\frac{\partial \dot{\boldsymbol{r}}}{\partial \dot{\boldsymbol{q}}} (\boldsymbol{\zeta} - \dot{\boldsymbol{\xi}}) = \frac{d}{dt} \left(\frac{\partial \dot{\boldsymbol{r}}}{\partial \dot{\boldsymbol{q}}} \right) \delta \boldsymbol{q} = \frac{\partial \dot{\boldsymbol{r}}}{\partial \boldsymbol{q}} \delta \boldsymbol{q} \tag{2.11}$$

Hence

$$\frac{\partial T}{\partial \boldsymbol{u}} (\boldsymbol{\zeta} - \dot{\boldsymbol{\xi}}) = \int_B \frac{\partial}{\partial \dot{\boldsymbol{r}}} \left(\frac{1}{2} \dot{\boldsymbol{r}}^{\mathrm{T}} dm \dot{\boldsymbol{r}} \right) \frac{\partial \dot{\boldsymbol{r}}}{\partial \dot{\boldsymbol{q}}} (\boldsymbol{\zeta} - \dot{\boldsymbol{\xi}}) = \frac{\partial T}{\partial \boldsymbol{q}} \delta \boldsymbol{q} \tag{2.12}$$

and Lagrange's equation of motion appears.

Next consider a the spacecraft/manipulator system. The generalized velocity for body k is written

$$\boldsymbol{u}_k = \begin{pmatrix} \boldsymbol{v}_k^k \\ \boldsymbol{\omega}_k^k \end{pmatrix} \in se(3) \tag{2.13}$$

which are the velocity and angular velocity vectors in body-fixed coordinates with respect to a body-fixed reference point P. The virtual displacement vector of body k is written

$$\boldsymbol{\xi}_k = \begin{pmatrix} \boldsymbol{\chi}_k \\ \boldsymbol{\theta}_k \end{pmatrix} \in se(3) \tag{2.14}$$

The kinetic energy is

$$T = \sum_{k=0}^{6} T_k, \ \ T_k = \frac{1}{2}\boldsymbol{u}_k^{\mathrm{T}} \boldsymbol{D}_k \boldsymbol{u}_k \tag{2.15}$$

where

$$\boldsymbol{D}_k = \begin{pmatrix} m_k \boldsymbol{I} & m_k \boldsymbol{S}(\boldsymbol{d}_{k,k^*}^k) \\ m_k \boldsymbol{S}(\boldsymbol{d}_{k,k^*}^k)^{\mathrm{T}} & \boldsymbol{M}_k^k \end{pmatrix} \tag{2.16}$$

is the inertia matrix in body-fixed coordinates with respect to a body-fixed reference point P, where m_k is the mass of body k, $\boldsymbol{d}_{k,k^*}^k$ is the offset from the origin in frame k to the center of mass k^* in body k, and $\boldsymbol{M}_k^k$ is the inertia matrix in k coordinates of body k referenced to the origin of frame k. $\boldsymbol{S}(\boldsymbol{a})$ denotes the skew-symmetric form of a general vector $\boldsymbol{a} = \begin{pmatrix} a_1 & a_2 & a_3 \end{pmatrix}^{\mathrm{T}}$ which is given by

$$\boldsymbol{S}(\boldsymbol{a}) = \begin{pmatrix} 0 & -a_3 & a_2 \\ a_3 & 0 & -a_1 \\ -a_2 & a_1 & 0 \end{pmatrix}. \tag{2.17}$$

The equation of motion then follows straightforwardly from (2.9) using the fact that $(\boldsymbol{\zeta}_k - \dot{\boldsymbol{\xi}}_k) = [\boldsymbol{u}_k, \boldsymbol{\xi}_k]$, which is the Lie derivative of $\boldsymbol{\xi}_k$ with respect to $\boldsymbol{u}_k$. Let

$$\boldsymbol{A}(\boldsymbol{\xi}_k) = \begin{pmatrix} \boldsymbol{S}(\boldsymbol{\theta}_k) & \boldsymbol{\chi}_k \\ 0 & 0 \end{pmatrix}, \ \ \boldsymbol{A}(\boldsymbol{u}_k) = \begin{pmatrix} \boldsymbol{S}(\boldsymbol{\omega}_k^k) & \boldsymbol{v}_k^k \\ 0 & 0 \end{pmatrix} \tag{2.18}$$

denote the associated 4×4 matrix representations of $se(3)$. Then the Lie bracket in $se(3)$ is found from the matrix commutator as follows:

$$\begin{aligned} \boldsymbol{A}([\boldsymbol{u}, \boldsymbol{\xi}]) = [\boldsymbol{A}\boldsymbol{u}, \boldsymbol{A}\boldsymbol{\xi}] &= \boldsymbol{A}(\boldsymbol{\xi}_k)\boldsymbol{A}(\boldsymbol{u}_k) - \boldsymbol{A}(\boldsymbol{u}_k)\boldsymbol{A}(\boldsymbol{\xi}_k) \\ &= -\begin{pmatrix} \boldsymbol{S}[(\boldsymbol{S}\boldsymbol{\omega}_k^k)\boldsymbol{\theta}_k] & \boldsymbol{S}(\boldsymbol{\omega}_k^k)\boldsymbol{\chi}_k + \boldsymbol{S}(\boldsymbol{v}_k^k)\boldsymbol{\theta}_k \\ 0 & 0 \end{pmatrix} \end{aligned}$$

It it seen that

$$\boldsymbol{\zeta}_k - \dot{\boldsymbol{\xi}}_k = -\begin{pmatrix} \boldsymbol{S}(\boldsymbol{\omega}_k^k) & \boldsymbol{S}(\boldsymbol{v}_k^k) \\ 0 & \boldsymbol{S}(\boldsymbol{\omega}_k^k) \end{pmatrix} \boldsymbol{\xi}_k \tag{2.19}$$

Finally, the motion constraints of the joints are accounted for. The independent generalized velocities of the complete system is $\boldsymbol{u}$, and $\boldsymbol{u}_k$ of body k satisfies

$$\boldsymbol{u}_k = \frac{\partial \boldsymbol{u}_k}{\partial \boldsymbol{u}} \boldsymbol{u} \tag{2.20}$$

while the virtual displacements of body k satisfies

$$\boldsymbol{\xi}_k = \frac{\partial \boldsymbol{u}_k}{\partial \boldsymbol{u}} \boldsymbol{\xi} \tag{2.21}$$

where $\boldsymbol{\xi}$ are the independent virtual displacements of the system. The equation of motion for the system then becomes

$$\sum_{k=0}^{6} \left\{ \left[\frac{d}{dt}\frac{\partial T_k}{\partial \boldsymbol{u}_k} + \frac{\partial T_k}{\partial \boldsymbol{u}_k} \begin{pmatrix} \boldsymbol{S}(\boldsymbol{\omega}_k^k) & \boldsymbol{S}(\boldsymbol{v}_k^k) \\ 0 & \boldsymbol{S}(\boldsymbol{\omega}_k^k) \end{pmatrix} - \boldsymbol{\tau}_k^{\mathrm{T}} \right] \frac{\partial \boldsymbol{u}_k}{\partial \boldsymbol{u}} \right\} \boldsymbol{\xi} = 0 \tag{2.22}$$

which, in view of the components of $\boldsymbol{\xi}$ being arbitrary, gives

$$\sum_{k=0}^{6} \left\{ \frac{\partial \boldsymbol{u}_k}{\partial \boldsymbol{u}}^{\mathrm{T}} \left[\frac{d}{dt}\frac{\partial T_k}{\partial \boldsymbol{u}_k}^{\mathrm{T}} + \begin{pmatrix} \boldsymbol{S}(\boldsymbol{\omega}_k^k) & 0 \\ \boldsymbol{S}(\boldsymbol{v}_k^k) & \boldsymbol{S}(\boldsymbol{\omega}_k^k) \end{pmatrix} \frac{\partial T_k}{\partial \boldsymbol{u}_k}^{\mathrm{T}} \right] \right\} = \boldsymbol{\tau} \tag{2.23}$$

where

$$\boldsymbol{\tau} = \sum_{k=0}^{6} \frac{\partial \boldsymbol{u}_k}{\partial \boldsymbol{u}}^{\mathrm{T}} \boldsymbol{\tau}_k \tag{2.24}$$

Define the symmetric positive definite inertia matrix

$$\boldsymbol{M}(\boldsymbol{\theta}) = \sum_{k=0}^{6} \boldsymbol{P}_k^{\mathrm{T}}(\boldsymbol{\theta}) \boldsymbol{D}_k \boldsymbol{P}_k(\boldsymbol{\theta}), \tag{2.25}$$

and the matrix

$$\boldsymbol{C}(\boldsymbol{\theta}, \boldsymbol{u}) = \sum_{k=0}^{6} [\boldsymbol{P}_k^{\mathrm{T}}(\boldsymbol{\theta}) \boldsymbol{D}_k \dot{\boldsymbol{P}}_k(\boldsymbol{\theta}) - \boldsymbol{P}_k^{\mathrm{T}}(\boldsymbol{\theta}) \boldsymbol{P}_k(\boldsymbol{\theta})] \tag{2.26}$$

where

$$\boldsymbol{W}_k(\boldsymbol{\theta}, \boldsymbol{u}) = \begin{pmatrix} \mathbf{0} & \boldsymbol{S}(\frac{\partial T_k}{\partial \boldsymbol{v}_k^k}^T) \\ \boldsymbol{S}(\frac{\partial T_k}{\partial \boldsymbol{v}_k^k}^T) & \boldsymbol{S}(\frac{\partial T_k}{\partial \boldsymbol{\omega}_k^k}^T) \end{pmatrix} \tag{2.27}$$

Clearly, $\dot{\boldsymbol{M}} - 2\boldsymbol{C}$ is skew symmetric. Then the model can be written in the form

$$\boldsymbol{M}(\boldsymbol{\theta})\dot{\boldsymbol{u}} + \boldsymbol{C}(\boldsymbol{\theta}, \boldsymbol{u})\boldsymbol{u} = \boldsymbol{\tau}. \tag{2.28}$$

which resembles the equations of motion for a fixed-base robot [35].

3. Total System Momentum

The development in this section is based on [22] where additional details and computational aspects are found. The position of the center of mass of the total system is

$$\boldsymbol{r}^{*I} = \frac{\sum_{k=0}^{6} m_k \, \boldsymbol{r}_{k^*}^I}{\sum_{k=0}^{6} m_k} \tag{3.1}$$

where m_k is the mass of link k and $\boldsymbol{r}_{k^*}^I$ is the position of the center of mass k^* in link k relative to the inertial frame I. The linear momentum is

$$\boldsymbol{p}^I = \sum_{k=0}^{6} m_k \, (\dot{\boldsymbol{r}}^*)^I. \tag{3.2}$$

The angular momentum around the center of mass of the total system is

$$\boldsymbol{h}^I = \sum_{k=0}^{6} (\boldsymbol{M}_{k^*,k}^I \, \boldsymbol{\omega}_k^I + m_k \, \boldsymbol{S}(\boldsymbol{s}_k{}^I) \, \dot{\boldsymbol{s}}_k^I) \tag{3.3}$$

where $\boldsymbol{M}_{k^*,k}^I = \boldsymbol{R}_k^I \, \boldsymbol{M}_{k^*,k}^k \, \boldsymbol{R}_I^k$ is the inertia matrix of body k around its center of mass k^*. The superscript denotes that the matrix is decomposed in the I frame. $\boldsymbol{s}_k = \boldsymbol{r}_{k^*} - \boldsymbol{r}^*$ is the position of link center of mass k^* relative to the system center of mass.

The only external force acting on the system is $\boldsymbol{\tau}_1$, which gives the following equation of motion for the total system:

$$\begin{pmatrix} \dot{\boldsymbol{p}}^I \\ \dot{\boldsymbol{h}}^I \end{pmatrix} = \boldsymbol{E}_0^I \boldsymbol{\tau}_1 \tag{3.4}$$

where

$$\boldsymbol{E}_0^I = \begin{pmatrix} \boldsymbol{R}_0^I & \boldsymbol{0} \\ \boldsymbol{0} & \boldsymbol{R}_0^I \end{pmatrix} \tag{3.5}$$

is a 6×6 transformation matrix, and $\boldsymbol{R}_0^I \in SO(3)$ is the 3×3 rotation matrix from frame I to frame 0. Obviously, $\boldsymbol{E}_0^I$ is orthogonal with det $\boldsymbol{E}_0^I = 1$ and $(\boldsymbol{E}_0^I)^{-1} = (\boldsymbol{E}_0^I)^{\mathrm{T}}$.

4. Velocity Kinematics and Jacobians

The end-effector linear and angular velocity is given by

$$\boldsymbol{u}_e := \boldsymbol{u}_6 = \begin{pmatrix} \boldsymbol{v}_6^6 \\ \boldsymbol{\omega}_6^6 \end{pmatrix} \tag{4.1}$$

and is expressed in terms of the generalized velocity vector $\boldsymbol{u} = (\boldsymbol{u}_0^{\mathrm{T}} \, \dot{\boldsymbol{\theta}}^{\mathrm{T}})^{\mathrm{T}}$ according to

$$\boldsymbol{u}_e = \boldsymbol{J}(\boldsymbol{\theta})\boldsymbol{u} = \boldsymbol{J}_0(\boldsymbol{\theta})\boldsymbol{u}_0 + \boldsymbol{J}_\theta(\boldsymbol{\theta})\dot{\boldsymbol{\theta}} \tag{4.2}$$

where the Jacobians are found in the same way as for fixed base manipulators.

The linear and angular momentum can be written

$$\begin{pmatrix} \boldsymbol{p}^I \\ \boldsymbol{h}^I \end{pmatrix} = \boldsymbol{P}_0(\boldsymbol{\theta})\boldsymbol{u}_0 + \boldsymbol{P}_\theta(\boldsymbol{\theta})\dot{\boldsymbol{\theta}} \tag{4.3}$$

If the linear and angular momentum is assumed to be zero, then

$$\boldsymbol{P}_0(\boldsymbol{\theta})\boldsymbol{u}_0 + \boldsymbol{P}_\theta(\boldsymbol{\theta})\dot{\boldsymbol{\theta}} = 0 \tag{4.4}$$

and the satellite velocity vector can be found from

$$\boldsymbol{u}_0 = -\boldsymbol{P}_0^{-1}(\boldsymbol{\theta})\boldsymbol{P}_\theta(\boldsymbol{\theta})\dot{\boldsymbol{\theta}} \tag{4.5}$$

In this case the end-effector velocity vector is found to be

$$\boldsymbol{u}_e = \boldsymbol{J}_g(\boldsymbol{\theta})\dot{\boldsymbol{\theta}} \tag{4.6}$$

where $\boldsymbol{J}_g = -\boldsymbol{J}_0\boldsymbol{P}_0^{-1}\boldsymbol{P}_\theta + \boldsymbol{J}_\theta$ was termed the generalized Jacobian matrix in [37]. The singularities of $\boldsymbol{J}_g$ were termed dynamic singularities in [27]. A fixed-base manipulator with joint variables $\boldsymbol{q}$ and Jacobian $\boldsymbol{J}_g(\boldsymbol{q})$ was specified in [38] where such a manipulator was termed the virtual manipulator corresponding to the system with zero linear and angular momentum.

5. Control Deviation in Rotation

Let $\boldsymbol{R} := \boldsymbol{R}_B^A$ denote the actual rotation matrix, while the desired rotation matrix is denoted $\boldsymbol{R}_d$. The kinematic differential equation for $\boldsymbol{R}$ in terms of the body-fixed angular velocity $\boldsymbol{\omega}^B$ is written $\dot{\boldsymbol{R}} = \boldsymbol{R}\boldsymbol{S}(\boldsymbol{\omega}^B)$. The attitude deviation is described by the rotation matrix $\tilde{\boldsymbol{R}}_B := \boldsymbol{R}_d^{\mathrm{T}}\boldsymbol{R} \in SO(3)$. The desired angular velocity vector $\boldsymbol{\omega}_d$ is defined by

$$\dot{\boldsymbol{R}}_d = \boldsymbol{S}(\boldsymbol{\omega}_d^A)\boldsymbol{R}_d. \tag{5.1}$$

Then time differentiation of $\tilde{\boldsymbol{R}}_B$ gives

$$\dot{\tilde{\boldsymbol{R}}}_B = \tilde{\boldsymbol{R}}_B\boldsymbol{S}(\tilde{\boldsymbol{\omega}}) \tag{5.2}$$

where $\tilde{\boldsymbol{\omega}} = \boldsymbol{\omega} - \boldsymbol{\omega}_d$ is the deviation in angular velocity. It is seen that the proposed definition of $\tilde{\boldsymbol{R}}$, $\tilde{\boldsymbol{\omega}}$ and $\boldsymbol{\omega}_d$ is consistent with the usual kinematic differential equation on $SO(3)$.

6. Euler Parameters

A rotation matrix $\boldsymbol{R}$ can be parameterized by a rotation ϕ around the unit vector $\boldsymbol{k}$ [12, 23]. The Euler parameters (ϵ, η) corresponding to $\boldsymbol{R}$ are written

$$\epsilon = \sin(\frac{\phi}{2})\boldsymbol{k}, \quad \eta = \cos(\frac{\phi}{2}). \tag{6.1}$$

The rotation matrix can then be written

$$\boldsymbol{R} = (\eta^2 - \epsilon^{\mathrm{T}}\epsilon)\boldsymbol{I} + 2\epsilon\epsilon^{\mathrm{T}} + 2\eta\boldsymbol{S}(\epsilon),$$

and it is seen that $\epsilon = \mathbf{0} \Leftrightarrow \tilde{\boldsymbol{R}} = \boldsymbol{I}$.

The kinematic differential equation for the rotation matrix $\boldsymbol{R}$ is $\dot{\boldsymbol{R}} = \boldsymbol{R}\boldsymbol{S}(\boldsymbol{\omega})$. while the associated kinematic differential equations for the Euler parameters are given by

$$\dot{\epsilon} = \frac{1}{2}[\eta\boldsymbol{I} + \boldsymbol{S}(\epsilon)]\boldsymbol{\omega} \tag{6.2}$$

$$\dot{\eta} = -\frac{1}{2}\epsilon^{\mathrm{T}}\boldsymbol{\omega} \tag{6.3}$$

7. Passivity Properties

From Eq. (6.3) a number of passivity results can easily be derived for the rotational kinematic equations. The following three-dimensional parameterizations of the rotation matrix will be studied.

$$\epsilon = \boldsymbol{k}\sin\frac{\phi}{2} \quad \text{(Euler parameter vector)} \tag{7.1}$$

$$\boldsymbol{e} = 2\eta\epsilon = \boldsymbol{k}\sin\phi \quad \text{(Euler rotation vector)} \tag{7.2}$$

$$\boldsymbol{\rho} = \frac{\epsilon}{\eta} = \boldsymbol{k}\tan\frac{\phi}{2} \quad \text{(Rodrigues vector)} \tag{7.3}$$

$$\boldsymbol{d} = 2\arccos|\eta|\boldsymbol{k} = \phi\boldsymbol{k} \quad \text{(angle-axis vector)} \tag{7.4}$$

Then, the following expressions hold:

$$\frac{d}{dt}[2(1-\eta)] = -2\dot{\eta} = \epsilon^{\mathrm{T}}\boldsymbol{\omega} \tag{7.5}$$

$$\frac{d}{dt}[2(1-\eta^2)] = -4\eta\dot{\eta} = 2\eta\epsilon^{\mathrm{T}}\boldsymbol{\omega} = \boldsymbol{e}^{\mathrm{T}}\boldsymbol{\omega} \tag{7.6}$$

$$\frac{d}{dt}[-2\ln|\eta|] = -2\frac{\dot{\eta}}{\eta} = \frac{\epsilon^{\mathrm{T}}}{\eta}\boldsymbol{\omega} = \boldsymbol{\rho}^{\mathrm{T}}\boldsymbol{\omega} \tag{7.7}$$

$$\frac{d}{dt}[\frac{1}{2}\phi^2] = -2\phi\frac{\dot{\eta}}{\sqrt{1-\eta^2}} = \phi\boldsymbol{k}^{\mathrm{T}}\boldsymbol{\omega} \tag{7.8}$$

It follows that the mappings $\boldsymbol{\omega} \mapsto \epsilon$, $\boldsymbol{\omega} \mapsto \boldsymbol{e}$, $\boldsymbol{\omega} \mapsto \boldsymbol{\rho}$, and $\boldsymbol{\omega} \mapsto \phi\boldsymbol{k}$ are all passive with the indicated storage functions. This is useful in attitude controller design [7, 39].

8. Coordination of Motion

If only the desired end effector motion is specified, the spacecraft motion is an internal motion, and the system can be viewed as a redundant manipulator system. The spacecraft/manipulator system is a relatively simple kinematic architecture as it is a serial structure of two six-degree-of-freedom mechanisms. Because of this it is possible to use positional constraints on the internal motion as in [6]. Redundancy can then be solved simply by assigning a constant nominal configuration $\boldsymbol{\theta}_n$ to the manipulator so that there are no singularities or joint limits close to $\boldsymbol{\theta}_n$. The selection of $\boldsymbol{\theta}_n$ can be based on engineering judgement. Then the satellite reference in SE_3 can be computed in real time solving

$$\boldsymbol{T}_{ee,d} = \boldsymbol{T}_{sat,d}\boldsymbol{T}_{man}(\boldsymbol{\theta}_n) \tag{8.1}$$

with respect to $\boldsymbol{T}_{sat,d}$, where

$$\boldsymbol{T}_{man} = \begin{pmatrix} \boldsymbol{R}_6^0 & \boldsymbol{r}_6^0 - \boldsymbol{r}_0^0 \\ \boldsymbol{0} & 1 \end{pmatrix} \in SE(3) \tag{8.2}$$

is computed from the forward kinematics of the manipulator. Thus redundancy is eliminated by specifying the remaining 6 degrees of freedom.

To achieve energy-efficient control it may be a good solution to control the end-effector tightly, while using less control effort on the remaining degrees of freedom. Then accurate end-effector control is achieved even when the control deviation for the vehicle is significant, thus eliminating the need for high control energy for vehicle motion. Further details are found in [8]. Related work is found in [15] where reaction wheels are used to counteract torques from the manipulator, while the translation of the vehicle is not controlled. In [1] the end-effector is accurately controlled using manipulator torques, while the external forces and torques are set to zero.

9. Nonholonomic Issues

The dynamics of underactuated free-floating robotic systems can be written in the form

$$\boldsymbol{M}(\boldsymbol{\theta})\dot{\boldsymbol{u}} + \boldsymbol{C}(\boldsymbol{\theta}, \boldsymbol{u})\boldsymbol{u} = \begin{bmatrix} 0 \\ \boldsymbol{\tau} \end{bmatrix} \tag{9.1}$$

where $\dim(\boldsymbol{\tau}) = m < n = \dim(\boldsymbol{u})$. The underactuation leads to a constraint on the acceleration of the system, given by the first $n - m$ equations of (9.1). [26] has given conditions under which this acceleration constraint can be integrated to a constraint on the velocity or a constraint on the configuration. If the acceleration constraint is not integrable, it is called a second-order nonholonomic constraint, and the system is called a second-order nonholonomic system. If the constraint can be integrated to a velocity constraint

$$g(\boldsymbol{\theta}, \boldsymbol{u}) = 0 \tag{9.2}$$

it is called a (first-order) nonholonomic constraint, while a constraint on the configuration $\boldsymbol{\theta}$

$$h(\boldsymbol{\theta}) = 0 \tag{9.3}$$

is called a holonomic constraint. For the case of an n-DOF manipulator on a vehicle that has no external forces or torques acting on it, it is shown in [24] that the linear momentum conservation equation is a holonomic constraint, while the angular momentum conservation equation is nonholonomic. If, in addition, not all the joints of the manipulator are actuated, this leads to second-order nonholonomic constraints [21].

It is shown in [26] that the system (9.1) does not satisfy Brockett's necessary condition [3]. So even if the system if controllable, it cannot be asymptotically stabilized by a state feedback control law $\boldsymbol{\tau} = \boldsymbol{\alpha}(\boldsymbol{\theta}, \boldsymbol{u})$ that is a continuous function of the state. This is a common feature of both first- and second-order nonholonomic systems, and different strategies have been proposed to evade this negative result. One approach has been to use continuous state feedback laws that asymptotically stabilizes an equilibrium manifold of the closed-loop system, instead of an equilibrium point. This approach is used for an underactuated free-flying system in [21]. Another approach has been the use of feedback control laws that are discontinuous functions of the state, in the attempt to asymptotically stabilize an equilibrium point. However, [5] has shown that affine systems, i.e. systems in the form

$$\dot{\boldsymbol{x}} = \boldsymbol{f}_0(\boldsymbol{x}) + \sum_{i=1}^{m} \boldsymbol{f}_i(x) u_i \tag{9.4}$$

which do not satisfy Brockett's necessary condition, cannot be asymptotically stabilized by discontinuous state feedback either. (This is under the assumption that one considers Filippov solutions for the closed-loop system, as proposed by [10].) As the system (9.1) is affine in the control $\boldsymbol{\tau}$, this result applies to the free-floating robotic systems. Typically, discontinuous feedback laws may give convergence to the desired configuration, without providing stability for the closed-loop system. Such a feedback control law is proposed for a free-flying space robot in [24], using a bidirectional approach. An important advantage of continuous over discontinuous feedback laws, is that continuous feedback control laws do not give chattering or the problem of physical realization of infinitely fast switching.

Another approach to evade Brockett's negative stabilizability result has been the use of continuous time-varying feedback laws $\boldsymbol{\tau} = \boldsymbol{\beta}(\boldsymbol{\theta}, \boldsymbol{u}, t)$. [36] proved that any one-dimensional nonlinear control system which is controllable, can be asymptotically stabilized by means of time-varying feedback control laws. The approach was first applied for nonholonomic systems by [33] who showed how continuous time-varying feedback laws could asymptotically stabilize a nonholonomic cart. To obtain faster convergence, [16] proposed to use continuous time-varying feedback laws that are non-differentiable at the equilibrium point that is to be stabilized. The feedback control laws are in [16] derived using averaging. However, the stability analysis of non-differentiable systems is nontrivial. For C^2 systems it is possible to infer exponential stability of the original system from that of the averaged system [14]. However, this result is not generally applicable to non-differentiable systems. In [16] an averaging result is developed for non-differentiable systems, under the condition that the system has certain homogeneity properties. The definitions of homogeneity are as follows: For any $\lambda > 0$ and any set of real parameters $r_1, \ldots, r_n > 0$, a dilation operator $\delta_\lambda^r : \mathcal{R}^{n+1} \to \mathcal{R}^{n+1}$ is defined by

$$\delta_\lambda^r(x_1, \ldots, x_n, t) = (\lambda^{r_1} x_1, \ldots, \lambda^{r_n} x_n, t) \tag{9.5}$$

A differential system $\dot{\boldsymbol{x}} = \boldsymbol{f}(\boldsymbol{x}, t)$ (or a vector field $\boldsymbol{f}$) with $\boldsymbol{f} : \mathcal{R}^n \times \mathcal{R} \to \mathcal{R}^n$ continuous, is homogeneous of degree $\sigma \geq 0$ with respect to the dilation δ_λ^r if its ith coordinate f^i satisfies the equation

$$f^i(\delta_\lambda^r(\boldsymbol{x}, t)) = \lambda^{r_i + \sigma} f^i(\boldsymbol{x}, t) \quad \forall \lambda > 0 \quad i = 1, \ldots, n \tag{9.6}$$

The origin of a system $\dot{\boldsymbol{x}} = \boldsymbol{f}(\boldsymbol{x}, t)$ is said to be exponentially stable with respect to the dilation δ_λ^r if there exists two strictly positive constants K and a such that along any solution $\boldsymbol{x}(t)$ of the system the following inequality is satisfied:

$$\rho_p^r(\boldsymbol{x}(t)) \leq K e^{-at} \rho_p^r(\boldsymbol{x}(0)) \tag{9.7}$$

where $\rho_p^r(\boldsymbol{x})$ is a homogeneous norm associated with the dilation δ_λ^r:

$$\rho_p^r(\boldsymbol{x}) = \left(\sum_{i=1}^{n} |x_i|^{\frac{p}{r_i}}\right)^{\frac{1}{p}} \quad \text{with} \quad p > 0 \tag{9.8}$$

For each set of parameters $r_1, \ldots, r_n$, all the associated norms are equivalent. The use of homogeneity properties of a system was first proposed by [13] and [11] for autonomous systems, and extended to time-varying systems by [30]. Under the assumption that a C^0 time-varying system is homogeneous of degree zero with respect to a given dilation, [16] proved that asymptotic stability of the averaged system implies exponential stability of the original system with respect to the given dilation. The use of periodic time-varying feedback laws that render the closed-loop system homogeneous, has proved to be a useful approach for exponential stabilization of nonholonomic systems. Further analysis tools have been developed. In [30] a converse theorem is presented, establishing the existence of homogeneous Lyapunov functions for time-varying asymptotically stable systems which are homogeneous of degree zero with respect to some dilation. [18, 20] present a solution to the problem of "adding integrators" for homogeneous time-varying C^0 systems. In order to avoid cancellation of dynamics, an extended version of this result is presented in [29]. Moreover, in [18, 20] also a perturbation result is presented for this class of systems. The result states that if a locally exponentially stable system that is homogeneous of degree zero, is perturbed by a vector field homogeneous of degree strictly positive with respect to the same dilation, the resulting system is still locally exponentially stable. Together, these results constitute a useful set of tools for developing exponentially stabilizing feedback control laws for nonholonomic systems. Amongst other, continuous time-varying feedback laws have been developed that exponentially stabilize nonholonomic systems in power form, including mobile robots, [16, 17, 30, 19], underactuated rigid spacecraft [18, 20, 4], surface vessels [29] and autonomous underwater vehicles (AUVs) [28]. Typically, the feedback control laws involve periodic time-varying terms of the form $\sin(t/\epsilon)$, where ϵ is a small positive parameter. Oscillations in the actuated degrees of freedom act together to provide a change in the unactuated degrees of freedom. The resulting behaviour for an AUV which has no forces along the body-fixed $y-$ and $z-$axis available, only control force in the body-fixed $x-$direction and control torques around the body-fixed $y-$ and $z-$axis, is shown in Fig. 9.1.

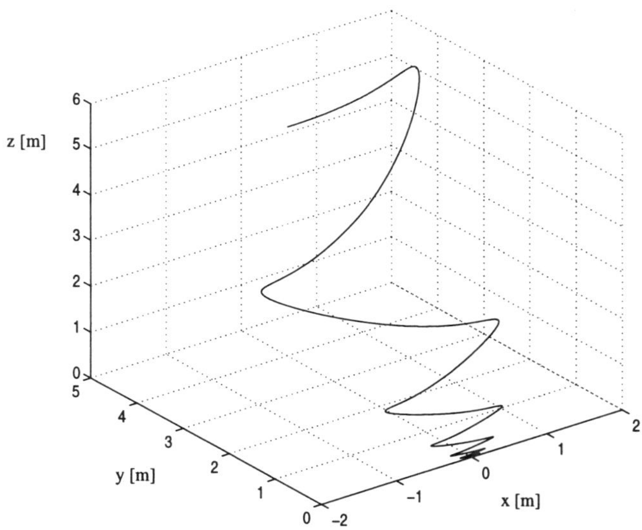

Fig. 9.1. The trajectory of the AUV in the xyz space

An interesting direction of research will be to apply these tools to develop exponentially stabilizing feedback laws for free-floating robotic systems. However, for the tools to apply the system must have the appropriate homogeneity properties. In [32] a continuous time-varying feedback law is proposed for a free-floating robotic system. The feedback control law is developed using averaging. [32] show that the averaged system is globally exponentially stable and that the trajectories of the original and the averaged system stay close within an $O(\sqrt{\epsilon})$ neighbourhood for all time. However, the system is not C^2 nor does it have the appropriate homogeneity properties, and therefore it is to the authors' best knowledge no rigorous theory available to conclude exponential stability of the original system from that of the averaged system. Therefore, it is an open question how to develop continuous time-varying feedback control laws that are proved to exponentially stabilize free-floating robotic systems. One approach to solve this problem may be to use the fact that the homogeneity properties of a system are coordinate dependent. One may thus seek to find a coordinate transformation that give system equations with the appropriate homogeneity properties. This is for instance done in [28, 29]. Another approach may be to develop analysis tools without the demand for homogeneity, for time-varying systems that are C^∞ everywhere except at the equilibrium point that we want to stabilize, where they are only C^0.

References

[1] Alexander H L, Cannon R H Jr 1987 Experiments on the control of a satellite manipulator. In: *Proc 1987 American Control Conf.* Seattle, WA

[2] Bremer H 1988 Über eine Zetralgleichung in der Dynamik. *Z Angew Math Mech.* 68:307–311

[3] Brockett R W 1983 Asymptotic stability and feedback stabilization. In: Brockett R W, Millmann R S, Sussman H J (eds) *Differential Geometric Control Theory.* Birkhäuser, Boston, MA, pp 181–208

[4] Coron J-M, Kerai E-Y 1996 Explicit feedbacks stabilizing the attitude of a rigid spacecraft with two control torques. *Automatica.* 32:669–677

[5] Coron J-M, Rosier L 1994 A relation between continuous time-varying and discontinuous feedback stabilization. *J Math Syst Estim Contr.* 4:67–84

[6] Egeland O 1987 Task-space tracking with redundant manipulators. *IEEE J Robot Automat.* 3:471–475

[7] Egeland O, Godhavn J-M 1994 Passivity-based attitude control of a rigid spacecraft. *IEEE Trans Automat Contr.* 39: 842–846

[8] Egeland O, Sagli J R 1993 Coordination of motion in a spacecraft/manipulator system. *Int J Robot Res.* 12:366–379

[9] Goldstein H 1980 *Classical mechanics.* (2nd ed) Addison Wesley, Reading, MA

[10] Hermes H 1967 Discontinuous vector fields and feedback control. In: Hale J K, LaSalle J P (eds) *Differential Equations and Dynamical Systems.* Academic Press, New York, pp 155-165

[11] Hermes H 1991 Nilpotent and high-order approximation of vector field systems. *SIAM Rev.* 33:238–264

[12] Hughes P C 1986 *Spacecraft Attitude Dynamics.* Wiley, New York

[13] Kawski M 1990 Homogeneous stabilizing feedback laws. *Contr Theo Adv Tech.* 6:497–516

[14] Khalil H K 1996 *Nonlinear systems.* (2nd ed) Prentice-Hall, Englewood Cliffs, NJ

[15] Longman R W, Lindberg R E, Zedd M F 1987 Satellite-mounted robot manipulators — New kinematics and reaction moment compensation. *Int J Robot Res.* 6(3):87–103

[16] M'Closkey R T, Murray R M 1993 Nonholonomic systems and exponential convergence: Some analysis tools. In: *Proc 32nd IEEE Conf Decision Contr.* San Antonio, TX, pp 943–948

[17] M'Closkey R T, Murray R M 1997 Exponential stabilization of driftless nonlinear control systems using homogeneous feedback. *IEEE Trans Automat Contr.* 42:614–628

[18] Morin P, Samson C 1995 Time-varying exponential stabilization of the attitude of a rigid spacecraft with two controls. In: *Proc 34th IEEE Conf Decision Contr.* New Orleans, LA, pp 3988–3993

[19] Morin P, Samson C 1996 Time-varying exponential stabilization of chained form systems based on a backstepping technique. In: *Proc 35th IEEE Conf Decision Contr.* Kobe, Japan, pp 1449–1454

[20] Morin P, Samson C 1997 Time-varying exponential stabilization of a rigid spacecraft with two control torques. *IEEE Trans Automat Contr.* 42:528–534

[21] Mukherjee R, Chen D 1992 Stabilization of free-flying under-actuated mechanisms in space. In: *Proc 1992 Amer Contr Conf.* Chicago, IL, pp 2016–2021

[22] Mukherjee R, Nakamura Y 1992 Formulation and efficient computation of inverse dynamics of space robots. *IEEE Trans Robot Automat.* 8: 400–406

[23] Murray R M, Li Z, Sastry, S S 1994 A Mathematical Introduction to Robotic Manipulation. CRC Press, Boca Raton, FL

[24] Nakamura Y, Mukherjee R 1991 Nonholonomic path planning of space robots via a bidirectional approach. *IEEE Trans Robot Automat.* 7:500–514

[25] Olver P J 1993 *Applications of Lie Groups to Differential Equations.* (2nd ed) Springer-Verlag, New York

[26] Oriolo G, Nakamura Y 1991 Control of mechanical systems with second-order nonholonomic constraints: underactuated manipulators. In: *Proc 30th IEEE Conf Decision Contr.* Brighton, UK, pp 2398–2403

[27] Papadopoulos E, Dubowsky S 1991 On the nature of control algorithms for free-floating space manipulators. *IEEE Trans Robot Automat.* 7:750–758

[28] Pettersen K Y, Egeland O 1996 Position and attitude control of an underactuated autonomous underwater vehicle. In: *Proc 35th IEEE Conf Decision Contr.* Kobe, Japan, pp 987–991

[29] Pettersen K Y, Egeland O 1997 Robust control of an underactuated surface vessel with thruster dynamics. In: *Proc 1997 Amer Contr Conf.* Albuquerque, New Mexico

[30] Pomet J-B, Samson C 1993 Time-varying exponential stabilization of nonholonomic systems in power form. Tech Rep 2126, INRIA

[31] Rosenberg R M 1977 *Analytical Dynamics of Discrete Systems.* Plenum Press, New York

[32] Rui C, Kolmanovsky I, McClamroch N H 1996 Feedback reconfiguration of underactuated multibody spacecraft. In: *Proc 35th IEEE Conf Decision Contr.* Kobe, Japan, pp 489–494

[33] Samson C 1991 Velocity and torque feedback control of a nonholonomic cart. In: Canudas de Wit C (ed) *Advanced Robot Control.* Springer-Verlag, London, UK, pp 125-151

[34] Samson C, Le Borgne M, Espiau B 1991 *Robot Control: The Task Function Approach.* Clarendon Press, Oxford, UK

[35] Sciavicco L, Siciliano B 1996 *Modeling and Control of Robot Manipulators.* McGraw-Hill, New York

[36] Sontag E, Sussmann H 1980 Remarks on continuous feedback. In: *Proc 19th IEEE Conf Decision Contr.* Albuquerque, NM, pp 916-921

[37] Umetani Y, Yoshida K 1987 Continuous path control of space manipulators. *Acta Astronaut.* 15:981–986

[38] Vafa Z, Dubowsky S. 1987 On the dynamics of manipulators in space using the virtual manipulator approach. In: *Proc 1987 IEEE Int Conf Robot Automat.* Raleigh, NC, pp 579–585

[39] Wen J T-Y, Kreutz-Delgado K 1991 The attitude control problem. *IEEE Trans Automat Contr.* 36:1148–1162

Underactuated Mechanical Systems

Mark W. Spong

Coordinated Science Laboratory, University of Illinois at Urbana–Champaign, USA

In this chapter we discuss the control of underactuated mechanical systems. Underactuated mechanical systems have fewer control inputs than degrees of freedom and arise in applications, such as space and undersea robots, mobile robots, flexible robots, walking, brachiating, and gymnastic robots. The Lagrangian dynamics of these systems may contain feedforward nonlinearities, non-minimum phase zero dynamics, nonholonomic constraints, and other properties that place this class of systems at the forefront of research in nonlinear control [22, 15]. A complete understanding of the control of these systems is therefore lacking. We will discuss the application of geometric nonlinear control, as well as methods based on passivity and energy for stabilization and tracking control. We will survey some of the existing results and point to open research problems.

1. Introduction

A mechanical system may be "underactuated" in several ways. The most obvious way is from intentional design as in the brachiation robot of Fukuda [31], the passive walker of McGeer [23], the Acrobot [5], or the Pendubot [39]. Underactuated systems also arise in mobile robot systems, for example, when a manipulator arm is attached to a mobile platform, a space platform, or an undersea vehicle [45]. A third way that underactuated systems arise is due to the mathematical model used for control design as, for example, when joint flexibility is included in the model [35]. It is also interesting to note that certain control problems for fully actuated redundant robots are similar to those for underactuated robots [7]. The class of underactuated mechanical systems is thus rich in both applications and control problems.

The class of underactuated mechanical systems is far too broad to survey in a single chapter. For fully actuated systems there are a number of control results that apply to the entire class, such as feedback linearization and passivity-based adaptive control [40]. By contrast, with the exception of the collocated partial feedback linearization result discussed below, there are few results that are applicable to the entire class of underactuated mechanical systems. For example, the control problems for flexible joint robots require somewhat different tools for analysis and controller design than the control problems for gymnastic robots like the Acrobot.

In this chapter we will confine our discussion primarily to control problems for serial link robots containing both active and passive joints, such as the Acrobot [13, 4, 37] and Pendubot [39, 1]. Our ultimate goal for studying such systems is to understand problems of balance and locomotion in both biological systems and in robotic systems. The reader is referred to the literature for treatment of other classes of underactuated systems, such as flexible link robots [3], flexible joint robots [35], space robots [10], mobile robots [27] or underwater robots [11].

The techniques we will discuss for control are mainly based on ideas of passivity and control of energy. Passivity-based control has a long and rich heritage having its roots in passive network synthesis and entering the control field via the Popov Criterion and the Kalman-Yakubovich-Popov Lemma [18]. Passivity in Lagrangian systems is equivalent to the now familiar skew-symmetry property [29], long known in classical mechanics and whose rediscovery in robot control led to breakthroughs in adaptive control of fully actuated manipulators [34]. In the nonlinear control field the exploitation of passivity has led to dramatic advances in controller design, with the appearance of concepts such as backstepping [19] and more recently forwarding [22, 32]. These methods are not yet generally applicable to all underactuated mechanical systems, but can be applied in special cases. In the area of robot locomotion, energy and passivity methods have already achieved some success. Indeed, the work of McGeer and others on passive walking [23] shows that stable limit cycle walking can be achieved by the natural tradeoff between kinetic and potential energy without feedback control of any kind. This work is fundamental, since, for example, there is considerable experimental evidence that a great part of the swing phase in human locomotion is passive. The muscles of the human leg are active primarily during the double support period, when the initial conditions on the angles and velocities of each of the limb segments are being established, after which they essentially turn off and allow the leg to swing through like a jointed pendulum [24]. This use of inertia and gravity coupled with the elastic energy stored and recovered from tendons, muscles, and bones, helps to account for the efficiency of animal locomotion.

2. Lagrangian Dynamics

For fully actuated mechanical systems a broad range of powerful techniques were developed in the last decade for the design of optimal, robust, adaptive, and learning controllers [40]. These techniques are possible because fully actuated systems possess a number of strong properties that facilitate control design, such as feedback linearizability, passivity, matching conditions, and linear parametrizability. For underactuated systems one or more of the above structural properties are usually lost. Moreover, undesirable properties such as higher relative degree and nonminimum phase behavior are manifested. For these reasons, control design becomes much more difficult and there are correspondingly fewer results available.

Consider the Lagrangian formulation of the dynamics of an n–degree-of-freedom mechanical system

$$D(q)\ddot{q} + C(q,\dot{q})\dot{q} + g(q) = B(q)\tau \tag{2.1}$$

where $q \in R^n$ is the vector of generalized coordinates, $\tau \in R^m$ is the input generalized force ($m < n$), and $B(q) \in R^{n\times m}$ has full rank for all q.

For a suitable partition of the vector q of generalized coordinates as $q^T = (q_1^T, q_2^T)$, where $q_1 \in R^{n-m}$ and $q_2 \in R^m$ we may write the system (2.1) as

$$\begin{aligned} d_{11}\ddot{q}_1 + d_{12}\ddot{q}_2 + h_1(q_1,\dot{q}_1,q_2,\dot{q}_2) + \phi_1(q_1,q_2) &= 0 & (2.2)\\ d_{12}\ddot{q}_1 + d_{22}\ddot{q}_2 + h_2(q_1,\dot{q}_1,q_2,\dot{q}_2) + \phi_2(q_1,q_2) &= b(q_1,q_2)\tau & (2.3)\end{aligned}$$

where h_i include Coriolis and centrifugal terms, and ϕ_i contains the terms derived from the potential energy, such as gravitational and elastic generalized forces. The $m \times m$ matrix $b(q_1, q_2)$ is assumed to be invertible.

Example 2.1. **Two-link robot.** Consider the two-link robot shown in Fig. (2.1):

$$\begin{aligned} d_{11}\ddot{q}_1 + d_{12}\ddot{q}_2 + h_1 + \phi_1 &= \tau_1 & (2.4)\\ d_{12}\ddot{q}_1 + d_{22}\ddot{q}_2 + h_2 + \phi_2 &= \tau_2 & (2.5)\end{aligned}$$

where

$$\begin{aligned} d_{11} &= m_1\ell_{c1}^2 + m_2(\ell_1^2 + \ell_{c2}^2 + 2\ell_1\ell_{c2}\cos(q_2)) + I_1 + I_2\\ d_{22} &= m_2\ell_{c2}^2 + I_2\\ d_{12} &= m_2(\ell_{c2}^2 + \ell_1\ell_{c2}\cos(q_2)) + I_2\\ h_1 &= -m_2\ell_1\ell_{c2}\sin(q_2)\dot{q}_2^2 - 2m_2\ell_1\ell_{c2}\sin(q_2)\dot{q}_2\dot{q}_1\\ h_2 &= m_2\ell_1\ell_{c2}\sin(q_2)\dot{q}_1^2\\ \phi_1 &= (m_1\ell_{c1} + m_2\ell_1)g\cos(q_1) + m_2\ell_{c2}g\cos(q_1+q_2)\\ \phi_2 &= m_2\ell_{c2}g\cos(q_1+q_2)\end{aligned}$$

If $\tau_1 = 0$ this system represents the Acrobot [13, 5], while if $\tau_2 = 0$ the system represents the Pendubot [39]. In addition, with $\phi_1 = 0 = \phi_2$ and $\tau_2 = 0$ one has the underactuated manipulator system considered in by several authors, such as [28, 9, 2].

Example 2.2. **Cart-pole system.** The cart-pole system is one of the classic examples and yet it still holds some interesting challenges from the standpoint of global nonlinear control. Referring to Fig. (2.2) the dynamics are given by:

$$\begin{aligned} (m_p + m_c)\ddot{x} + m_p\ell\cos\theta\ddot{\theta} - m_c\dot{\theta}^2\sin\theta &= F\\ m_p\ell\cos\theta\ddot{x} + m_p\ddot{\theta} - m_p\ell g\sin\theta &= 0\end{aligned}$$

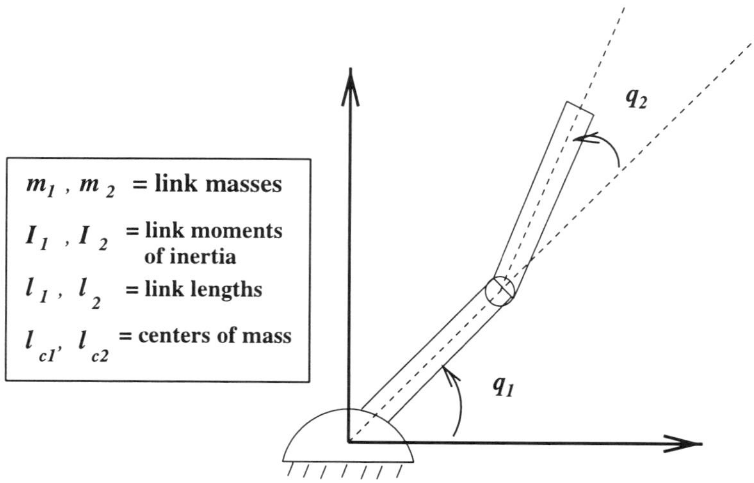

Fig. 2.1. Two link robot

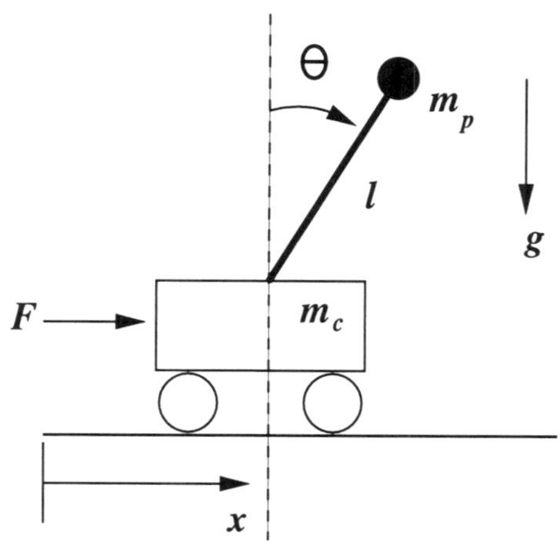

Fig. 2.2. Cart-pole system

For simplicity we normalize all constants to unity. To put the system in standard form, we set $q_1 = \theta$, $q_2 = x$, $\tau = F$, and write the equations as

$$\ddot{q}_1 + \cos q_1 \ddot{q}_2 - \sin q_1 = 0 \tag{2.6}$$
$$\cos q_1 \ddot{q}_1 + 2\ddot{q}_2 - \dot{q}_1^2 \sin q_1 = \tau \tag{2.7}$$

2.1 Equilibrium Solutions and Controllability

The nature of the fixed points of (2.2), (2.3) is closely tied to the controllability of the system. Let $\tau = \bar{\tau}$ = constant. Then, since the terms h_i are quadratic in the velocities $\dot{q}_i$, the equilibrium solutions satisfy

$$\phi_1(q_1, q_2) = 0 \tag{2.8}$$
$$\phi_2(q_1, q_2) = b(q_1, q_2)\bar{\tau} \tag{2.9}$$

and may either be isolated fixed points for each fixed $\bar{\tau}$, as in the case of the Acrobot, and Pendubot or they may be higher dimension as happens (for $\bar{\tau} = 0$) in systems without potential terms. For example, in the absence of gravity, the Pendubot dynamics satisfies

$$\phi_i(q_1, q_2) = 0 \qquad \text{for } i = 1, 2 \tag{2.10}$$

for all $(q_1, q_2) \in Q$, where Q denotes the two dimensional configuration space.

In the first case, systems with potential terms are linearly controllable around (almost all) fixed points, i.e. the Taylor series linearization is a controllable linear system. Systems without potential terms are generally not linearly controllable. Their local controllability properties are therefore more subtle to determine.

We may interpret Eq. (2.2) as a (dynamic) constraint on the accelerations of the generalized coordinates. It is then interesting to ask whether these constraints are holonomic, i.e. integrable. For many of the most interesting cases, including underactuated robot manipulators [28], the Acrobot [37] and Pendubot [39], the PVTOL system [43], the TORA system, and underwater robots, these constraints turn out to be completely nonintegrable as shown in [30]. An important consequence is that the system (2.2), (2.3) is (strongly) accessible, since nonintegrability of the second order constraint equations means that the dimension of the reachable set is not reduced.

Accessibility does not imply stabilizability of an equilibrium configuration using time-invariant continuous state feedback (either static or dynamic). In fact, for systems without potential terms it is known [30] that such stabilizability is not possible. The proof of this follows from an application of Brockett's Theorem [6]. The situation here is, therefore, quite similar to the case of control of nonholonomic mobile robots. Of course, systems with potential terms are exponentially stabilizable by linear time-invariant feedback.

3. Partial Feedback Linearization

An interesting property that holds for the entire class of underactuated mechanical systems is the so-called collocated partial feedback linearization property [36], which is a consequence of positive definiteness of the inertia matrix. A related property, the non-collocated partial feedback linearization, holds for a restricted class of underactuated systems.

3.1 Collocated Linearization

Collocated linearization refers to a control that linearizes the equations associated with the actuated degrees of freedom q_2. Equivalently, collocated linearization can be thought of as input/output linearization [14] with respect to an output $y = q_2$. The result states that the original system (2.2), (2.3) is feedback equivalent to the system

$$d_{11}\ddot{q}_1 + h_1 + \phi_1 = -d_{12}u \tag{3.1}$$

$$\ddot{q}_2 = u \tag{3.2}$$

with a suitable nonlinear feedback control

$$\tau = \alpha(q_1, \dot{q}_1, q_2, \dot{q}_2) + \beta(q_1, q_2)u \tag{3.3}$$

where u is a new control input to be determined. The derivation is straightforward and is contained in [36].

3.2 Non-collocated Linearization

Non-collocated linearization refers to linearizing the passive degrees of freedom and is possible under a special assumption on the inertia matrix of the robot.

Definition 3.1. *The system (2.2), (2.3) is (locally) Strongly Inertially Coupled if and only if*

$$rank(d_{12}(q)) = n - m \qquad \text{for all } q \in \mathcal{B}$$

where $\mathcal{B}$ is a neighborhood of the origin. The Strong Inertial Coupling is global if the rank condition holds for all $q \in Q$.

Note that Strong Inertial Coupling requires $m \geq n - m$, i.e. that the number of active degrees of freedom be at least as great as the number of passive degrees of freedom. Under the assumption of Strong Inertial Coupling we may compute a pseudo–inverse $d_{12}^{\dagger}$ for d_{12} as

$$d_{12}^{\dagger} = d_{12}^T(d_{12}d_{12}^T)^{-1}$$

and show the existence of a feedback control τ that transforms the system into the following feedback equivalent

$$\begin{aligned} \ddot{q}_1 &= u \\ \ddot{q}_2 &= -d_{12}^{\dagger}(d_{11}u + h_1 + \phi_1) \end{aligned}$$

The details are contained in [36]. We note that a system satisfying the (local) Strong Inertial Coupling Property is known as an *Internal/External Convertible* system in the terminology of Getz [12].

Example 3.1. **Cart-pole system.** The cart-pole system,

$$\begin{aligned} \ddot{q}_1 + \cos q_1 \ddot{q}_2 - \sin q_1 &= 0 \\ \cos q_1 \ddot{q}_1 + 2\ddot{q}_2 - \dot{q}_1^2 \sin q_1 &= \tau \end{aligned}$$

is strongly inertially coupled for $-\pi/2 < q_1 < \pi/2$ but not strongly inertially coupled globally. For collocated linearization it is easy to show that the control law

$$\tau = (2 - \cos^2 q_2)u + \cos q_1 \sin q_1 - \dot{q}_1^2 \sin q_1 \tag{3.4}$$

results in the feedback equivalent system

$$\begin{aligned} \ddot{q}_1 &= \sin q_1 - \cos q_1 u \\ \ddot{q}_2 &= u \end{aligned}$$

which is valid globally, while the control law

$$\tau = 2 \tan q_1 - \dot{q}_1^2 \sin q_1 - \frac{1 + \sin^2 q_1}{\cos q_1} u \tag{3.5}$$

results in the feedback equivalent system

$$\begin{aligned} \ddot{q}_1 &= u \\ \ddot{q}_2 &= \tan q_1 - \frac{1}{\cos q_1} u \end{aligned}$$

valid for $q_1 \in (-\pi/2, \pi/2)$.

4. Cascade Systems

The advantages of the first stage partial feedback linearization are both a conceptual and a structural simplification of the control problem. We can write the systems under consideration, after the first stage partial feedback linearization, as

$$\dot{x} = Ax + Bu \tag{4.1}$$

$$\dot{\eta} = w(\eta) + h(\eta, x) + g(\eta, x)u \tag{4.2}$$

with suitable definitions of all quantities, such that $h(\eta, 0) = 0$. The pair (A, B) is controllable since the linear system is a set of m double integrators and the expression

$$\dot{\eta} = w(\eta) \tag{4.3}$$

represents the zero dynamics [14]. If the control term u is chosen to be a function only of x, for example $u = -Kx$, then the system will be in cascade form

$$\dot{x} = \bar{A}x \tag{4.4}$$
$$\dot{\eta} = \bar{w}(\eta, x) \tag{4.5}$$

where $\bar{A} = A - BK$ is a Hurwitz matrix and $\bar{w}(\eta, x) = w(\eta) + h(\eta, x) - g(\eta, x)Kx$. There are a number of local and global stabilization results for special classes of such cascade systems. Both the nature of the equilibrium solution of the zero dynamics and the nature of the coupling between the x and η subsystems determines the type of results that can be proven. See [33] for a detailed treatment of the latest results.

4.1 Passivity and Energy Control

For general nonlinear systems of the form (4.4), (4.5) local asymptotic stability of the origin is guaranteed if the origin of the zero dynamics is locally asymptotically stable. Such systems are called minimum phase. Global stability requires consideration of issues such as peaking [42]. The systems considered here generally have multiple equilibrium points and are non-minimum phase in a neighborhood of a typical equilibrium point. Nevertheless we can utilize the special structure of the system (as a Lagrangian system) to show global stability in some cases. The crucial fact is the following,

Theorem 4.1. *Given the Lagrangian mechanical system (2.2), (2.3), the zero dynamics of the collocated feedback equivalent system (3.1), (3.2), equivalently (4.3), also defines a Lagrangian system, in particular, there exists a positive definite scalar (energy) function, $E(\eta)$, such that*

$$L_w E = 0 \tag{4.6}$$

The proof of this theorem is straightforward and is omitted. Mainly one needs to show that the kinetic energy of the original system is positive definite when restricted to the zero dynamics manifold $x = 0$ of (4.4), (4.5). It is interesting to note that, in the case of non-collocated linearization, the zero dynamics fails to be a Lagrangian system. However, in some cases a Lyapunov-like function E may still be found satisfying (4.6). The importance of this result is that it can be used to ensure stability of the interconnection.

Consider a slightly simplified (single input) system in the form

$$\dot{x} = Ax + Bu \tag{4.7}$$
$$\dot{\eta} = w(\eta) + g(\eta, x)u \tag{4.8}$$

Since $L_w E = 0$ we have

$$\dot{E} = L_g E u \tag{4.9}$$

which implies that (4.8) defines a passive system with respect to the input u and output $y_\eta = L_g E$. If we therefore choose $u = -Kx$ in (4.7) so that the transfer function $K(sI - A)^{-1}B$ is Strictly Positive Real (SPR), i.e. so that (4.1) is passive with respect to the output $y_1 = Kx$, then (4.7), (4.8) can be represented as a feedback interconnection of passive systems and is therefore passive. Stability of the interconnection follows from an additional (detectability) assumption [33]. The trajectory of the system will, in fact, converge to a particular energy level, which corresponds to a particular trajectory on the zero dynamics manifold. This idea has been used to design swingup controllers for systems like the cart-pole, Acrobot, and Pendubot [38].

Example 4.1. **Swingup control.** Applying this control to the cart-pole system results in the response shown in Fig. 4.1. Note that asymptotic stability

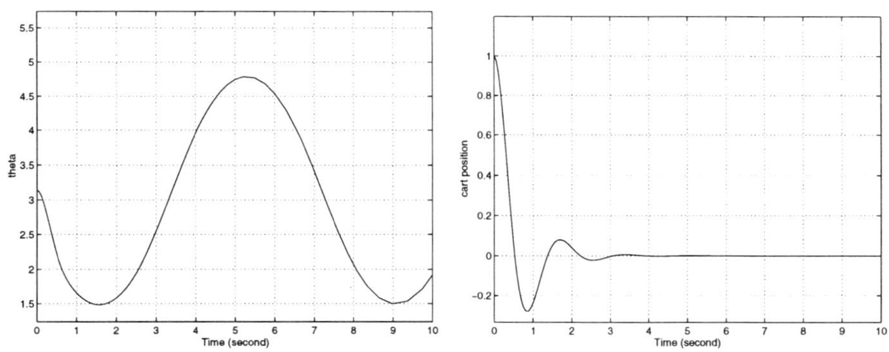

Fig. 4.1. Response of the cascade cart-pole system

is only guaranteed to a manifold and not to a fixed point. For this reason, control must eventually switch to a second control that locally stabilizes the equilibrium point. However, the control design is very simple and widely applicable as one way to overcome problems with feedforward nonlinearities and non-minimum phase zero dynamics [25, 37, 39].

4.2 Lyapunov Functions and Forwarding

Important extensions of the proceeding ideas are due to Teel [44], Mazenc and Praly [22], Sepulchre, Janković, and Kokotović [16, 32] and others who provide constructive procedures for global and semi-global asymptotic stabilization to a fixed point, rather than just to a manifold, for restricted subclasses of underactuated systems. A detailed discussion of these results is outside the scope of this chapter. The reader is referred to [33] and the references therein for details.

As a brief glimpse into one such approach we can illustrate the basic idea of the method of *forwarding* [32]. Suppose that a Lyapunov function V_0 for the zero dynamics (4.3) is known and satisfies (4.6). Since $\bar{A} = A - BK$ is Hurwitz, $V_1 = x^T Px$ defines a Lyapunov function for (4.4) where P satisfies a Lyapunov equation for $\bar{A}$. The techniques in [33] provide procedures for constructing cross terms $\Phi(\eta, x)$ such that

$$V(\eta, x) = V_0(\eta) + \Phi(\eta, x) + V_1(x) \tag{4.10}$$

defines a Lyapunov function for the system (4.4), (4.5). Calculating $\dot{V}$ along trajectories of (4.4), (4.5) gives

$$\dot{V} = L_w V_0 + L_{h-gKx} V_0 + \dot{\Phi} + L_{\bar{A}x} V_1 \tag{4.11}$$

The trick is now to show that there exists Φ satisfying

$$\dot{\Phi} = -L_{h-gKx} V_0 \tag{4.12}$$

that simultaneously guarantees the required properties for V to be a (radially unbounded) Lyapunov function for the system. If such an expression can be found then global stability of (4.4), (4.5) is assured. One may now go back and augment the control input u as

$$u = -Kx + v \tag{4.13}$$

and consider the system

$$\dot{x} = \bar{A}x + Bv \tag{4.14}$$

$$\dot{\eta} = \bar{w}(\eta, x) + g(\eta, x)v \tag{4.15}$$

This system is of the form

$$\dot{z} = F(z) + G(z)v \tag{4.16}$$

where a Lyapunov function V is known satisfying $L_F V \leq 0$. It then follows that a Jurdjevic-Quinn type of control [17]

$$v = -L_G V \tag{4.17}$$

can be used to achieve global stability and, under further restrictions, global asymptotic stability.

4.3 Hybrid and Switching Control

The ideas in the previous section are applicable only to restricted classes of mechanical systems. The same is true for other techniques, such as backstepping [20], the technique of adding integrators due to Mazenc and Praly [22], and the saturation approach of Teel [43]. In the case of backstepping the system state equations must have a lower triangular structure while for forwarding the state equations must have an upper triangular structure. Even when applicable these methods lead to designs which can be computationally difficult. For example, the computation of the cross term Φ in (4.10) is possible only in simple examples.

One way to avoid these computational difficulties is to consider a hybrid control architecture that switches among several controllers, each of which may be simpler to design. For example, the global stabilization of the inverted position of the Pendubot using a single smooth controller designed using integrator forwarding is currently not possible. However, since the Pendubot is linearly controllable in a neighborhood of the inverted configuration, one need only design a nonlinear controller so that the trajectory intersects a suitable neighborhood of the desired equilibrium (swingup control) and then switch to a linear controller to stabilize the system around the equilibrium (balance control).

As an added benefit, the design of switching controllers in the context of locomotion is likely to lead to an improved understanding of locomotion in biological systems. The problem of locomotion while maintaining balance encompasses the transition from standing to walking and back to standing, as well as the transition among various gaits of locomotion. For example, a human is constantly starting, stopping, performing tasks while standing, sitting down, standing up, etc. Such a complex behavior cannot be achieved with a single smooth controller but may be achieved by switching among multiple controllers.

Consider a supervisory control architecture shown in Fig. 4.2 for the problem of swingup and balance of a gymnastic robot. The Supervisor switches between a nonlinear Swing Up Controller and a linear Balance Controller when the trajectory enters the basin of attraction of the local balancing controller. This architecture also allows for robustness to disturbances. Since the balance controller is only local, a large disturbance is handled by switching back into the swing up mode and re-converging to the basin of attraction of the balance controller. A successful swingup and balance control for the Acrobot is shown in Fig. 4.3.

4.4 Nonholonomic Systems

Underactuated systems which are not linearly controllable, typically those without gravitational or elastic terms in the dynamics, are amenable to motion planning and control approaches similar to those that used for mobile

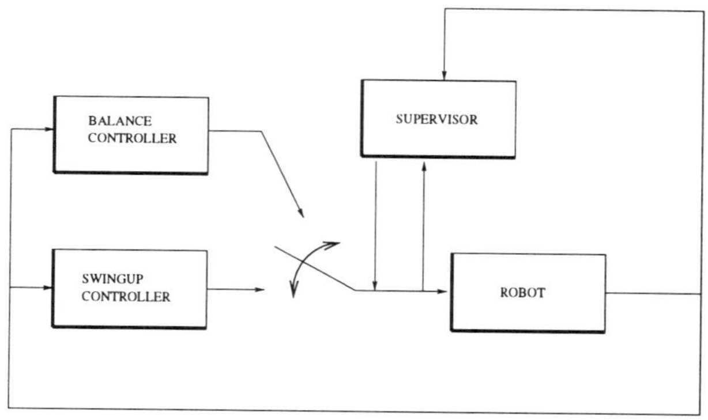

Fig. 4.2. Supervisory control architecture

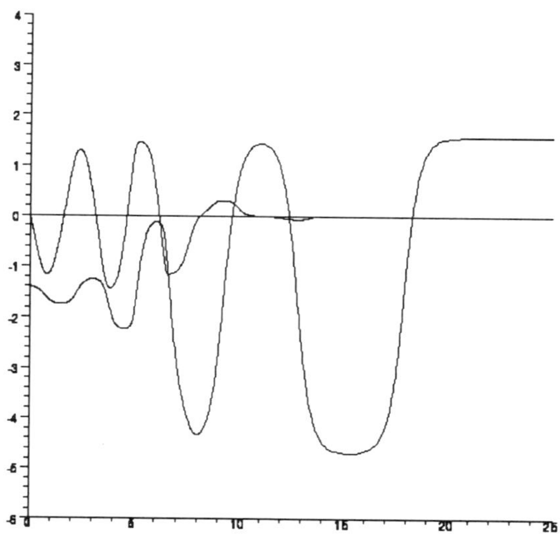

Fig. 4.3. Swingup and balance control for the Acrobot

robots. Consider the Pendubot in the absence of gravity, which after partial feedback linearization, can be expressed in the form [9]

$$\ddot{q}_1 = u \tag{4.18}$$
$$\ddot{q}_2 = -c \sin q_2 \dot{q}_1^2 - (1 + c \cos q_2) u \tag{4.19}$$

The main difference between systems, such as the above, with acceleration constraints and mobile robots with velocity constraints is the presence of the drift term in the equations of motion, which complicates the controllability analysis. For example, for systems without drift, accessibility (in the sense of full rank of the accessibility distribution [14]) implies controllability by Chow's Theorem [26]. This is no longer true for systems with drift.

A stronger notion of controllability for underactuated systems is the property of small time local controllability (STLC) [41]. A sufficient condition for STLC was given for underactuated mechanical systems in [8]. This result is important since it implies the existence of either discontinuous or time periodic feedback controllers to stabilize the system to a point. Algorithms for point to point control of the above system are given in [9, 30].

5. Conclusions

Underactuated mechanical systems of the type considered here present many challenging opportunities for future research.

1. **Robust and adaptive control:** It is well known that the Lagrangian dynamic equations of robots are linear in the inertia parameters and that this linearity is generally lost when the system is written in state space. For fully actuated systems, the passivity-based adaptive control [29] circumvents this difficulty. However, for underactuated systems, passivity is lost. The recursive design techniques of integrator backstepping and integrator forwarding are the proper extensions of passivity-based design techniques when they are applicable. At the present time these techniques are applicable for those systems that retain linearity in the parameters and which satisfy certain structural properties, growth conditions on the nonlinear coupling, etc. The extension of these methods for robust and adaptive control of larger classes of underactuated systems is thus a research problem of major importance.
2. **Saturation methods:** The method of Teel [43] using saturation functions is a powerful technique for achieving semi-global and global stabilization results. As in the case of backstepping and forwarding, results exist only for restricted classes of systems. Extending these methods to larger classes of underactuated systems is an important problem.
3. **Stability of hybrid systems:** The research problems in the use of hybrid and logic-based switching control for underactuated systems are

mainly at the supervisory level, i.e. determining when to switch and proving stability. This turns out to be highly non-trivial. Formal stability results exist only for limited classes of hybrid systems. It is, in fact, known that stability results for the general class of hybrid systems cannot be obtained. This is because the vocabulary for describing hybrid systems is too expressive to permit such strong results. One can embed a universal Turing machine into a hybrid system so that the stability question can reduce, in the worst case, to the *halting problem* [21], which is known to be undecidable. Thus one necessarily must focus on specific classes of hybrid systems in order to make progress.

References

[1] Álvarez-Gallegos Ja, Álvarez-Gallegos Jq, González-Hernández H G 1997 Analysis of the dynamics of an underactuated robot: The forced pendubot. Preprint
[2] Arai H, Tachi S 1991 Position control of a manipulator with passive joints using dynamic coupling. *IEEE Trans Robot Automat.* 8(4)
[3] Book W J 1982 Recursive Lagrangian dynamics of flexible manipulator arms via transformation matrices. In: *Proc IFAC Symp CAD Multivar Tech Syst.* W. Lafayette, IN, pp 5–17
[4] Bortoff S, Spong M W 1992 Pseudolinearization of the acrobot using spline functions. In: *Proc 31st IEEE Conf Decision Contr.* Tucson, AZ, pp 593–598
[5] Bortoff S A 1992 Pseudolinearization using spline functions with application to the Acrobot. PhD thesis, University of Illinois at Urbana–Champaign
[6] Brockett R W 1983 Asymptotic stability and feedback stabilization. In: Brockett R W, Millmann R S, Sussman H J (eds) *Differential Geometric Control Theory.* Birkhäuser, Boston, MA, pp 181–208
[7] De Luca A 1996 Nonholonomic behavior in redundant robot arms. Lecture Notes, Dutch Institute of Summer School
[8] De Luca A, Mattone R, Oriolo G 1996 Dynamic mobility of redundant robots using end-effector commands. In: *Proc 1996 IEEE Int Conf Robot Automat.* Minneapolis, MN, pp 1760–1767
[9] De Luca A, Mattone R, Oriolo G 1996 Control of underactuated mechanical systems: Application to the planar 2r robot. In: *Proc 35th IEEE Conf Decision Contr.* Kobe, Japan, pp 1455–1460
[10] Dubowsky S, Papadopoulos E 1993 The kinematics, dynamics, and control of free-flying and free-floating space robotic systems. *IEEE Trans Robot Automat.* 9(5)
[11] Fossen T I 1994 *Guidance and Control of Ocean Vehicles.* Wiley, Chichester, UK
[12] Getz N H 1995 Dynamic inversion of nonlinear maps with applications to nonlinear control and robotics. PhD thesis, University of California at Berkeley
[13] Hauser J, Murray R M 1990 Nonlinear controllers for non-integrable systems: the Acrobot example. In: *Proc 1990 Amer Contr Conf.* San Diego, CA, pp 669–671
[14] Isidori A 1989 *Nonlinear Control Systems.* Springer-Verlag, Berlin, Germany
[15] Janković M, Sepulchre R, Kokotović P V 1995 Global stabilization of an enlarged class of cascade nonlinear systems. Preprint

[16] Janković M, Sepulchre R, Kokotović P V 1996 Global adaptive stabilization of cascade nonlinear systems. In: *Proc 13th IFAC World Congr.* San Francisco, CA, pp 311–316
[17] Jurdjevic V, Quinn J P 1978 Controllability and stability. *J. Diff Eqs.* 28:381–389
[18] Khalil H K 1992 *Nonlinear Systems.* Macmillan, New York
[19] Kokotović P V, Krstic M, Kanellakopoulos I 1992 Backstepping to passivity: Recursive design of adaptive systems. In: *Proc 31st IEEE Conf Decision Contr.* Tucson, AZ, pp 3276–3280
[20] Krstic M, Kanellakopoulos I, Kokotović PV 1995 *Nonlinear and Adaptive Control Design.* Wiley, New York
[21] Lewis H R, Papadimitriou C H 1981 *Elements of the Theory of Computation.* Prentice-Hall, Englewood Cliffs, NJ
[22] Mazenc F, Praly L 1995 Adding an integration and global stabilization of feedforward systems. *IEEE Trans Automat Contr.* 40
[23] McGeer T 1990 Passive dynamic walking. *Int J Robot Res.* 9(2)
[24] McMahon T A 1984 *Muscles, Reflexes, and Locomotion.* Princeton University Press, Princeton, NJ
[25] Mukherjee R, Chen D 1993 Control of free-flying underactuated space manipulators to equilibrium manifolds. *IEEE Trans Robot Automat.* 9:561–570
[26] Murray R M, Li Z, Sastry S S 1994 *A Mathematical Introduction to Robotic Manipulation.* CRC Press, Boca Raton, FL
[27] Murray R M, Sastry S S 1993 Nonholonomic motion planning: Steering using sinusoids. *IEEE Trans Automat Contr.* 38:700–716
[28] Oriolo G, Nakamura Y 1991 Control of mechanical systems with second order nonholonomic constraints: Underactuated manipulators. In: *Proc 30th IEEE Conf Decision Contr.* Brighton, UK, pp 306–308
[29] Ortega R, Spong M W 1989 Adaptive motion control of rigid robots: a tutorial. *Automatica.* 25:877–888
[30] Reyhanoglu M, van der Schaft A J, McClamroch N H, Kolmanovsky I 1996 Nonlinear control of a class of underacturated systems. In: *Proc 35th IEEE Conf Decision Contr.* Kobe, Japan, pp 1682–1687
[31] Saito F, Fukuda T, Arai F 1993 Swing and locomotion control for two-link brachiation robot. In: *Proc 1993 IEEE Int Conf Robot Automat.* Atlanta, GA, vol 2, pp 719–724
[32] Sepulchre R, Janković M, Kokotović P V 1996 Integrator forwarding: A new recursive nonlinear robust design. In: *Proc 13th IFAC World Congr.* San Francisco, CA, pp 85–90
[33] Sepulchre R, Janković M, Kokotović P V 1997 *Constructive Nonlinear Control.* Springer-Verlag, London, UK
[34] Slotine J-J E, Li W 1991 *Applied Nonlinear Control.* Prentice-Hall, Englewood Cliffs, NJ
[35] Spong M W 1987 Modeling and control of elastic joint robots. *ASME J Dyn Syst Meas Contr.* 109:310–319
[36] Spong M W 1994 The control of underactuated mechanical systems. In: *1st Int Conf Mechatron.* Mexico City
[37] Spong M W 1995 The swingup control problem for the acrobot. *IEEE Contr Syst Mag.* 15(1):49–55
[38] Spong M W 1996 Energy based control of a class of underactuated mechanical system. In: *Proc 13th IFAC World Congr.* San Francisco, CA, vol F, pp 431–436

[39] Spong M W, Block D 1995 The pendubot: A mechatronic systems for control research and education. In: *Proc 34th IEEE Conf Decision Contr.* New Orleans, LA, pp 555–557

[40] Spong, M W, Lewis F L, Abdallah C T 1992 *Robot Control: Dynamics, Motion Planning, and Analysis.* IEEE Press, Piscataway, NJ

[41] Sussmann H J 1987 A general theorem on local controllability. *SIAM J Contr Opt.* 25:158–194

[42] Sussmann H, Kokotović P 1991 The peaking phenomenon and the global stabilization of nonlinear systems. *IEEE Trans Automat Contr.* 36:424–439

[43] Teel A 1992 Using saturation to stabilize a class of single-input partially linear composite systems. In: *Proc 3rd IFAC Symp Nonlinear Contr Syst.* Bordeaux, France, pp 24–26

[44] Teel A R, Praly L 1995 Tools for semi-global stabilization by partial state and output feedback. *SIAM J Contr Opt.* 33:1443–1488

[45] Wichlund K Y, Sørdalen O J, Egeland O 1995 Control of vehicles with second-order nonholonomic constraints: Underactuated vehicles. In: *Proc 3rd Euro Contr Conf.* Rome, Italy, pp 3086–3091

Trends in Mobile Robot and Vehicle Control

Carlos Canudas de Wit

Laboratoire d'Automatique de Grenoble, ENSIEG–INPG, France

This chapter presents an overview of some of the actual trends in control design for mobile robots and multibody wheeled vehicles. The control schemes are presented by area of application. Our discussion includes automatic parking, path following, vision-guided vehicle systems, multibody vehicle control. Approaches such as the follow-the-leader principle and other areas like car platooning in highways and transportation systems are also discussed.

1. Introduction

A lot of work has been done in the area of control of mobile robots and vehicles. Most of the literature in the mobile robot nonlinear control area has been motivated by intellectual curiosity driven by the stabilization obstruction indicated first by Brockett theorem [1]. This theorem shows that the nonlinear integrator, which is diffeomorphic equivalent to the unicycle kinematics, cannot be stabilized by time-invariant smooth feedback. Being established this difficulty, many control alternatives (time-varying, nonsmooth) have been proposed. The work in [10] provides a complete survey on different control designs for nonholonomic systems. A study on the structural properties of the nonholonomic models and the closed-loop properties of several feedback laws are also given in the last three chapters of the book [3].

Other problems, such as trajectory and path tracking, can be solved via smooth controllers provided that additional assumptions are made. For instance, in the trajectory tracking problem (the system states should follow a target model), the lost of stabilization (via smooth time-invariant feedback) is partially recovered as far as the target model is kept in permanent motion. In the path tracking problem, it is also well understood that convergence to the target path can be obtained in the spatial coordinates, see [3].

However, not all the mobile robot and vehicle control problems can be cast into the three categories of problems mentioned above. For instance, the problem of controlling a set of vehicles in platooning gives rise to problems like collision avoidance, string stability, etc. In some applications, the states used in the feedback laws are not necessarily directly measured from the existing sensors, but should be indirectly computed from other sensors, or estimated from state observers. For example, while controlling a platoon of vehicles, the variables such as a relative distance and orientations will be probably more easy to measure than the absolute position and altitude of each vehicle. Another example can be found in vision-guided systems; the "output" variables are not necessarily given in terms of the Cartesian coordinates and angles with respect to a fixed frame, but as a function of the internal TV camera coordinates —such as the perspective projection— in the moving vehicle frame.

It is thus often necessary to reformulate the original problem, in order to incorporate the available sensor information, and to redefine the control objectives which are not necessarily reachable by using the existing solutions proposed in the literature. One of the purposes of this chapter is to review some of these problems and to describe some of the adopted solutions by area of application. Our discussion here goes beyond the problem of mobile robot control, which seems to be a research area mainly developed at the universities and some government research laboratories, and addresses new control problems formulated by the car industry and transportation systems.

The chapter is organized by category of problems. After recalling well-known model properties, the following sections deal sequentially with the following problems: automatic parking, path following, visual-based system, multibody vehicle control, vehicle platooning (train-like vehicles and car platooning in highways and transportation systems).

2. Preliminaries

Through this chapter we will be concerned with nonholonomic systems, where each independent vehicle will be modeled as:

$$\dot{z} = G(z)u \tag{2.1}$$

$$\dot{u} = \nu \tag{2.2}$$

where z describes the generalized coordinates of dimension n, u is the generalized "velocity" vector of dimension[1] $m < n$, and ν is the "acceleration"(or torque) control input. In short, the matrix $G(z)$ describes the direction at which motion is possible at the given position z, i.e. $A(z)G^T(z) = 0$, where $A(z)\dot{z} = 0$ describes the nonholonomic constraints. With an abuse of language, the upper equation (2.1) is known as the kinematic model, where together with the cascade integrator (2.1), (2.2), the equations are known as the dynamic model. The reason for this nomenclature is that the Lagrange model of a class of nonholonomic system systems is feedback equivalent to the above simplified equations. Besides, most of the more important properties of this model are captured by the kinematic structure of matrix $G(z)$ (see [3]):

[1] In this chapter we will only consider vehicles with restricted mobility where the number of control variables is smaller than the dimension of the generalized coordinates.

- **Controllability**. If the input matrix $G(z)$ has full rank, and the involutive distribution $\bar{\Delta} = \text{inv span}\ \{\text{col}(G(z))\}$ has constant maximal dimension for all z (i.e. dimension m), then the kinematic model (2.1) is fully controllable. Indeed, for the kinematic driftless system (2.1), the strong accessibility condition coincides with the involutive distribution $\bar{\Delta} = \text{inv span}\ \{\text{col}(G(z))\}$, implying controllability. This means that a mobile robots can be driven from any initial configuration $z(0)$ to any final one $z(T)$ in a finite time, by manipulating the "velocity" control input $\nu(t)$. This property is preserved when the cascade integrator is added to the kinematic model. Hence the dynamic system (2.1), (2.2) having a drift term is small-time-locally controllable.
- **Stabilization**. The kinematic model (2.1) is not stabilizable by a continuous static time-invariant state feedback $u(z)$. Indeed, the so-called Brockett's necessary condition is not satisfied by a continuous static feedback, since the map $(z, u) \mapsto G(z)u$ is not onto in the neighborhood of the equilibria. A similar problem is also found with the dynamic model (2.1)–(2.2).
- **Linear approximation**. The linear approximation of the kinematic model (2.1) about the equilibrium point (i.e. $z = u = 0$) is not controllable. Indeed, the model $\dot{z} = G(0)u$ is neither controllable nor stabilizable. Notice, however, that the full nonlinear model $\dot{z} = G(z)u$ is controllable in spite of the lack of the smooth stabilizability.
- **Feedback linearizability** The kinematic and dynamic models are not full state linearizable by smooth static time-invariant state feedback.[2] For some special cases where only a subset of the coordinates z are of interest, it is possible to perform partial state linearization under permanent motion condition of the target coordinates (this problem is widely covered in Chapter 8 of [3]. Also, for some output functions relevant to the control problem at hand, output linearization (in the spatial coordinates) is also be possible. This last problem will be discussed in detail in the next sections.

For the sake of simplicity, in what follows we will only consider kinematic models.

3. Automatic Parking

Automatic parking is a typical application where both path planning and feedback control need to be combined. Some car builders are now studying automatic parking mechanisms that may be integrated in the next generation of commercial products. Industries of transportation buses and trucks, where this problem is more difficult to be performed manually, are also seeking

[2] This can be easily seen by observing that there is no way to locally invert the linear approximation $\dot{z} = G(0)u$, i.e. to find a change of coordinates in the input space so that $\bar{u} = G(0)u$ for all u and $\bar{u}$.

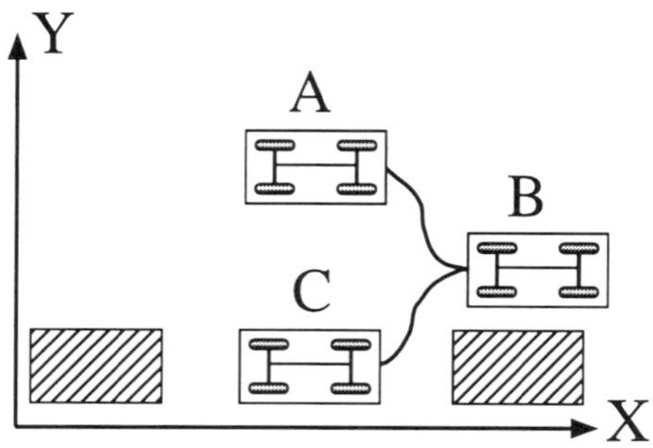

Fig. 2.1. A typical parking maneuver

for similar features. An automatic park maneuver is decomposed into two, somewhat separated, problems:

- motion planning, and
- stabilization along the planned trajectory.

Figure 2.1 shows a typical parking maneuver. The vehicle should be steered from point A to B, and then to point C with a final prescribed orientation. Since motion should be performed in clustered environments, the smooth curve $\mathcal{C}_{ABC}$ is defined before motion starts, using polynomial interpolation or splines, in order to account for possible geometric constraints. One way to characterize the smooth planar curve $\mathcal{C}_{ABC}$ is by using the Frénet frame: the curve $\mathcal{C}_{ABC}$ is expressed as a function of the length of the path s and the path curvature $\kappa(s)$. Regular Cartesian parameterization is also possible, $y = f(x)$, where (x, y) are the Cartesian coordinates, and $f(x)$ is a polynomial function. Smoothness of the full curve $\mathcal{C}_{ABC}$ imposes additional conditions on the path curvature of $f(x)$ (and its higher partial derivatives) at the boundary points of the curves $\mathcal{C}_{AB}$ and $\mathcal{C}_{BC}$. This may lead to polynomial curves of high degrees, but ensures the continuity and existence of bounded open-loop control $u(t)$. The degree of the polynomial curve to be planned depends on the type of problem to be considered. For instance, velocity control based on the kinematic model will require polynomial $y = f(x)$ of lower order than the ones obtained from a dynamic formulation.

Once the path $y = f(x)$ has been planned under the above mentioned smoothness requirements, assuming that the vehicle under study is located already on the curve $y = f(x)$, the complete motion planning problem consists in finding the smooth open-loop control $u(t)$ steering the system from its initial configuration (point A) to the final configuration (point C), along the curve $\mathcal{C}_{ABC}$. This problem has been extensively treated in connection with flat systems in papers by Rouchon and co-workers (see for instance [18, 6]). The idea is better explained through the following simple example.

Consider the unicycle kinematics:

$$\dot{x} = v\cos\theta \tag{3.1}$$
$$\dot{y} = v\sin\theta \tag{3.2}$$
$$\dot{\theta} = w \tag{3.3}$$

where v and w are the kinematic inputs: the translational velocity and the rotational velocity, respectively. The position in the plane is given by (x, y), and the orientation angle with respect to x is θ. Given the polynomial curve $y = f(x)$, our problem is to find w and v so as to stay on this curve. The system (3.2) is assumed to lie initially on the path, and to be oriented along the tangent of the curve. The ratio of change dy/dx, obtained from the first two equations of the unicycle kinematics, is used to define the desired angle θ_d, as a function of the curve $y = f(x)$, that makes the vehicle remain on the path:

$$\tan\theta_d = \left.\frac{dy}{dx}\right|_{y=f(x)} = \frac{d}{dx}f(x)$$

From (3.3) we get the desired rotational velocity w_d by setting $\dot{\theta}_d = w_d$, which substituted in the time-derivative of the above relation gives:

$$\frac{d\tan\theta_d}{d\theta_d}\dot{\theta}_d := \frac{df(x)}{dx}\dot{x} \tag{3.4}$$
$$\frac{1}{\cos^2\theta_d}w_d = v\cdot\frac{d^2f(x)}{dx^2}\cos\theta_d \tag{3.5}$$
$$w_d = v\cdot\frac{d^2f(x)}{dx^2}\cos^3\theta_d \tag{3.6}$$
$$= v\cdot\frac{d^2f(x)}{dx^2}\left[1+\left(\frac{dy}{dx}\right)^2\right]^{-3/2} \tag{3.7}$$

Remark 3.1. The degree of the polynomial $y(x)$ should be at least of order two. However, initial and final constraints on the curve endpoints may increase this order.

Remark 3.2. The desired rotational velocity w_d depends on the translational velocity. This is typical in path following problems, and is used to avoid control singularities: it regularizes the open-loop control w_d when the vehicle stops. The translational velocity v can be seen as an additional control degree of freedom that defines the actual position along the curve. The time-profile of $v(t)$ defines the time at which the end points of the curve are reached. v_d can be planned off-line, but it can also be defined by the driver during the parking maneuver.

Remark 3.3. In the spatial coordinate $s = \int_0^t |v|dt$, the unicycle equations together with the open-loop control developed above ($w = w_d$) can also be rewritten as:

$$\frac{dx}{ds} = \cos\theta \tag{3.8}$$

$$\frac{dy}{ds} = \sin\theta \tag{3.9}$$

$$\frac{d\theta}{ds} = w/v|_{w=w_d} = \frac{d^2 f(x)}{dx^2}\left[1+\left(\frac{dy}{dx}\right)^2\right]^{-3/2} \tag{3.10}$$

The open-loop control w_d/v is nothing else than the radius of curvature of $f(x)$ that is made to coincide with the instantaneous radius of curvature of the unicycle kinematics w/v. This solution corresponds to the inverse control problem in the sense that for a given trajectory specified in the system state coordinates, it is possible to invert the system and compute the system inputs that provide such a trajectory. As it will be shown next, this controller can be also seen as a partially linearizable state feedback that linearize the "output" $y(s)$ in the s-space.

Stabilization along the planned trajectory can be easily explained by first observing that the above "open-loop" control[3] linearizes the y coordinate in the s-space.

From (3.9), we see that the input w appears after computing the second partial derivative of y, i.e.

$$\frac{d^2y}{ds^2} = \cos\theta\frac{d\theta}{ds} = \cos\theta\cdot\frac{w}{v}$$

Linearization of y in the s coordinates implies selecting the ratio w/v as:

$$\frac{w}{v} = \frac{1}{\cos\theta}u \quad \rightarrow \quad \frac{d^2y}{ds^2} = u$$

where u is the new control variable to be designed. The condition needed for the invariance of the manifold $y - f(x) = 0$ is that $\frac{d^2y}{ds^2} = \frac{d^2f(x)}{ds^2}$. Hence $u = \frac{d^2f(x)}{ds^2}$. Since the path specifications are here given as a function of x and not as a function of s (although this may be possible), u should be rewritten only as a function of $f(x)$. After some calculation we can shown that

$$\frac{d^2f(x)}{ds^2} = \cos^2\theta\cdot\frac{d^2f(x)}{dx^2} - \frac{df(x)}{dx}\sin\theta\frac{w}{v}$$

which, after transformation back to the original control coordinates in $\frac{w}{v}$, gives

[3] It can be argued that w_d is not really an open-loop control since it explicitly depends on the state x via the first and second partial derivatives of $f(x)$. However, note that this control only ensures the invariance of the manifold $y - f(x) = 0$, but it does not ensure its attractibility. In that sense, w_d is considered here as an "open-loop" feedback.

$$\frac{w}{v} = \frac{1}{\cos\theta}u = \frac{1}{\cos\theta}\frac{d^2f(x)}{ds^2} \tag{3.11}$$

$$= \frac{d^2f(x)}{dx^2}\left[1+\left(\frac{dy}{dx}\right)^2\right]^{-3/2} \tag{3.12}$$

that is the same control obtained by imposing that the orientation angle of the vehicle coincides with the tangent of the curve to be tracked. Attractibility of this curve is now obtained by simply adding corrective terms in u as:

$$u = \frac{d^2f(x)}{ds^2} + k_v\left(\frac{df(x)}{ds} - \frac{dy}{ds}\right) + k_p\left(f(x) - y\right)$$

which in the spatial error coordinates $\tilde{y} := y_d - y = f(x) - y$ gives a second-order linear space invariant model

$$\frac{d^2\tilde{y}}{ds^2} + k_v\frac{d\tilde{y}}{ds} + k_p\tilde{y} = 0$$

An important issue here is that the attractibility of the curve is not a function of time, as in classical stability definitions, but as a function of the length of path. From the above equation, it can be observed that

$$|\tilde{y}(s)| \leq |\tilde{y}(0)|\exp\{-c\cdot s(t)\}$$

Since the length of the path in the parking maneuver is limited (i.e. $s(t) \in [0, L_{ABC}]$, asymptotic stability in the classical sense (where s is replaced by time) is not possible, but instead uniform ultimate stability is obtained. This implies that the error $\tilde{y}(s)$ will converge uniformly to a ball of radius r, that can be made arbitrarily small by increasing the constant c which depends on the controller gains k_p and k_v. For many applications concerning this class of problems, this kind of approach suffices for most practical purposes.

This type of idea has been applied to other more complex systems like multi-steering trailer systems, and systems moving in an environment with obstacles. Some of these works are mentioned and discussed in [6].

4. Path Following

The path following problem can be seen to some extent as a particular case of the automatic parking maneuver problem. However, as it will be discussed later, it may present some important differences motivating a separate discussion.

The main distinction between the path following problem and the motion planning problem (like the automatic parking maneuver) is that the explicit path planning of the curve $\mathcal{C}_{ABC}$ in Fig. 2.1 may not always be needed. If the vehicle is initially located close to the line or path to be tracked, it may not

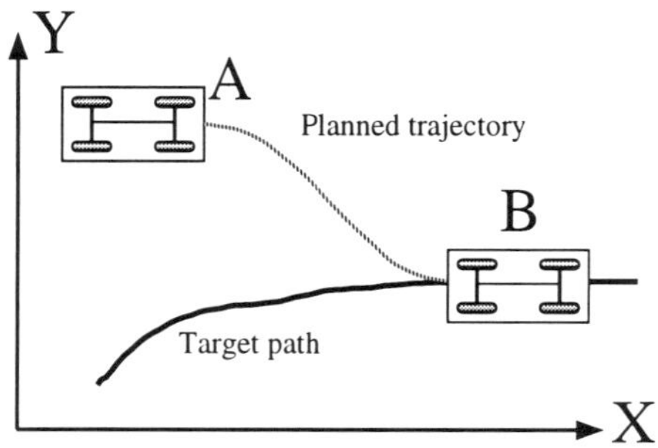

Fig. 3.1. Schematic view of the path following problem

be explicitly needed to characterize the trajectory along which the vehicle will reach the target path. The rejection properties of the used regulator will determine the way the initial error and disturbances will be canceled. The target path is given by the problem, in opposition to the automatic parking maneuver where only a target end-point posture is given and a curve is planned to reach such a point. Figure 3.1 shows a schematic view of the path following problem. An explicit planning of the trajectory from A to B may not be needed.

The other main difference with respect to the problem of automatic parking is that due to the long distances of the path to be tracked, it is not suitable to use fixed frames to formulate the vehicle kinematic and the regulation problem. It is thus suitable to use, instead, variables related to a frame attached to the moving vehicle and describing the error distance between the path and the vehicle. As will be discussed later, the selected variables are not always accessible from direct measurements, but it should be observed or computed from other indirect measures.

The path following problem is found in many applications ranging from mines to urban transportation systems, where a vehicle or a set of vehicles need to be guided along a path. The on-line information about the path is sometimes obtained via vision systems, magnetic fields or distance sensors. They provide information about the curve to be followed in term of distances, tangents, radius of curvature, etc. An example is shown in Figs. 4.1 and 4.2 corresponding to the CiVis concept recently proposed by Matra/Renault-VI where an articulated bus is aimed at following a path marked on the road. This system is proposed as a cheap alternative to urban trains running on railways, where the infrastructure may be too costly.

Except for the case of the unicycle kinematics where the solution is unique, in general the path following problem has not a unique formulation. Several choices for the variables (or points in the vehicle) to be regulated are possible. Figure 4.3 shows an example where the distances d_1 and d_2 are two possi-

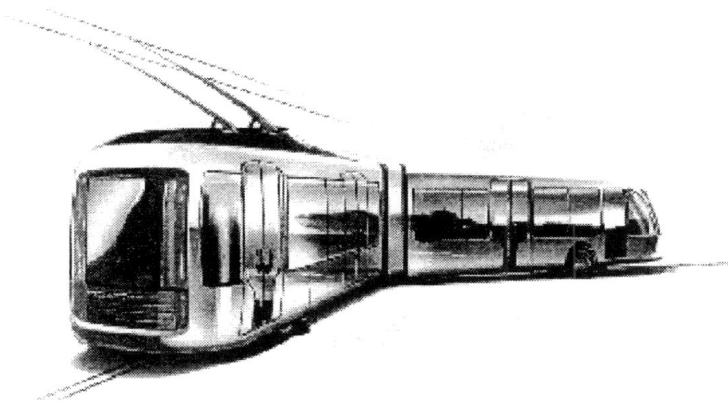

Fig. 4.1. The CiVis concept proposed by Matra and Renault: 3/4 rear-view of the articulated unit (courtesy of Renault-VI)

Fig. 4.2. The CiVis concept proposed by Matra and Renault-VI: crossing a square (courtesy of Renault-VI)

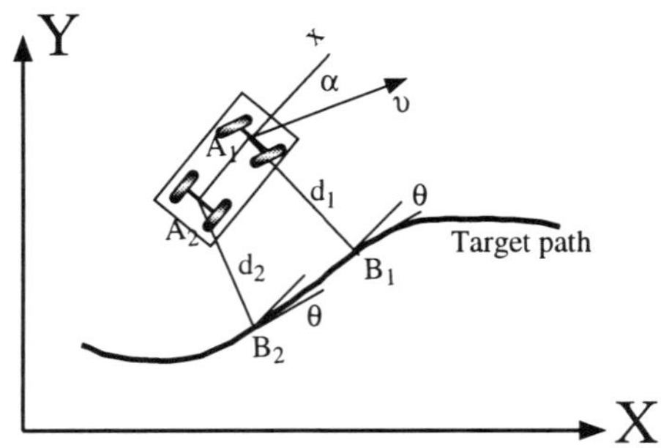

Fig. 4.3. Two possible coordinate sets for the path following problem

ble choices for the output variables, describing the shorter signed distance between the points A_1 and A_2 on the vehicle, and the points B_1 and B_2 on the path. The other important coordinate is θ that describes the orientation difference between the path tangent at point B_i and the axis x attached to the vehicle. To make the discussion simpler, assume that the steering angle α is the control input to be designed, and v is the translational velocity. Then, the simplified kinematic models expressed in the coordinates (d_i, θ) are:

$$\begin{cases} \dot{d}_1 &= v\sin(\theta+\alpha) \\ \dot{\theta} &= v/l\sin\alpha - \frac{v}{\kappa_1(s)-d_1}\cos(\theta+\alpha) \end{cases}$$

$$\begin{cases} \dot{d}_2 &= v\cos\alpha\sin\theta \\ \dot{\theta} &= v/l\sin\alpha - \frac{v}{\kappa_2(s)-d_2}\cos\alpha\cos\theta \end{cases}$$

where $\kappa_i(s)$ is the radium of curvature of the path at point i. These models are valid locally as far as $\kappa_i(s) - d_i > 0$. In many applications this hypothesis will hold since the radio of curvature of the path is likely to be large, and the initial vehicle position will be close to the path. The variables d_1 and d_2 are suitable choices because they correspond to system outputs that can be linearized in the spatial coordinates (i.e. flat outputs in the s-space). For instance, for the model expressed in the (d_1, θ) coordinates, the control

$$\alpha = \arcsin(-kd) - \theta$$

with the magnitude of kd smaller than one, yields the following linear spatial equation

$$d' + kd = 0$$

where $d' = \frac{\partial d}{\partial s}$. In the same sense as discussed before, d will decrease as s will increase. Since in the path following problem the length of the path may not be limited or small (as is the case for the parking maneuvers), the spatial convergence induced by the above equation is particularly well adapted for the problem at hand. When d is small, the control law indicates that α will tend to $-\theta$. Then the kinematic equation describing the variation of θ tends to

$$\dot{\theta} = -\frac{v}{l} \sin \alpha - \frac{v}{\kappa_1(s)}$$

which indicates that the front wheel axis of the vehicle is orthogonal to the tangent to the path at point B_1.

Remark 4.1. A similar study can be performed for the second model given above. In this case, the length of the path of the rear axis defined as $s = \int_0^t |v \cos \alpha| dt$ is used instead. The following spatial model is obtained:

$$\begin{aligned} d_2' &= \sin \theta \\ \theta' &= \frac{1}{l} \tan \alpha - \frac{1}{\kappa_2(s) - d} \cos \theta \end{aligned}$$

and the output d_2 is also linearizable. Since the control α appears only at the level of the variation of θ, the resulting linearized system will be of order two.

Remark 4.2. The above controllers were discussed assuming that the steering angle is directly controlled. Extensions to velocity and torque (or acceleration) control are possible by following the same development as in the automatic parking section.

Remark 4.3. Other nonlinear control designs are possible, not necessarily by resorting to linearization. An alternative is to define the control from a Lyapunov design. There is no consensus on which of the existing approaches may be preferable. Indeed, the distinction should probably be made at the level of the disturbance rejection properties. Robustness of these control schemes has not been studied enough. Clearly, this is a central problem since the control design is often done on the basis of the kinematic models as it has been demonstrated here. In practice, unmodeled dynamics and disturbances such as: additional system dynamics, actuator nonlinearities, lateral sliding due to the frictional contact and deformation of ties, mechanical flexibilities of the transmissions, asymmetries in the acceleration and deceleration vehicle characteristics, as well as many other factors, may degrade global system performance. Robustness will thus be one of the major issues when designing new feedback laws.

Remark 4.4. As opposed to the automatic parking problem, where the path planning trajectory is mandatory, the path following problem can be solved by control laws only depending on the characteristics of the path to be tracked, which in many cases is not known *a priori*. Indeed, one of the main difficulties, which is a problem *per se*, is to observe or to compute from indirect measurements the variables relevant for control. This problem is discussed next.

5. Visual-based Control System

Many of the problems mentioned in the previous chapter require the computation of variables such as d and θ. In many applications this information is not available from direct measures, and should be estimated using information from other sensors such as: TV cameras, radars, proximity sensors, and others. For instance, the lateral control designed in connection with the PATH program uses information collected from magnets placed along the road. Other applications use vision systems to detect the target path characteristics. The CiVis concept proposed by Renault, and the European PROMOTE program are examples of systems based on this idea. Examples of recent works in this area can be found in [21, 8, 12] among others. The problem is described next.

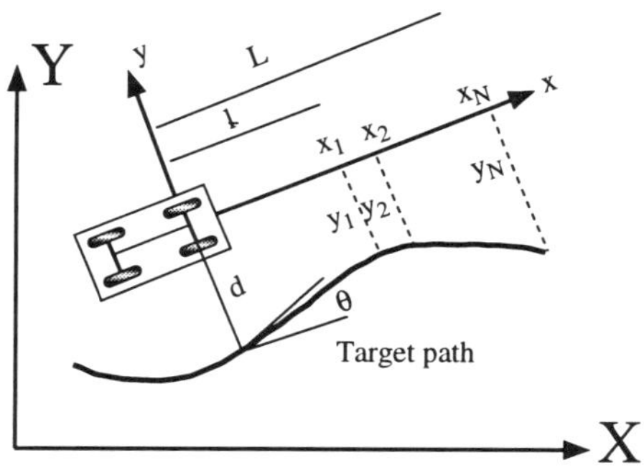

Fig. 5.1. Example of a system under vision guidance

Consider the system shown in Fig. 5.1. A vision system located on the vehicle provides measures from the x axis of the moving frame at different points placed at fixed distances $\{x_1, x_2, \dots, x_N\}$. In opposition to the problem stipulated in [8], these points do not necessarily span the positive x axis from zero to L. The reason is that the used TV camera may not necessarily cover a full angle of 180 degrees in the vertical axis. Therefore, some important measures needed for control (i.e. the distance d, the tangent angle θ, and eventually the radius of curvature at point A, see Fig. 5.1) are not directly measurable. This means that interpolation is not enough to characterize the contour of the target path as is the case in [8], but that some type of contour prediction is necessary to obtain d and θ. Note that at time instant t, the prediction of these variable will necessarily need past information of the contour expressed in some type of invariant coordinates (i.e. the radius of curvature, and its higher partial derivatives).

An important issue here is that in the moving vehicle frame, the contour is not time-invariant when represented in polynomial form. Indeed, the contour should be represented as

$$y = f(x,t), \quad \forall x \in [0,L]$$

Making the optical center of the camera to coincide with the center of the front wheel axis, the perspective projection y/x can be measured up to a Gaussian noise. Hence, given the points $\{x_1, x_2, \ldots, x_N\}$ in the area $[l, L]$, and the corresponding measured perspective projections, the sets of points $(x_i, y_i), \forall i = 1, 2 \ldots, N$ in the region $[l, L]$ can be computed.

Frezza and co-workers [8] formulated the problem for the unicycle kinematics (see the previous section) in the vehicle's moving frame, i.e.

$$\begin{aligned} \dot{x} &= wy + v \\ \dot{y} &= -wx \end{aligned}$$

and proposed a local representation of the contour around $x = 0$ using the following coordinates:

$$\begin{aligned} d(t) &= f(0,t) \\ \theta &= \frac{\partial f}{\partial x}(0,t) \\ \kappa(t) &= \frac{\partial^2 f}{\partial x^2}(0,t) \end{aligned}$$

where d is the distance between the vehicle (unicycle) and the path, θ is the relative vehicle and path orientation, and κ is the radius of curvature. The differential nonlinear equation in the above coordinates is

$$\begin{aligned} \dot{d} &= \theta(v - wd) \\ \dot{\theta} &= \kappa(v - wd) - w(\theta^2 + 1) \end{aligned}$$

The above system can be linearized in d and θ, via v and w, by a static feedback function of (d, θ, κ). To compute such a feedback, it is thus necessary to have the information about $f(0,t)$, and its first and second partial derivatives. For this, it is necessary to estimate the function $f(x,t)$. In [8], they propose to model the contour by a set of cubic B-splines $y(x,t) = \phi(x)^T a(t)$, where $a(t)$ is the time-varying polynomial coefficient vector, and $\phi(x)$ is the base function vector. All the points y_i can be represented in that form and organized in a vector representation

$$Y(x,t) = \Phi(x)a(t) + \epsilon$$

with ϵ being a measurement noise. A model for the variation of $a(t)$ can be derived using the above expression in the partial differential equation governing the evolution of the surface. This equation has the form

$$\dot{a}(t) = g(w, v, x, a)$$

and thus the vector $a(t)$ becomes an additional unmeasured state of the system. An observer-based control scheme needs to be designed, and stability of the complete system need to be studied. Frezza and co-workers proposed to use an extended Kalman filter to estimate $a(t)$, but they do not provide stability analysis of the resulting closed-loop system. The problem becomes even more complex when only a set of measures in the interval $[l, L]$ is available, as was mentioned at the beginning of this section.

Although some fundamental problems related to the controllability of the moving contour and the characterization of the steerable variables of the systems have been recently investigated [12], many problems are still open and clearly deserve more attention.

6. Multibody Vehicle Control

The control of groups of transportation units is nowadays a domain of intensive research. Car platooning is probably the most important industrial driving force for research in this area. Many studies in this domain have been carried out by programs such as PATH (California Partners for Advanced Transit and Highways [22]), as well as for other programs involving automated or semi-automated city cars (i.e. the French PRAXITEL program [5]), heavy transportation vehicles (i.e. the European PROMOTE program), and optimization of urban transports (i.e. the CiVis concept proposed by the Matra and Renault partnership). Other domains of interest are the air traffic management and unmanned submarine vehicles.

In terms of robotic applications, the concept of multibody train-like vehicles has been proposed [9] to face issues of heavy-duty applications in cluttered indoor environments. Among the possible concepts of such multibody system motions, the "follow-the-leader" behaviour is considered to be the most relevant. The follower vehicles should be controlled to track the leader car signature that need to be reconstructed on-line. The ideas can be applied to multibody-train vehicles [2], as well as to the problem of car platooning [16]. In the latter case, the inter-space distance between vehicles can also be controlled. These two classes of applications will be discussed next.

6.1 Multibody Train Vehicles

Application of the "follow-the-leader" principle to multibody train vehicles implies that the surface swept by the whole train will be equal to the surface swept by the first vehicle. Hence, this behaviour is particularly useful in application where the multibody system is required to move in clustered environments such as mines or nuclear power-stations. Figure 6.1 shows an

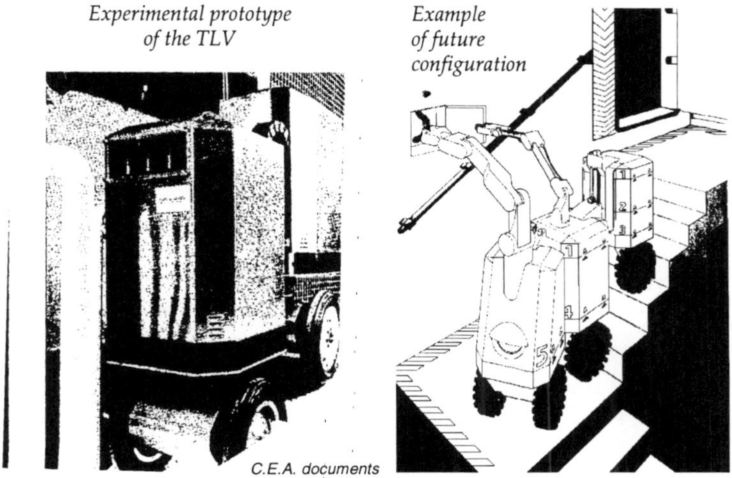

Fig. 6.1. Left side: the 2-cart experimental TLV descending an inclined plane; right side: a multi-cart configuration (courtesy of CEA)

example of a TLV configuration provided by CEA (French Center of Atomic Energy).

The kinematic model of this system can be derived in the relative angle coordinates (α_i, β_i) shown in Fig. 6.2, which provides a schematic view of the TLV system. The model is given by:

$$\begin{aligned} \dot{\alpha}_i &= v_{i-1} f(\alpha_i, \beta_i) + w_{i-1} \\ \dot{\beta}_i &= v_i f(\alpha_i, \beta_i) + w_i \end{aligned}$$

where

$$f(\alpha_i, \beta_i) = \frac{1}{l} \frac{\sin(\beta_i - \alpha_i)}{\sin \beta_i}$$

A peculiarity of this system with respect to multi-steered vehicles is that each vehicle wheel axis is independently controlled by the rotational control variable w_i. However, some generic singularity problems can be predicted when the wheel axis is oriented 90 degrees with respect to the pulling bar direction (i.e. $\beta_i = \pi/2$).

A problem to be solved priori to the control design is the reconstruction of the leading vehicle path signature so as to define a suitable set of reference angles for the variables (α, β). An algorithm for the path reconstruction is given in [2]. The way to define the reference angles is not unique. Two alternatives are:

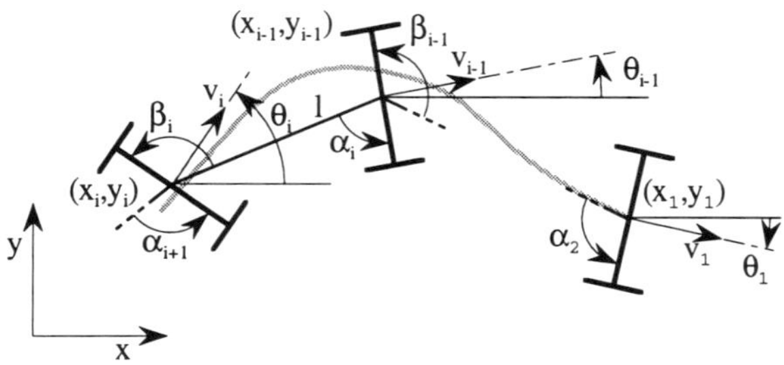

Fig. 6.2. Illustration of the train coordinates

- a virtual train reference placed along the leader path, or
- a set of individual car references placed as close as possible to the path.

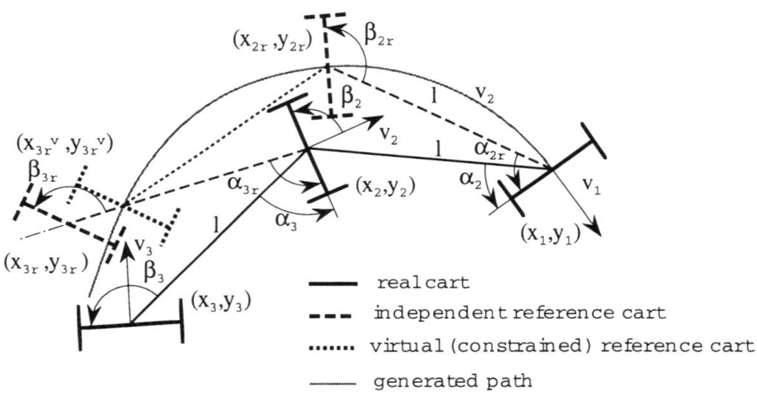

Fig. 6.3. The reference carts

Figure 6.3 sketches these two possibilities. The second solution is found to be more suitable for control design because it has the advantage of reducing the error propagation improving the transient responses.

Having defined the set of references $(\alpha_{ir}, \beta_{ir})$, a nonlinear control law based on backstepping ideas can be designed. The control provides boundedness of the error variables and convergence of such variables to a compact set with arbitrarily small radius. As in the case of path following, the difference between $(\alpha_{ir}, \beta_{ir})$ and (α_i, β_r) approaches zero as the curvilinear distance of the leader path increases.

An experimental test carried out along nontrivial trajectories is shown in Fig. 6.4. The tracking error d shown in this figure describes the Euclidean distance between the desired position (obtained by the reference model generation algorithm) and the real position of the second cart at each sampled time. Issues of control saturation and singularities are also here considered. Note that this trajectory includes motions along singular configurations yielding acceptable small peaks on d, due to singularity crossing. Details of these experiments are further described in [13].

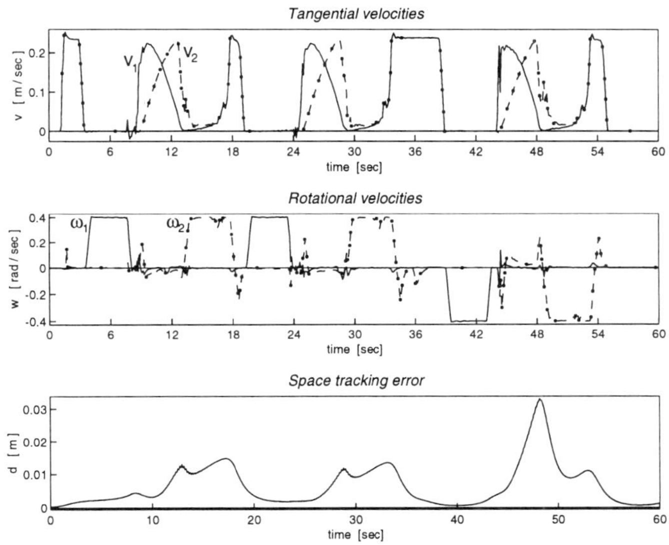

Fig. 6.4. Test of the 2-cart experimental TLV on a trajectory with singularities

Similar ideas can be applied to a set of vehicles without mechanical links. The extension of the control design described before to a class independent vehicles with different degrees of steerability and mobility has been studied in [16]. The studied class of vehicles includes differential steering cars, front-wheel-driven and steered cars, as well as 4-wheel-steered vehicles. Figures 6.5 and 6.6 show a typical motion of a 3-car platoon under feedback control obtained from the control design in [16] (see also [14]). The proposed control improves over other existing approaches in the sense that it can handle leader path signatures with arbitrarily small radius of curvature. Other control schemes for vehicle platooning found in the literature, often deal with approximated linear models only valid for small deviations (i.e. see [5, 20, 4] among others). They are mainly designed for application where the radius of curvature of the leader path signature is high, like in highways. The general problem of car platooning in highways and transportation systems is indeed more complex than just controlling a single platoon by regulating the lateral and longitudinal deviations. The next section discusses the generality of this problem.

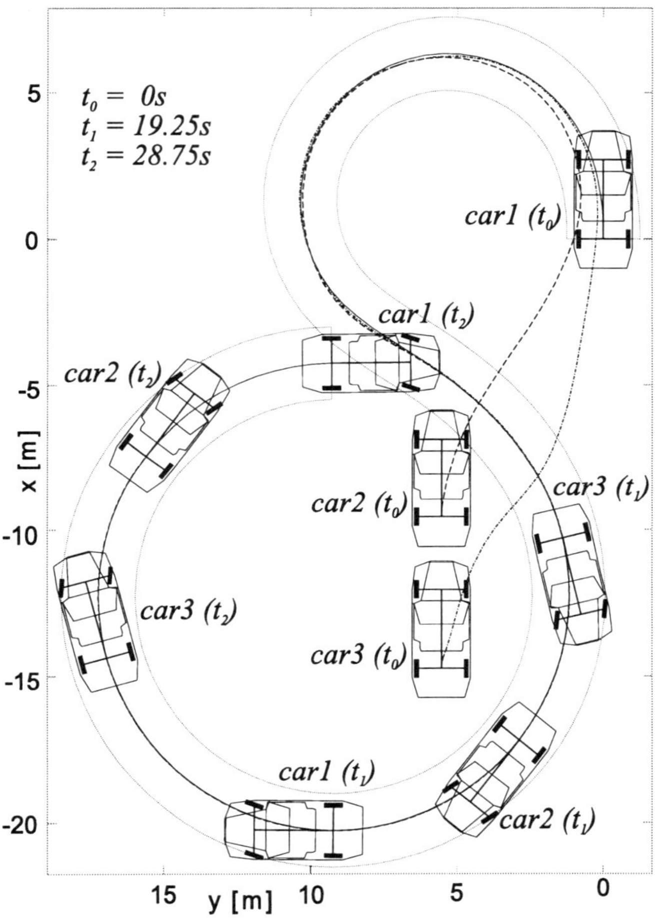

Fig. 6.5. Simulation of a 3-car platoon: motion

6.2 Car Platooning in Highways and Transportation Systems

The two major areas of applications for control of vehicles with high complexity will be probably concentrated in the next years along the following two applications area:

– automated highway systems, and
– heavy vehicles and urban transportation systems.

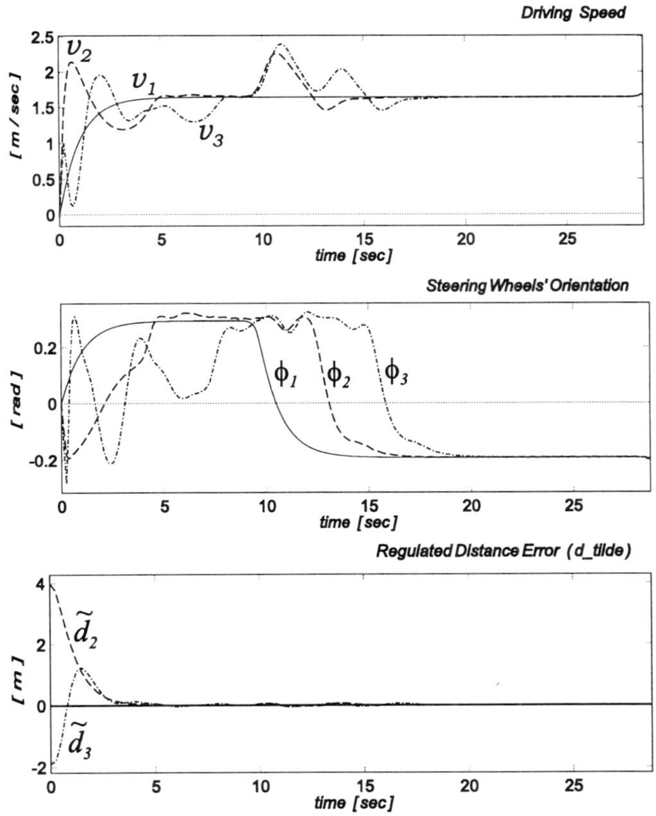

Fig. 6.6. Simulation of a 3-car platoon: main curves

An automated highway system is aimed at improving the capacity and safety of existing highways by platooning vehicles so as to decrease the inter-vehicle distance among the vehicles. Since the inter-vehicle distance are expected to be quite short (1 or 2 meters), human drives cannot react quickly enough as automated systems will do.

Control design of automated highway systems, as stipulated in the PATH program [22], is structured in a five-layer architecture:

– network
– link
– planning and coordination

– regulation
– vehicle dynamics

The network and the link layers belong to the roadside system, whereas the coordination, the regulation and the vehicle dynamics layers belong to the vehicle system. Each of these layers gives rise to nontrivial and quite complicated control problems. Some of them are briefly described next, see [11, 22] for further discussion.

– **Vehicle dynamic control layer.** This layer receives reference values from the regulation layer in terms of acceleration profiles, and vehicle steering angles. These references should be tracked in spite of the unmodeled dynamics, and nonlinearities of the actuators and interfaces such as: the hydraulic break system, air/fuel injection systems, flexibilities in the mechanical transmissions, automatic transmission, etc. This layer gives rise to "standard" control problems in the sense that the interfaces may be reasonably well modeled by differential equations coming from physics laws, and the corresponding control designs may be based on existing (linear or nonlinear) control methods. Most of these are well posed control problems.
– **Regulation layer.** Its goal is to perform the maneuvers defined by the higher layers. Most of the problems here are formulated at the kinematic level. Lateral and longitudinal control are examples of typical tasks to be performed. The control scheme based on the following-the-leader principle described in the previous section is also an example of a task to be performed in the regulation layer. Issues such as string stability and platooning error propagation are also part of the considered problems. Restrictions imposed by the coordination layer on the control actions (velocity and acceleration), so as to prevent collisions within and with other platoons, need also to be integrated. The dynamics owned by the used sensors (vision, proximity, etc.) need to be considered as well.
– **Coordination layer.** Its role is to ensure that the maneuvers defined by the link layer are performed safely. A typical example is to ensure that collision may not occur within the vehicles of a given platoon. The leading vehicle of this platoon should also consider the state conditions of the leading and rear platoons to avoid collision among them. This problem of collision avoidance has been recently studied via min-max optimization where each vehicle is modeled by a simple double integrator. An analytic solution was found, and safety acceleration and deceleration regions were determined. Therefore, if the regulation layer can preserve the vehicles within this region then a safe vehicle motion is ensured, see [7].
– **Link layer.** This layer establishes traffic conditions (in terms of density and flow profiles) so as to realize the capacity of a single highway or a stretch of highways. It determines when and how the platoons should be splitting, joining or changing lines. Works in this area are more scarce. Models used for control are derived at macroscopic level based on mass

conservation laws for highways. These models are described by partial differential equations. An example of a model describing a one-lane highway is given in [11]

$$\frac{\partial}{\partial t}K(x,t) = -\frac{\partial}{\partial x}\{(K(x,t)V(x,t)\}$$

where $K(x,t)$ is the density of the highway, $V(x,t)$ is the traffic velocity. The vehicle flow rate is given by: $\phi(x,t) = K(x,t)V(x,t)$. Control objectives are formulated in terms of tracking a desired density and flow profile, by commanding the traffic velocity; see for example [11] where this problem is extensively studied.
- **Network layer.** This layer determines the vehicles routing within the highway system so as to optimize the total time needed to go from the initial to the final destination. A time-optimal stochastic control problem may be formulated here. Other types of distribution laws, as the ones mentioned in the link layer, can also be considered.

Full automation may be questionable from reasons other than technical. For instance, safety may be an important issue that strongly demands for a semi-automated system, in which the driver safety becomes a priority. It will not be surprising that future programs in automated highways are re-oriented towards systems where the three lower layers are kept fully automated, whereas the two higher layers (network and link) are let to the driver consideration. Some additional information may be also accessible from modern navigation systems providing the information to the driver about some possible optimal routing. If this tendency is confirmed, then new control problems will be considered where the interaction between driver and control system will be dominant. In heavy vehicles platooning and in urban transportation systems, there is no such an aim for global flow and density optimization as for the intelligent highway systems. Some of the programs concerning platooning of heavy-duty vehicles have mainly been launched by European consortia. The motivation is different from the American IHS program. Fuel economy is one of them. In Europe the fuel cost on the global budget of the transportation industry is high, and the benefits of reducing the fuel budget may be very important. Expectation in fuel cost reduction while platooning heavy vehicles is between 10% and 15% of economy due to the reduction on the aerodynamic forces. One example of such a program is PROMOTE (a continuation of the PROMETEUS program) which is conducted by a large consortium involving car builders, vehicle research centers and traffic regulation offices (Mercedes-Benz, Fiat Research Center, Saab-Scania, Volvo and Dal), car equipment manufacturers (Bosch, EMI-CRL Thorn, Daimler-Benz, Iveco, Wabco, ZF and TåV), and transportation companies from several European countries. Apart from the control design aspects, the project also looks for solutions to social aspects such as the insurance responsibility, and considers the possible driver unions point of view, that estimates that such systems may jeopardize their activity. This last point may not be critical since the project aim is not to avoid the driver, but to equip the vehicle with an automatic mode allowing to perform platooning up to a short enough inter-vehicle distance, which is not reachable by the driver alone due to the inherent low bandwidth of the human reactions. However, the driver role in this programs is to perform maneuvers and actions corresponding to the higher layers of the IHV programs.

In such transportation programs, part of the tasks defined in the coordination layer, such as the transition task during the splitting and joining the platoon, may be confined to the driver, whereas the lower layers of regulation and vehicle dynamics control will be automated. Most of the safety action will be ensured by the driver with the help of diagnostic and supervisory algorithms.

The control problems found at the two lowest layers are of the same nature of the ones in the IHV systems, except that its complexity will increase. Some of the reasons are:

- the kinematic of the articulated heavy duty vehicles is more complex than simple cars,
- the complexity of vehicle actuators is certainly higher: break systems are more involved, the flexibilities of the mechanical transmissions are more important, etc.,
- the closed-loop bandwidth of the regulation layer may be lower due to the substantial load carried by the vehicles.

7. Conclusions

This chapter has presented an application-oriented overview of some of the actual trends in control design for mobile robots and multibody wheeled vehicles, including transportation systems and intelligent highways. The presentation is not exhaustive, and many other areas and problems have not been discussed. The following discussion summarizes some of the more important issue treated here:

- **Systems.** Discussion here has concerned mobile robots but also vehicles and transportation systems. Apart from some applications mentioned in connection with nuclear applications, mobile robots are scarcely found in industry. Space application may be a clear area for them, but space research programs are limited. Most of the prototypes of mobile robots have been developed and studied by universities and some government research centers. Nevertheless, they have been a great source of inspiration for studying new control strategies (nonlinear, periodic controllers, discontinuous feedback) and understanding their inherent model properties (nonholonomy, controllability, steerability, stabilizability, etc.). As opposed to them, vehicles and transportation systems have a well identified industry driving force which concerns many areas of our society (cars, urban and heavy duty transports, etc). Car industry is seeking to introduce new automatic features in the next generation of their products. This starts to have a great impact in the research directions taken by the universities. Many new control problems, at many levels, are still open.

- **Control strategies.** Before proceeding to the feedback law design, it is necessary to formulate the control problem to be solved. It has been shown that for a given task like platooning, different control strategies can be taken. For instance, if the "follow-the-leader" principle is adopted, then there are at least two different ways to define the reference variables to be tracked. Similarly, in vision-guided systems, the control problem can be directly formulated in the TV camera coordinates or in the Cartesian space. The way the problem is formulated is thus not just a matter of personal choices, but it can substantially change the problem complexity, and also the control properties of the considered system. For the problems discussed in this chapter, there is no consensus on what formulation is preferable. This clearly depends on the system at hand, and other technological factors such as sensor location and bandwidth, actuator capacity, etc.
- **Feedback laws.** The problem of regulation and tracking of kinematic models considered here is quite well understood. Although, many approaches for the posture stabilization problem have been proposed in the literature (periodic and discontinuous feedbacks), few of them are of practical use. For instance even if periodic time-varying controllers may provide exponential stability, it would be difficult to accept in urban traffic systems a transient behaviour with large oscillations. It would be better accepted to use feedback laws with guaranteed damped transients, even if they are not asymptotically stable in the classic way. Typically, a spatial convergence (in terms of the path length) will be obtained, and it is particularly well adapted to the problem at hand.
- **Future challenges.** Up to now, most of the mobile robot systems, and some of the mentioned intelligent highway systems, are seeking for fully automated operation. For different reasons such as algorithm complexity, safety, and impossibility of modelling the full environment surrounding the vehicle, fully automated system may be abandoned as a future challenge. Instead, semi-automated systems combining human operation with some automatic features, will be more promising. In robotics, human presence may be introduced by teleoperation links, which will avoid dealing with highly complex and unsolved control problems (decision trees) for which human skill is better adapted. In automobile control and transportation systems, their presence is tautological. For both areas, one of the most important challenges will be the understanding in how to put the human into the control loop.

Acknowledgement. Thanks are due to R. Horowitz for discussion on the PATH program and associated problems. A grateful acknowledgement also goes to the research group of Renault-VI for the enlightened discussion on some of the new problems concerning the transportation and urban system and by providing the illustrations of the CiVis concept.

References

[1] Brockett R W 1983 Asymptotic stability and feedback stabilization. In: Brockett R W, Millmann R S, Sussman H J (eds) *Differential Geometric Control Theory.* Birkhäuser, Boston, MA, pp 181–208

[2] Canudas de Wit C, NDoudi-Likoho A, Micaelli A 1997 Nonlinear control for a train-like vehicle. *Int J Robot Res.* 16:300–319

[3] Canudas de Wit C, Siciliano B, Bastin G (eds) 1997 *Theory of Robot Control.* Springer-Verlag, London, UK

[4] Chee W, Tomizuka M 1994 Lane change maneuver of automobiles for the intelligent vehicle and highway systems. In: *Proc 1994 Amer Contr Conf.* Baltimore, MD, pp 3586–3589

[5] Daviet P, Parent M 1993 Contrôle longitudinal d'un train de véhicules. In: *Automatique pour les Véhicules Terrestres.* pp 187–196

[6] Fliess M, Lévine J, Martin P, Rouchon P 1995 Flatness and defect of nonlinear systems: introductory theory and examples. *Int J Contr.* 61:1327–1361

[7] Frankel J, Alvarez L, Horowitz R, Li P Y 1995 Safety-oriented maneuvers for IVHS. In: *Proc 1995 Amer Contr Conf.* Seattle, WA, pp 668–672

[8] Frezza R, Soatto S, Picci G 1997 Visual path following by recursive spline updating. Preprint

[9] Hirose S, Morishima A 1990 Design and control of mobile robot with an articulated body. *Int J Robot Res.* 9:99–114

[10] Kolmanovsky I, McClamroch N H 1995 Developments in nonholonomic control problems. *IEEE Contr Syst Mag.* 15(6):20–36

[11] Li P Y, Horowitz R, Alvarez L, Frankel J, Robertson A M 1997 An automated highway system link layer controller for traffic flow stabilization. *Transp Res C.* 5(1):11–37

[12] Ma Y, Košecká J, Sastry S 1997 Vision guided navigation for nonholonomic mobile robot. Preprint

[13] Micaelli A, Louveau F, Sabourin D, Canudas de Wit C, Ndoudi-Likoho A 1997 Follow-the-leader control for a train-like vehicle: implementation and experimental results. Int Rep CEA

[14] NDoudi-Likoho A D 1997 Commande non-linéaire des véhicules multi-corps à roues. PhD thesis, Laboratoire d'Automatique de Grenoble, INPG

[15] NDoudi-Likoho A, Canudas de Wit C 1995 Dynamic nonlinear control for a platoon of cars. In: *Prepr 1st IFAC Work Advances Automot Contr.* Ascona, Switzerland, pp 197–202

[16] NDoudi-Likoho A, Canudas de Wit C 1997 Nonlinear control of multibody wheeled vehicles. *36th IEEE Conf Decision Contr.* San Diego, CA

[17] Nogami J, Nakasuka S, Tanabe T 1996 Concept of future air traffic management and its real-time decision support utilizing concept learning. In *Prepr 13th IFAC World Congr.* San Francisco, CA, vol 8, pp 351–356

[18] Rouchon P, Fliess M, Lévine J, Martin P 1993 Flatness and motion planning: the car with n-trailers. In: *Proc 2nd Euro Contr Conf.* Groningen, NL, pp 1518–1522

[19] Sheikholeslam S, Desoer C A 1992 Combined longitudinal and lateral control of a platoon of vehicles. In: *Proc 1992 Amer Contr Conf.* Chicago, IL, pp 1763–1767

[20] Shladover S, Desoer C A, Hedrick J K, Tomizuka M, Walrand J, Zhang W B, McMahon D, Peng H, Sheikholeslam S, McKeown N 1991 Automatic vehicle control developments in the PATH program. *IEEE Trans Vehic Tech.* 40:114–130

[21] Tsakiris D, Rives P, Samson C 1997 Applying visual servoing techniques to control non-holonomic mobile robots. In: *Proc IEEE/RSJ/INRIA Work New Trends Image-based Robot Servoing.* Grenoble, France

[22] Varaiya P 1993 Smart cars on smart road: Problems of control. *IEEE Trans Automat Contr.* 32:195–207

Vision-based Robot Control

Peter I. Corke[1] and Gregory D. Hager[2]

[1] CSIRO Manufacturing Science and Technology, Australia
[2] Department of Computer Science, Yale University, USA

The topic of vision-based robot control has been investigated for more than 20 years and over that time several major, and well understood, approaches have evolved. This chapter describes the fundamental principles of these methods, and discusses their relative strengths and weaknesses. The discussion emphasizes the interdependence of vision and control, for example, the vision system provides input to the robot control loop, but the vision system may utilize control techniques to track the target. We also discuss issues such as dynamic performance, approaches to image feature extraction, the impact of current technology trends, future applications and research challenges.

1. Introduction

A great many tasks routinely performed by humans (for example machine control, driving, assembly, or fruit picking) are based on visually perceived information. In order for robots to perform such tasks, without extensive instrumentation or re-engineering of the environment, they must also have the ability to perceive and act upon visual information. Computer vision is therefore an important sensor for robotic systems since it mimics the human sense of vision and allows for non-contact measurement of the environment.

Limited vision capability has been available in commercial robot controllers for many years now. It is used for tasks such as inserting parts with respect to fiducial marks on printed circuit boards, or for grasping unorganized parts moving on conveyor belts. Typically these systems adopt a 'look' then 'move' strategy — a well calibrated camera and vision system determines the desired robot end-effector pose and the robot system is commanded to make the appropriate motion. The accuracy of the resulting motion clearly depends directly on the quality of the camera calibration and the accuracy of the robot. The systems in operation today are able to achieve the necessary precision using high-quality and expensive components and good system engineering.

An alternative approach to increasing the performance of the overall system is to use a vision system to continuously guide, or steer, the robot end-effector toward the target. Such a closed-loop position control structure for a robot end-effector is referred to as *visual servoing* system. A visual-feedback control loop, like any feedback control system, will increase closed-loop accuracy and robustness to error in the sensor or the robot — allowing the accuracy of the vision system and robot to be relaxed while maintaining overall accuracy.

We can more formally define visual servoing as the use of visual information to control the *pose* of the robot relative to a target object or a set of target features. The term robot here will encompass not only conventional manipulator arms, but also mobile robots, cars, aircraft or underwater vehicles. Such systems may include more than one camera, and the cameras can be placed either on the robot observing the target (the so-called 'eye in hand' configuration), or in the world observing the robot and the target. The term target refers to the object(s) relative to which the robot is being positioned.

The field of vision-based robot control began with the work of Shirai and Inoue [34] in 1973 and significant, albeit slow, progress was made up to around 1990. Progress in that era was hindered largely by technological issues, in particular extracting information from video data streams. Since 1990 there has been a marked rise in interest in this field, largely fueled by personal computing power crossing the threshold which allows analysis of scenes at a sufficient rate to 'servo' a robot manipulator.

The reported use of visual information to guide robots is now quite extensive and encompasses applications as diverse as manufacturing (part mating, alignment), vehicle control (cars, planes, underwater, space), teleoperation, tracking cameras and fruit picking as well as emulating human dynamic skills in diverse areas such as ping-pong, air hockey, juggling, catching and balancing.

The remainder of this chapter will provide a brief introduction to the principles behind visual servoing and highlight the control issues involved. Section 2. introduces some important concepts and defines the notation that we will use. Section 3. discusses the common approaches to using vision in control, both kinematics and dynamics. Section 4. then describes the application of control and estimation methods to the vision problems of image Jacobian and pose estimation, image feature extraction, and camera control. For greater detail the reader is referred to [5, 20] (which have extensive bibliographies), and [14] for a discussion of the vision issues related to visual servoing.

2. Fundamentals

2.1 Camera Imaging and Geometry

Before being able to apply visual sensors to robot control it is essential to have an understanding of the sensor, typically a CCD camera. A camera contains a lens that forms a 2D projection of the scene on the image plane where the sensor is located. In practice the lens is not ideal and introduces a number of distortions [6].

Using homogeneous coordinates the lens perspective transformation may be expressed in linear form for an arbitrary camera location. Using matrix representation the point $(x,\ y,\ z)$ is transformed

$$\begin{bmatrix} U \\ V \\ W \end{bmatrix} = \begin{bmatrix} \alpha_x & 0 & X_0 & 0 \\ 0 & \alpha_y & Y_0 & 0 \\ 0 & 0 & 1 & 0 \end{bmatrix} \begin{bmatrix} 1 & 0 & 0 & 0 \\ 0 & 1 & 0 & 0 \\ 0 & 0 & -1/f & 1 \\ 0 & 0 & 0 & 1 \end{bmatrix} \left({}^0\mathbf{T}_c\right)^{-1} \begin{bmatrix} x \\ y \\ z \\ 1 \end{bmatrix} \tag{2.1}$$

$$= \mathbf{C} \begin{bmatrix} x & y & z & 1 \end{bmatrix}^T \tag{2.2}$$

where the *intrinsic* parameters are the X- and Y- axis scaling factor in pixels/mm, α_x and α_y, image plane offset in pixels (X_0, Y_0), focal length f, and the *extrinsic* parameters representing the camera position in world coordinates ${}^0\mathbf{T}_c$. The image plane coordinates in pixels are then expressed in terms of the homogeneous coordinates as

$$u = \frac{U}{Z}, \quad v = \frac{V}{Z} \tag{2.3}$$

in units of pixels. The camera calibration matrix $\mathbf{C}$ encapsulates the intrinsic and extrinsic parameters and is typically determined by a calibration procedure for which there is a considerable literature (for example [36]).

This projection (2.1) causes direct depth information to be lost so that each point on the image plane corresponds to a ray in 3D space. Therefore, some additional information is needed to determine the 3D coordinates corresponding to an image plane point, a precursor to 3D robot pose control. This additional information may come from multiple cameras (for example stereo vision), multiple views with a single camera, or knowledge of the geometric relationship between several feature points on the target.

2.2 Image Features and the Image Feature Parameter Space

In the computer vision literature, an *image feature* is any structural feature than can be extracted from an image (e.g. an edge, a corner or a distinctive region). Typically, an image feature will correspond to the projection of a physical feature of some object (e.g. the robot tool) on the camera image plane. A good image feature is one that can be located unambiguously in different views of the scene, such as a hole in a gasket [10] or a contrived pattern [9].

We define an *image feature parameter* to be any real-valued quantity that can be calculated from one or more image features. Image feature parameters that have been used for visual servo control include the image plane coordinates of points in the image [4, 9, 16, 35], the distance between two points in the image plane and the orientation of the line connecting those two points [10], perceived edge length, the area of a projected surface and the relative areas of two projected surfaces [32], the centroid and higher order moments of a projected surface [32], the parameters of lines in the image plane [9], and the parameters of an ellipse in the image plane.

In order to perform visual servo control, we select a set of image feature parameters. Once we have chosen a set of k image feature parameters, we can define an image feature parameter vector $\mathbf{f} = [f_1 \cdots f_k]^T$. Since each f_i is a (possibly bounded) real valued parameter, we have $\mathbf{f} = [f_1 \cdots f_k]^T \in \mathcal{F} \subseteq \Re^k$, where $\mathcal{F}$ represents the *image feature parameter space.*

2.3 Camera Sensor

In a visual servo system the camera performs the function of the sampler. The sample rate for most visual servo systems is dictated by the frame rate of the camera, and this is defined by broadcast television standards to be 30 Hz in the US and Japan and 25 Hz elsewhere. The standard was not driven by control requirements, but rather by the need for low transmission bandwidth, ease of decoding, and minimal human perception of flicker. This latter requirement also led to the adoption of interlacing where each frame is transmitted as two sequential fields containing respectively all the even picture lines and all the odd lines. For computer vision applications interlacing introduces undesirable artifacts since the two fields are exposed at different points in time which leads to blurring or tearing of rapidly moving objects. It is therefore quite common for vision-based control systems to process the fields individually, treating them as images with reduced vertical resolution, and doubling the visual sample rate.

A camera's output reflects the integrated intensity over the exposure interval which is typically the same as the frame time. An ideal visual sampler would capture the instantaneous state of the scene and in practice this can only be approximated by using a short exposure interval (by means of a mechanical, or more commonly an 'electronic' shutter) but this is at the expense of small integrated charge in the sensor and consequently a poor signal to noise ratio.

Table 2.1. Taxonomy of visual control structures according to Sanderson and Weiss

	Error signal defined in task space	Error signal defined in image plane
No joint level feedback	Position-based visual servo	Image-based visual servo
Joint level feedback	Dynamic position-based look-and-move	Dynamic image-based look-and-move

3. Vision in Control

Sanderson and Weiss [32] introduced a taxonomy of visual servo systems, into which all subsequent visual servo systems can be categorized. The four categories in this taxonomy are given in Tab. 2.1 and are shown schematically in Figs. 3.1 and 3.2. It should be noted that, by common usage, all such systems are today referred to as visual servo systems, but the distinction regarding the presence or absence of joint-level feedback is an important one. For several reasons, nearly all implemented systems adopt joint-level feedback. Firstly, the relatively low sampling rates available from vision makes direct control of a robot end-effector with complex, nonlinear dynamics an extremely challenging control problem. Using internal feedback with a high sampling rate generally presents the visual controller with idealized axis dynamics [6]. Secondly, many robots already have an interface for accepting Cartesian velocity or incremental position inputs to the internal position controller.

The second major classification of systems, the columns in Tab. 2.1, distinguishes *position-based* control from *image-based* control. In position-based control, features are extracted from the image and used in conjunction with a geometric model of the target and the known camera model to estimate the pose of the target with respect to the camera. Feedback is computed by reducing errors in estimated pose space. In image-based servoing, control values are computed on the basis of image features directly. However this presents a significant challenge to controller design since the plant (relationship between robot motion and image features) is non-linear and highly coupled.

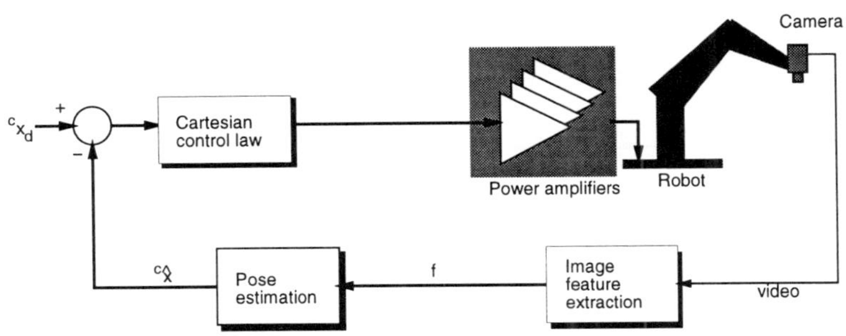

Fig. 3.1. Position-based visual servo (PBVS) structure

Another axis in the taxonomy is to distinguish between systems which only observe the target object, *endpoint open-loop* (EOL), and those which observe both the target object and the robot end-effector *endpoint closed-loop* (ECL). The primary difference is that EOL system must rely on an explicit hand-eye calibration when translating a task specification into a visual servoing algorithm. The relative merits of these two methods, in particular robustness with respect to calibration errors is discussed in [20, 17, 13, 12].

3.1 Position-based Approach

In position-based visual servoing image feature parameters extracted from the image are used, with an *a priori* known geometric object model, to estimate the pose of the target with respect to the camera. A considerable literature exists on the problem of pose estimation [19].

The problem is now one of reducing the error between the current and the desired pose of the robot, both defined in the task space. This control structure neatly separates the control problem from the estimation problems involved in computing pose from visual data.

It is useful to consider the positioning task [31] as being *fulfilled* with the end-effector in pose $\mathbf{x}_e$ if $\mathbf{E}(\mathbf{x}_e) = \mathbf{0}$. Once a suitable kinematic error function, $\mathbf{E}(.)$, has been defined, a regulator is created which reduces the estimated value of the kinematic error function to zero. This regulator produces, at every time instant, a desired end-effector velocity screw, $\dot{\mathbf{r}}$, which is sent to the robot control subsystem. This velocity can be transformed to required joint velocity using a technique such as resolved-rate motion control.

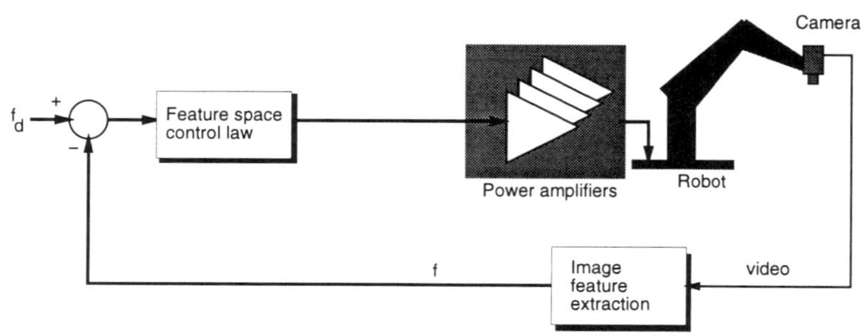

Fig. 3.2. Image-based visual servo (IBVS) structure

3.2 Image-based Approach

Image-based visual servo control uses the location of features on the image plane directly for feedback, avoiding the pose estimation step. For example, consider Fig. 3.3, where it is desired to move the robot so that the camera's view changes from the initial to the final view, and the feature parameter vector from $\mathbf{f}_0$ to $\mathbf{f}_d$. Implicit in $\mathbf{f}_d$ is that the robot is normal to, and centered over the plane at the desired distance. Many tasks can be described in terms of the motion of image features, for instance aligning visual cues in the scene, edge following, or catching.

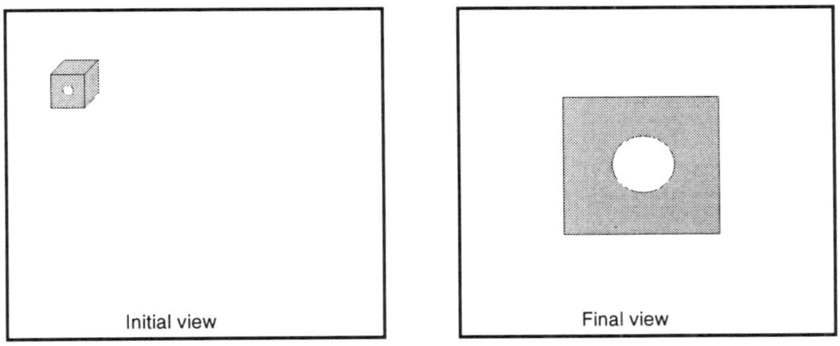

Fig. 3.3. Example of initial and desired view of a cube

In general the relationship between relative pose and feature position is non-linear and cross-coupled such that motion of one end-effector DOF will result in the complex motion of many features. The differential relationship can be represented by the image Jacobian, $\mathbf{J}_v \in \Re^{k \times m}$,

$$\dot{\mathbf{f}} = \mathbf{J}_v \dot{\mathbf{r}} \tag{3.1}$$

where $\mathbf{r}$ represents the coordinates of the end-effector in some parameterization of the m-dimensional task space, $\dot{\mathbf{r}}$ represent the corresponding end-effector velocity, $\mathbf{f}$ represent a vector of image feature parameters and $\dot{\mathbf{f}}$ the corresponding vector of k image feature parameter velocities.

$$J_v(\mathbf{r}) = \left[\frac{\partial \mathbf{f}}{\partial \mathbf{r}}\right] = \begin{bmatrix} \frac{\partial f_1(\mathbf{r})}{\partial r_1} & \cdots & \frac{\partial f_1(\mathbf{r})}{\partial r_m} \\ \vdots & & \vdots \\ \frac{\partial f_k(\mathbf{r})}{\partial r_1} & \cdots & \frac{\partial f_k(\mathbf{r})}{\partial r_m} \end{bmatrix}. \tag{3.2}$$

The dimension of the image Jacobian will vary depending on the task and the number of feature parameters. The image Jacobian is also referred to as the *feature sensitivity matrix*, the *interaction matrix* [9] and the $\mathbf{B}$ matrix [27].

For the case of a single point at (x, y, z) with velocity $[\dot{T}_x\, \dot{T}_y\, \dot{T}_z\, \omega_x\, \omega_y\, \omega_z]'$, both with respect to the camera, the image plane velocity is given by [20]

$$\begin{bmatrix} \dot{u} \\ \dot{v} \end{bmatrix} = \begin{bmatrix} \frac{\lambda}{z} & 0 & \frac{-u}{z} & \frac{-uv}{\lambda} & \frac{\lambda^2+u^2}{\lambda} & -v \\ 0 & \frac{\lambda}{z} & \frac{-v}{z} & \frac{-\lambda^2-v^2}{\lambda} & \frac{uv}{\lambda} & u \end{bmatrix} \begin{bmatrix} T_x \\ T_y \\ T_z \\ \omega_x \\ \omega_y \\ \omega_z \end{bmatrix} \tag{3.3}$$

The Jacobian for the case of multiple points is simply obtained by stacking the Jacobians for each pair of image point coordinates. Image Jacobians for other features such as lines, circles and ellipses can also be derived.

Visual servo control applications typically require the computation of $\dot{\mathbf{r}}$, given as input $\dot{\mathbf{f}}$. When $k = m$ and $\mathbf{J}_v$ is nonsingular, $\mathbf{J}_v^{-1}$ exists and

$$\dot{\mathbf{r}} = \mathbf{J}_v^{-1}\dot{\mathbf{f}} \tag{3.4}$$

which is the approach used by Feddema [10].

When $k \neq m$, $\mathbf{J}_v^{-1}$ does not exist. In the case $k > m$ there are more feature parameters than task degrees of freedom. Typically this will result in a set of inconsistent equations, since the k visual features parameters are noisy estimates obtained from a computer vision system. The appropriate pseudo-inverse is given by

$$\mathbf{J}_v^{+} = (\mathbf{J}_v^T\mathbf{J}_v)^{-1}\mathbf{J}_v^T. \tag{3.5}$$

and

$$\dot{\mathbf{r}} = \mathbf{J}_v^{+}\dot{\mathbf{f}}. \tag{3.6}$$

This approach has been demonstrated [16, 22, 9] and has been shown to increase robustness, particularly with respect to singularities [22].

When $k < m$, the system is under-constrained which implies that we are not observing enough features to uniquely determine the object motion $\dot{\mathbf{r}}$, i.e. there are certain components of the object motion that can not be observed. In this case, the appropriate pseudo-inverse is given by

$$\mathbf{J}_v^{+} = \mathbf{J}_v^T(\mathbf{J}_v\mathbf{J}_v^T)^{-1}. \tag{3.7}$$

The null space of $\mathbf{J}_v$ contains those components of the object velocity that are unobservable, for example the motion of a point along a projection ray. The null space of the image Jacobian plays a significant role in hybrid methods, in which some degrees of freedom are controlled using visual servo, while the remaining degrees of freedom are controlled using some other modality [4]. The condition number of the image Jacobian gives an indication of *resolvability* [26], that is how well task space motion can be perceived on the image plane. Feddema describes an algorithm [10] to automatically select a subset of the measurable features so as to minimize the condition number of the image Jacobian.

In general the image Jacobian, see (3.1), is a function of target distance, z, which is often not known. Common approaches are to compute the Jacobian based on a fixed estimate of distance [9] or an estimate based on full or partial pose estimation [16, 10].

3.3 Dynamics

The discussion above has ignored the dynamics of the robot manipulator and the visual sensing system, but these are critical since they affect the overall performance and stability of the closed-loop system. The dynamics of robot systems have been studied extensively. A well designed robot generally uses fast high-gain axis controllers to provide idealized position or velocity tracking and rejection of disturbances such as inertial forces and payload variation. The robot system may also introduce latency in its command interpreter, control implementation or communications protocol.

What is rather less obvious is that the vision system and camera also have dynamics — a significant latency, perhaps one or more video sample intervals, caused by:

1. The transport time of pixels from the camera to the vision system. It takes almost a full frame time to transfer all pixels from the camera.
2. Finite exposure time of the camera. Unless a fast shutter is used, the image is the integrated intensity over the frame time. For a moving object it can be shown that this introduces a lag in the estimate of the object's centroid [6].
3. Finite computation time for the computer vision algorithm. This time is a function of the algorithm as well as the computing hardware. Simple hardware blob tracking can be very fast and completely overlapped with pixel transport from the camera, while whole frame adaptive segmentation, optical flow or stereo matching algorithms may take many frame times to execute.

The latency, a pure time delay, is represented by e^{-sT} and its Pade approximation of

$$e^{-sT} \approx \frac{1 - sT/2}{1 + sT/2} \tag{3.8}$$

allows classical analysis techniques to be used to investigate the effect of delay [33, 6]. This shows that systems will tend toward instability as loop gain or loop delay is increased. Most reported visual servo systems use proportional feedback but with the loop gain sufficiently 'detuned' so as to ensure stability. These systems are generally of Type I (position feedback and a velocity controlled actuator), and typically show a settling time to a step target demand of anything from 0.5 to 10 seconds. However for moving targets, as expected, they show considerable error or lag.

This level of performance is considerably less than would be expected even given the low sample rate imposed by the camera and vision system. If we assume a vision system operating at 60 samples per second (RS170 field rate), as many systems do, then the common 'rule of thumb' would lead to an expected closed-loop bandwidth of 6 to 12 Hz, or in time domain terms a rise time of between 60 and 120 ms. If the system dynamics were simple then most laboratory systems should easily achieve this level of performance, even through *ad hoc* tuning.

In order to try and achieve performance closer to what is achievable classic techniques can be used such as increasing the loop gain and/or adding some series compensator (which may also raise the system's Type). These approaches have been investigated [6] and while able to dramatically improve performance are very sensitive to plant parameter variation, and a high-performance specification can lead to the synthesis of unstable compensators which are unusable in practice. Predictive control, in particular the Smith Predictor, is often cited [3, 33, 6] but it too is very sensitive to plant parameter variation.

Corke [6] has shown that estimated velocity feedforward can provide a greater level of performance, and increased robustness, than is possible using feedback control alone. Similar conclusions have been reached by others for visual [8] and force [7] control. Utilizing feedforward changes the problem from one of control system design to one of estimator design. The duality between controllers and estimators is well known, and the advantage of changing the problem into one of estimator design is that the dynamic process being estimated, the target, generally has simpler linear dynamics than the robot and vision system. While a predictor can be used to 'cover' an arbitrarily large latency, predicting over a long interval leads to poor tracking of high-frequency unmodeled target dynamics.

The problem of delay in vision-based control has also been solved by nature. The eye is capable of high-performance stable tracking despite total open-loop delay of 130 ms due to perceptual processes, neural computation and communications. Considerable neurophysiological literature [11, 30] is concerned with establishing models of the underlying control process which is believed to be both non-linear and variable structure.

4. Control and Estimation in Vision

The discussion above has considered a structure where image feature parameters provided by a vision system provide input to a control system, but we have not addressed the hard question about how image feature parameters are computed or how image features are reliably located within a changing image. The remainder of this section discusses how control and estimation techniques are applied to the problem of image feature parameter calculation and image Jacobian estimation.

4.1 Image Feature Parameter Extraction

The fundamental vision problem in vision-based control is to extract information about the position or motion of objects at a sufficient rate to close a feedback loop with reasonable performance. The challenge, then, is to process a data stream of about 7 Mbyte/sec (monochrome) or 30 Mbyte/sec (color).

There are two general broad classes of image processing algorithms used for this task: full-field image processing followed by segmentation and matching, and localized feature detection. Many tracking problems can be solved using either approach, but it is clear that the data-processing requirements for the solutions vary considerably. Full-frame algorithms such as optical flow calculation or region segmentation tend to lead to data intensive processing using specialized hardware to extract features. More recently the active vision paradigm has been adopted. In this approach, feature-based algorithms which concentrate on spatially localized areas of the image are used. Since image processing is local, high data bandwidth between the host and the digitizer is not needed. The amount of data that must be processed is also greatly reduced and can be handled by sequential algorithms operating on standard computing hardware. Since there will be only small changes from one scene to the next, once the feature location has been initialized, the feature location is predicted from its previous position and estimated velocity [37, 29, 8]. Such systems are cost-effective and, since the tracking algorithms reside in software, extremely flexible and portable.

The features used in control applications are typically variations on a very small set of primitives: simple "blobs" computed by segmenting based on gray value color, "edges" or line segments, corners based on line segments, or structured patterns of texture. For many reported systems tracking is not the focus and is often solved in an *ad hoc* fashion for the purposes of a single demonstration.

Recently, a freely available package XVision[1] implementing a variety of specially optimized tracking algorithms has been developed. The key in XVision is to employ *image warping* to geometrically transform image windows so that image features appear in a canonical configuration. Subsequent processing of the warped window can then be simplified by assuming the feature is in or near this canonical configuration. As a result, the image processing algorithms used in feature-tracking can focus on the problem of accurate *configuration adjustment* rather than general-purpose feature detection.

On a typical commodity processor, for example a 120 MHz Pentium, XVision is able to track a 40×40 blob (position), a 40×40 texture region (position) or a 40 pixel edge segment (position and orientation) in less than a millisecond. It is able to track 40×40 texture patch (translation, rotation, and scale) in about 2 ms. Thus, it is easily possible to track 20 to 30 features of this size and type at frame rate.

Although fast and accurate image-level performance is important, experience has shown that tracking them is most effective when geometric, physical, and temporal *models* from the surrounding task can be brought to bear on the tracking problem. Geometric models may be anything from weak assump-

[1] http://www.cs.yale.edu/html/yale/cs/ai/visionrobotics/yaletracking.html

tions about the form of the object as it projects to the camera image (e.g. contour trackers) to full-fledged three-dimensional models with variable parameters [2]. The key problem in model-based tracking is to integrate simple features into a consistent whole, both to predict the configuration of features in the future and to evaluate the accuracy of any single feature.

Another important part of such reliable feature trackers is the filtering process to estimate target state based on noisy observations of the target's position and a dynamic model of the target's motion. Filters proposed include tracking filters [23], $\alpha - \beta - \gamma$ filters [1], Kalman filters [37, 8], AR, ARX or ARMAX models [28].

4.2 Image Jacobian Estimation

The image-based approach requires an estimate, explicit or implicit, of the image Jacobian. Some recent results [21, 18] demonstrate the feasibility of online image Jacobian estimation. Implicit, or learning methods, have also been investigated to learn the non-linear relationships between features and manipulator joint angles [35] as have artificial neural techniques [24, 15].

The problem can also be formulated as an adaptive control problem where the image Jacobian represents a highly cross-coupled multi-input multi-output (MIMO) plant with time varying gains. Sanderson and Weiss [32] proposed independent single-input single-output (SISO) model-reference adaptive control (MRAC) loops rather than MIMO controllers. More recently Papanikolopoulos [27] has used adaptive control techniques to estimate the depth of each feature point in a cluster.

4.3 Other

Pose estimation, required for position-based visual servoing, is a classic computer vision problem which has been formulated as an estimation problem [37] and solved using an extended Kalman filter. The filter state is the relative pose expressed in a convenient parameterization. The observation function performs the perspective transformation of the world point coordinates to the image plane, and the error is used to update the filter state.

Control loops are also required in order to optimize image quality and thus assist reliable feature extraction. Image intensity can be maintained by adjusting exposure time and/or lens aperture, while other loops based on simple ranging sensors or image sharpness can be used to adjust camera focus setting. Field of view can be controlled by an adjustable zoom lens. More complex criteria such as resolvability and depth of field constraints can also be controlled by moving the camera itself [25].

5. The Future

5.1 Benefits from Technology Trends

The fundamental technologies required for visual servoing are image sensors and computing. Fortunately the price to performance ratios of both technologies are improving due to continuing progress in microelectronic fabrication density (described by Moore's Law), and the convergence of video and computing driven by consumer demands. Cameras may become so cheap as to become ubiquitous, rather than using expensive robots to position cameras it may be cheaper to add large numbers of cameras and switch between them as required.

Early and current visual servo systems have been constrained by broadcast TV standards, with limitations discussed above. In the last few years non-standard cameras have come onto the market which provide progressive scan (non-interlaced) output, and tradeoffs between resolution and frame rate. Digital output cameras are also becoming available and have the advantage of providing more stable images and requiring a simpler computer interface. The field of electro-optics is also booming, with phenomenal developments in laser and sensor technology. Small point laser rangefinders and scanning laser rangefinders are now commercially available. The outlook for the future is therefore bright. While progress prior to 1990 was hampered by technology, the next decade offers an almost overwhelming richness of technology and the problems are likely to be in the areas of integration and robust algorithm development.

5.2 Research Challenges

The future research challenges are in three different areas. One is robust vision, which will be required if systems are to work in complex real-world environments rather than black velvet draped laboratories. This includes not only making the tracking process itself robust, but also addressing issues such as initialization, adaptation, and recovery from momentary failure. Some possibilities include the use of color vision for more robust discrimination, and non-anthropomorphic sensors such as laser rangefinders mentioned above which eliminate the need for pose reconstruction by sensing directly in task space.

The second area is concerned with control and estimation and the following areas are suggested:

- Robust image Jacobian estimation from measurements made during task execution, and proofs of convergence.
- Robust or adaptive controllers for improved dynamic performance. Current approaches [6] are based on known constant processing latency, but more sophisticated visual processing may have significant variability in latency.

- Establishment of performance measures to allow quantitative comparison of different vision based control techniques.

A third area is at the level of systems and integration. Specifically, a vision-based control system is a complex entity, both to construct and to program. While the notion of programming a stand-alone manipulator is well-developed there no equivalent notions for programming a *vision-based* system. Furthermore, adding vision as well as other sensing (tactile, force, etc.) significantly adds to the *hybrid* modes of operation that needs to be included in the system. Finally, vision-based systems often need to operate in different *modes* depending on the surrounding circumstances (for example a car may be following, overtaking, merging etc.). Implementing realistic vision-based system will require some integration of discrete logic in order to respond to changing circumstances.

6. Conclusion

Both the science and technology vision-based motion control have made rapid strides in the last 10 years. Methods which were laboratory demonstrations requiring a technological tour-de-force are now routinely implemented and used in applications. Research is now moving from demonstrations to pushing the frontiers in accuracy, performance and robustness.

We expect to see vision-based systems become more and more common. Witness, for example, the number of groups now demonstrating working vision-based driving systems. However, research challenges, particularly in the vision area, abound and are sure to occupy researchers for the foreseeable future.

References

[1] Allen P, Yoshimi B, Timcenko A 1991 Real-time visual servoing. In: *Proc 1991 IEEE Int Conf Robot Automat.* Sacramento, CA, pp 851–856

[2] Blake A, Curwen R, Zisserman A 1993 Affine-invariant contour tracking with automatic control of spatiotemporal scale. In: *Proc Int Conf Comp Vis.* Berlin, Germany, pp 421–430

[3] Brown C 1990 Gaze controls with interactions and delays. *IEEE Trans Syst Man Cyber.* 20:518–527

[4] Castano A, Hutchinson S A 1994 Visual compliance: Task-directed visual servo control. *IEEE Trans Robot Automat.* 10:334–342

[5] Corke P 1993 Visual control of robot manipulators — A review. In: Hashimoto K (ed) *Visual Servoing.* World Scientific, Singapore, pp 1–31

[6] Corke P I 1996 *Visual Control of Robots: High-Performance Visual Servoing.* Research Studies Press, Taunton, UK

[7] De Schutter J 1988 Improved force control laws for advanced tracking applications. In: *Proc 1988 IEEE Int Conf Robot Automat.* Philadelphia, PA, pp 1497–1502

[8] Dickmanns E D, Graefe V 1988 Dynamic monocular machine vision. *Mach Vis Appl.* 1:223–240

[9] Espiau B, Chaumette F, Rives P 1992 A new approach to visual servoing in robotics. *IEEE Trans Robot Automat.* 8:313–326

[10] Feddema J, Lee C, Mitchell) 1991 Weighted selection of image features for resolved rate visual feedback control. *IEEE Trans Robot Automat.* 7:31–47

[11] Goldreich D, Krauzlis R, Lisberger S 1992 Effect of changing feedback delay on spontaneous oscillations in smooth pursuit eye movements of monkeys. *J Neurophys.* 67:625–638

[12] Hager G D 1997 A modular system for robust hand-eye coordination. *IEEE Trans Robot Automat.* 13:582–595

[13] Hager G D, Chang W-C, Morse A S 1994 Robot hand-eye coordination based on stereo vision. *IEEE Contr Syst Mag.* 15(1):30–39

[14] Hager G D, Toyama K 1996 XVision: Combining image warping and geometric constraints for fast visual tracking. In: *Proc 4th Euro Conf Comp Vis.* pp 507–517

[15] Hashimoto H, Kubota T, Lo W-C, Harashima F 1989 A control scheme of visual servo control of robotic manipulators using artificial neural network. In: *Proc IEEE Int Conf Contr Appl.* Jerusalem, Israel, pp TA-3–6

[16] Hashimoto K, Kimoto T, Ebine T, Kimura H 1991 Manipulator control with image-based visual servo. In: *Proc 1991 IEEE Int Conf Robot Automat.* Sacramento, CA, pp 2267–2272

[17] Hollinghurst N, Cipolla R 1994 Uncalibrated stereo hand eye coordination. *Image Vis Comp.* 12(3):187–192

[18] Hosoda K, Asada M 1994 Versatile visual servoing without knowledge of true Jacobian. In: *Proc IEEE Int Work Intel Robot Syst.* pp 186–191

[19] Huang T S, Netravali A N 1994 Motion and structure from feature correspondences: A review. *IEEE Proc.* 82:252–268

[20] Hutchinson S, Hager G, Corke P 1996 A tutorial on visual servo control. *IEEE Trans Robot Automat.* 12:651–670

[21] Jägersand M, Nelson R 1996 On-line estimation of visual-motor models using active vision. In: *Proc ARPA Image Understand Work.*

[22] Jang W, Bien Z 1991 Feature-based visual servoing of an eye-in-hand robot with improved tracking performance. In: *Proc 1991 IEEE Int Conf Robot Automat.* Sacramento, CA, pp 2254–2260

[23] Kalata P R 1984 The tracking index: A generalized parameter for $\alpha - \beta$ and $\alpha - \beta - \gamma$ target trackers. *IEEE Trans Aerosp Electron Syst.* 20:174–182

[24] Kuperstein M 1988 Generalized neural model for adaptive sensory-motor control of single postures. In: *Proc 1988 IEEE Int Conf Robot Automat.* Philadelphia, PA, pp 140–143

[25] Nelson B, Khosla P K 1993 Increasing the tracking region of an eye-in-hand system by singularity and joint limit avoidance. In: *Proc 1993 IEEE Int Conf Robot Automat.* Atlanta, GA, vol 3, pp 418–423

[26] Nelson B, Khosla P 1994 The resolvability ellipsoid for visual servoing. In: *Proc IEEE Conf Comp Vis Patt Recog.* pp 829–832

[27] Papanikolopoulos N P, Khosla P K 1993 Adaptive robot visual tracking: theory and experiments. *IEEE Trans Automat Contr.* 38:429–445

[28] Papanikolopoulos N, Khosla P, Kanade T 1991 Vision and control techniques for robotic visual tracking. In: *Proc 1991 IEEE Int Conf Robot Automat.* Sacramento, CA, pp 857–864

[29] Rizzi A, Koditschek D 1991 Preliminary experiments in spatial robot juggling. In: Chatila R, Hirzinger G (eds) *Experimental Robotics II.* Springer-Verlag, London, UK.

[30] Robinson D 1987 Why visuomotor systems don't like negative feedback and how they avoid it. In: Arbib M, Hanson A (eds) *Vision, Brain and Cooperative Behaviour.* MIT Press, Cambridge, MA

[31] Samson C, Le Borgne M, Espiau B 1992 *Robot Control: The Task Function Approach.* Clarendon Press, Oxford, UK

[32] Sanderson A, Weiss L 1983 Adaptive visual servo control of robots. In: Pugh A (ed) *Robot Vision.* Springer-Verlag, Berlin, Germany, pp 107–116

[33] Sharkey P, Murray D 1996 Delays versus performance of visually guided systems. *IEE Proc Contr Theo Appl.* 143:436–447

[34] Shirai Y, Inoue H 1973 Guiding a robot by visual feedback in assembling tasks. *Patt Recogn.* 5:99–108

[35] Skaar S, Brockman W, Hanson R 1987 Camera-space manipulation. *Int J Robot Res.* 6(4):20–32

[36] Tsai R 1986 An efficient and accurate camera calibration technique for 3D machine vision. In: *Proc IEEE Conf Comp Vis Patt Recogn.* pp 364–374

[37] Wilson W 1994 Visual servo control of robots using Kalman filter estimates of robot pose relative to work-pieces. In: Hashimoto K (ed) *Visual Servoing.* World Scientific, Singapore, pp 71–104

Sensor Fusion

Thomas C. Henderson[1], Mohamed Dekhil[1], Robert R. Kessler[1], and Martin L. Griss[2]

[1] Department of Computer Science, University of Utah, USA
[2] Hewlett Packard Labs, USA

Sensor fusion involves a wide spectrum of areas, ranging from hardware for sensors and data acquisition, through analog and digital processing of the data, up to symbolic analysis all within a theoretical framework that solves some class of problem. We review recent work on major problems in sensor fusion in the areas of theory, architecture, agents, robotics, and navigation. Finally, we describe our work on major architectural techniques for designing and developing wide area sensor network systems and for achieving robustness in multisensor systems.

1. Introduction

Multiple sensors in a control system can be used to provide:

- *more information*,
- *robustness*, and
- *complementary information*.

In this chapter, we emphasize the first two of these. In particular, some recent work on wide area sensor systems is described, as well as tools which permit empirical performance analysis of sensor systems.

By *more information* we mean that the sensors are used to monitor wider aspects of a system; this may mean over a wider geographical area (e.g. a power grid, telephone system, etc.) or diverse aspects of the system (e.g. air speed, attitude, acceleration of a plane). Quite extensive systems can be monitored, and thus, more informed control options made available. This is achieved through a higher level view of the interpretation of the sensor readings in the context of the entire set.

Robustness has several dimensions to it. First, statistical techniques can be applied to obtain better estimates from multiple instances of the same type sensor, or multiple readings from a single sensor [15]. Fault tolerance is another aspect of robustness which becomes possible when replacement sensors exist. This brings up another issue which is the need to monitor sensor activity and the ability to make tests to determine the state of the system (e.g. camera failed) and strategies to switch to alternative methods if a sensor is compromised.

As a simple example of a sensor system which demonstrates all these issues, consider a fire alarm system for a large warehouse. The sensors are widely dispersed, and, as a set, yield information not only about the existence of a fire, but also about its origin, intensity, and direction of spread. Clearly, there is a need to signal an alarm for any fire, but a high expense is incurred for false alarms. Note that complementary information may lead to more robust systems; if there are two sensor types in every detector such that one is sensitive to particles in the air and the other is sensitive to heat, then potential non-fire phenomena, like water vapor or a hot day, are less likely to be misclassified.

There are now available many source materials on multisensor fusion and integration; for example, see [1, 3, 14, 17, 24, 28, 29, 32, 33], as well as the biannual IEEE Conference on Multisensor Fusion and Integration for Intelligent Systems.

The basic problem studied by the discipline is to satisfactorily exploit multiple sensors to achieve the required system goals. This is a vast problem domain, and techniques are contingent on the sensors, processing, task constraints, etc. Since any review is by nature quite broad in scope, we will let the reader peruse the above mentioned sources for a general overview and introduction to multisensor fusion.

Another key issue is the role of control in multisensor fusion systems. Generally, control in this context is understood to mean control of the multiple sensors and the fusion processes (also called the multisensor fusion architecture). However, from a control theory point of view, it is desirable to understand how the sensors and associated processes impact the control law or system behavior. In our discussion on robustness, we will return to this issue and elaborate our approach. We believe that robustness at the highest level of a multisensor fusion system requires adaptive control.

In the next few sections, we will first give a review of the state of the art issues in multisensor fusion, and then focus on some directions in multisensor fusion architectures that are of great interest to us. The first of these is the revolutionary impact of networks on multisensor systems (e.g. see [45]), and Sect. 3. describes a framework that has been developed in conjunction with Hewlett Packard Corporation to enable enterprise wide measurement and control of power usage. The second major direction of interest is robustness in multisensor fusion systems and we present some novel approaches for dealing with this in Sect. 4. As this diverse set of topics demonstrates, multisensor fusion is getting more broadly based than ever!

2. State of the Art Issues in Sensor Fusion

In order to organize the disparate areas of multisensor fusion, we propose five major areas of work: theory, architectures, agents, robotics, and navigation. These cover most of the aspects that arise in systems of interest, although there is some overlap among them. In the following subsections, we highlight topics of interest and list representative work.

2.1 Theory

The theoretical basis of multisensor fusion is to be found in operations research, probability and statistics, and estimation theory. Recent results include methods that produce a minimax risk fixed-size confidence set estimate [25], select minimal complexity models based on Kolmogorov complexity [23], use linguistic approaches [26], tolerance analysis [22, 21], define information invariance [11], and exploit genetic algorithms, simulated annealing and tabu search [6]. Of course, geometric approaches have been popular for some time [12, 18]. Finally, the Dempster-Shafer method is used for combining evidence at higher levels [30].

2.2 Architecture

Architecture proposals abound because anybody who builds a system must prescribe how it is organized. Some papers are aimed at improving the computing architecture (e.g. by pipe-lining [42]), others aim at modularity and scalability (e.g. [37]). Another major development is the advent of large-scale networking of sensors and requisite software frameworks to design, implement and monitor control architectures [9, 10, 34]. Finally, there have been attempts at specifying architectures for complex systems which subsume multisensor systems (e.g. [31]). A fundamental issue for both theory and architecture is the conversion from signals to symbols in multisensor systems, and no panacea has been found.

2.3 Agents

A more recent approach in multisensor fusion systems is to delegate responsibility to more autonomous subsystems and have their combined activity result in the required processing. Representative of this is the work of [5, 13, 40].

2.4 Robotics

Many areas of multisensor fusion in robotics are currently being explored, but the most crucial areas are force/torque [44], grasping [2], vision [39], and haptic recognition [41].

2.5 Navigation

Navigation has long been a subject dealing with multisensor integration. Recent topics include decision theoretic approaches [27], emergent behaviors [38], adaptive techniques [4, 35], frequency response methods [8]. Although the majority of techniques described in this community deal with mobile robots, there is great interest in applying these approaches to riverboats, cars, and other modes of transportation.

We will now present some very specific work relating to wide area sensor networks and robustness of multisensor systems.

3. Wide Area Sensor Networks

In collaboration with Hewlett-Packard Laboratories (HPL), we have been developing an experimental distributed measurement software framework that explores a variety of different ways to make the development of Distributed Measurement and Control (DMC) systems easier. Our technology offers the ability to rapidly develop, deploy, tune, and evolve complete distributed measurement applications. Our solution makes use of transducers attached to the HP Vantera Measurement and Control Nodes for DMC [7]. The software technology that we have developed and integrated into our testbed includes distributed middleware and services, visual tools, and solution frameworks and components. The problem faced here is that building robust, distributed, enterprise-scale measurement applications using wide area sensor networks has high value, but is intrinsically difficult. Developers want enterprise-scale measurement applications to gain more accurate control of processes and physical events that impact their applications.

A typical domain for wide area sensor networks is energy management. When utilities are deregulated, more precise management of energy usage across the enterprise is critical. Utilities will change utility rates in real-time, and issue alerts when impending load becomes critical. Companies can negotiate contracts for different levels of guaranteed or optional service, permitting the utility to request equipment shut off for lower tiers of service. Many Fortune 500 companies spend tens of millions of dollars each year on power, which could change wildly as daily/hourly rates start to vary dynamically across the corporation. Energy costs will go up by a factor of 3 to 5 in peak load periods. Measurement nodes, transducers and controllers distributed across sites and buildings, will be attached to power panels and which enable energy users to monitor and control usage. Energy managers at multiple corporate sites must manage energy use and adjust to and balance cost, benefit and business situation. Site managers, enterprise workflow and measurement agents monitor usage, watch for alerts and initiate load shifting and shedding (see Fig. 3.1).

The complete solution requires many layers. Data gathered from the physical processes is passed through various information abstraction layers to provide strategic insight into the enterprise. Likewise, business level decisions are passed down through layers of interpretation to provide control of the processes. Measurement systems control and access transducers (sensors and actuators) via the HP Vantera using the HP Vantera Information Backplane publish/subscribe information bus. Some transducers are self-identifying and provide measurement units and precision via Transducer Electronic Data Sheets (TEDS - IEEE 1451.2).

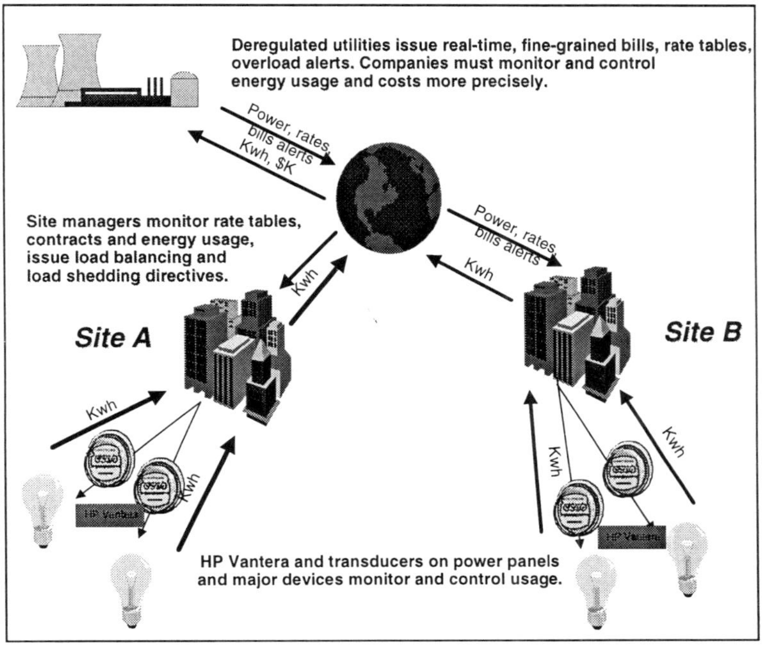

Fig. 3.1. Energy management

3.1 Component Frameworks

How are applications like this built? Enterprise and measurement applications are built from a set of components, collectively called frameworks. Each framework defines the kinds of components, their interfaces, their services, and how these components can be interconnected. Components can be constructed independently, but designed for reuse. In a distributed environment, each component may be able to run on a different host, and communicate with others [16].

For this testbed, we have developed the (scriptable) Active Node measurement oriented framework, using the Utah/HPL CWave visual programming framework as a base [34]. This framework defines three main kinds of measurement components: (1) Measurement Interface Nodes that provide gateways between the enterprise and the measurement systems, (2) Active Nodes that provide an agency for measurement abstraction, and (3) Active Leaf Nodes that act as proxies to monitor and control transducers. Each of these nodes communicates with the other types and the measurement devices.

The key component for the CWave measurement agency component model is the Active Node. Active nodes allow a measurement engineer to write scripts which communicate with other nodes via the HP Vantera publish/subscribe information bus and thus control component interactions. The scripts run via the Microsoft ActiveX Scripting engine.

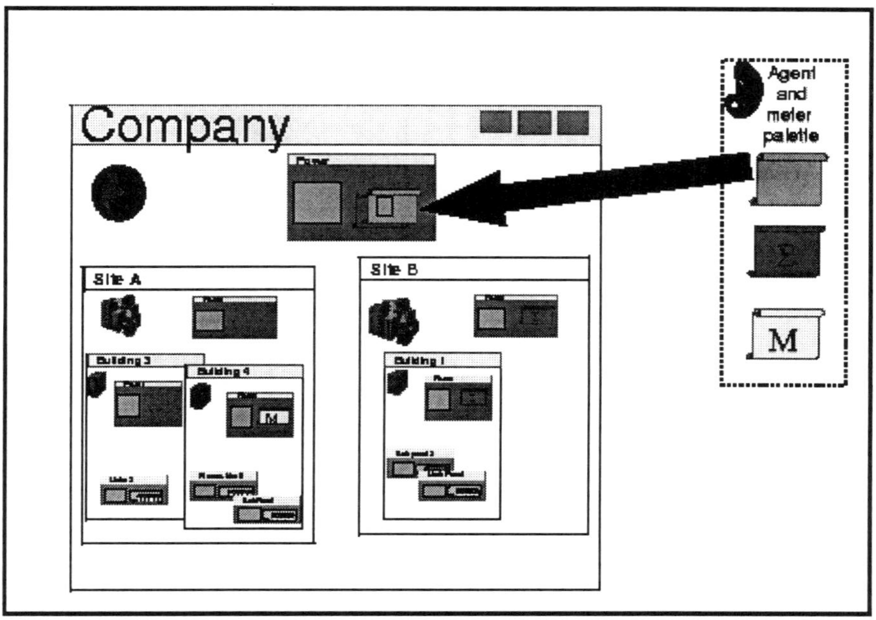

Fig. 3.2. Basic structure of agent and analysis components

One or more agent scripts, written in VBScript, JavaScript, or Perl can be downloaded into the component (by drag and drop). Once downloaded, each script runs its own execution thread and can access the full ActiveX scripting engine. These scripts can publish new topics, subscribe to subjects, and interact with the CWave environment. The Measurement Interface Nodes (MIN) act as gateways between the HP Vantera Information Bus and the enterprise world.

CWave is a uniquely *visual* framework, consisting of components and tools that are targeted for quickly building, evolving, testing and deploying distributed measurement systems. CWave features include extensibility, visual development of scripts, downloaded programs, multi-machine visualization and control, drag-and-drop, and wiring. Active Nodes are instances of a CWave component that has been specially interfaced to the HP Vantera environment, and make heavy use of the publish/subscribe capabilities.

In Fig. 3.2 CWave components are shown which represent buildings and sites. This was built by dragging components from the palette, customizing the names, and setting properties from the property sets. Detail can be suppressed or revealed via zooming and selective viewing. Nesting of components provides a natural namespace for components that is important in controlling the scope of publish/subscribe. We also show a palette of *agent scripts* that can be dragged and dropped onto an Active Node or Active Leaf Node. The scripts can then be propagated to children. The CWave environment can be used by engineers to develop components and skeletal applications, by installers to configure and customize a system, or by the end user engineers to adjust measurements, to monitor processing, etc.

The deployed set of agent scripts work together to collect the power data from several sensors, and combine these into an average energy measurement over the interval. These measurements are propagated up, with averaging, and tests for missing measurements to ensure robustness of the result. Other agent scripts can be deployed to monitor or log data at various Active Nodes, or to test for conditions and issue alerts when certain (multi-sensor) data conditions arise.

CWave provides visual programming, threads and distributed management. The addition of the Active Node framework provides a flexible and customizable model for combining and integrating the inputs and control of a multitude of sensors.

4. Robustness

Multisensor fusion techniques have been applied to a wide range of problem domains, including: mobile robots, autonomous systems, object recognition, navigation, target tracking, etc. In most of these applications, it is necessary that the system perform even under poor operating conditions or when subcomponents fail. We have developed an approach to permit the developer to obtain performance information about various parts of the system, from theory, simulation or actual execution of the system. We have developed a semantic framework which allows issues to be identified at a more abstract level and then monitored at other levels of realization of the system.

Our overall goal is shown in Fig. 4.1. A system is comprised of software, hardware, sensors, user requirements, and environmental conditions. A system model can be constructed in terms of models of these components. An analysis can be done at that level, but is usually quite abstract and of a worst-case kind of analysis. We want to use such models to gain insight into where in a simulation or actual system to put taps and monitors to obtain empirical performance data. These can be used for several purposes:

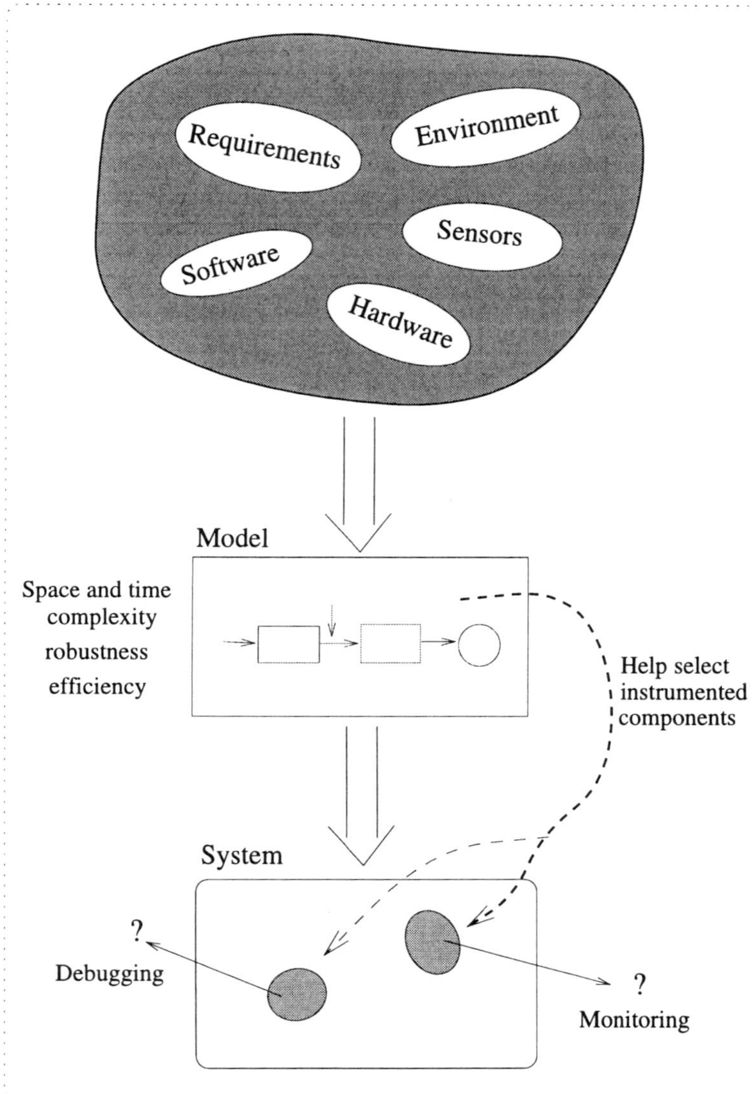

Fig. 4.1. The proposed modeling approach

- *Debugging*: the designer can interactively watch important aspects of the system as it runs.
- *Design Choices*: the designer may want to measure the performance of alternative solutions over various domains (time, space, error, etc.).
- *Adaptive Response*: the designer may want to put in place monitors which watch the performance of the system, or its context, and change techniques during execution.

This allows robustness to be built into a system, and the user can do so on well-founded information. For a more detailed discussion of this approach, see [9, 10] where we applied the framework to the performance comparison of a visual technique and a sonar technique for indoor wall pose estimation.

4.1 Instrumented Sensor Systems

Since we are putting in functions to monitor the data passing through a module and its actions, we call our approach *Instrumented Sensor Systems*. Figure 4.2 shows the components of our framework. As shown in the figure,

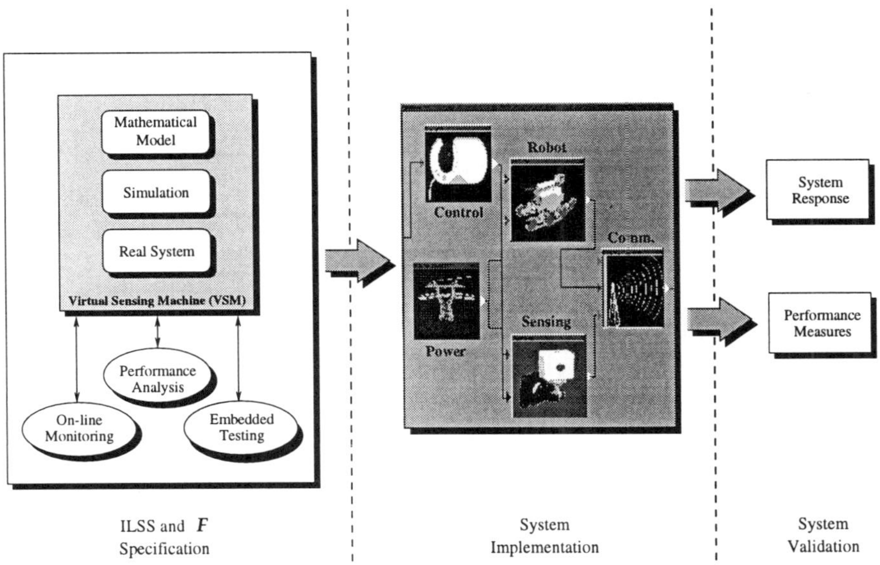

Fig. 4.2. The instrumented logical sensor system components

this approach allows for a specification, an implementation, and the validation of results. This is an extension of our previous work on *Logical Sensor Specifications* [19, 20]. A logical sensor is basically a system object with a name, characteristic (typed) output, and a select function which chooses between various alternate methods for producing the desired data. An *Instrumented Logical Sensor* has the following additional functions:

- *Taps*: Provide for a trace of the flow of data along a component connection path,
- *Tests*: Run the currently selected method on a known input/output pair and check for correctness.
- *Monitors*: Check for failure or adaptive mode conditions (monitors may run tests as part of their monitoring activity).

We have applied these ideas to the development of mobile robot systems. (Also, see [46] which influenced our work.)

4.2 Adaptive Control

Another area we are currently exploring is adaptive control. Figure 4.3 shows a basic feedback control loop. In a stable environment such an approach is

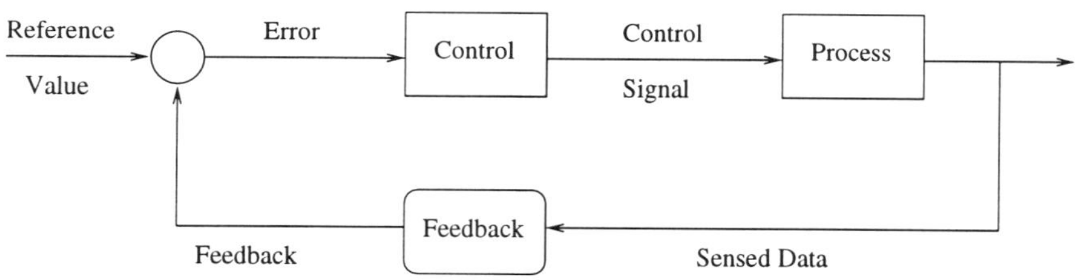

Fig. 4.3. Simple feedback control loop

feasible and many standard techniques can be applied (e.g. Kalman filtering is often used in navigation). Our concern is with a higher level of robustness — namely, when the context changes or assumptions fail, methods must be provided so the system can detect and handle the situation (see Fig. 4.4).

Our view is that sensor data can be used to alter the parameters of the control function. For example, consider a control function which solves for a variable (say, V which is a voltage to be applied to a motor), at each instant by determining where the following equation is 0:

$$\frac{dV}{dt} = a * (1 - \frac{V}{b}) - \frac{V}{(1 + V^2)} = f(V) \tag{4.1}$$

The parameters a and b represent features of the environment, and if sensors are used to determine them, then our function is really $f(V; a, b)$. In this case when the parameters change value, the solution space may change

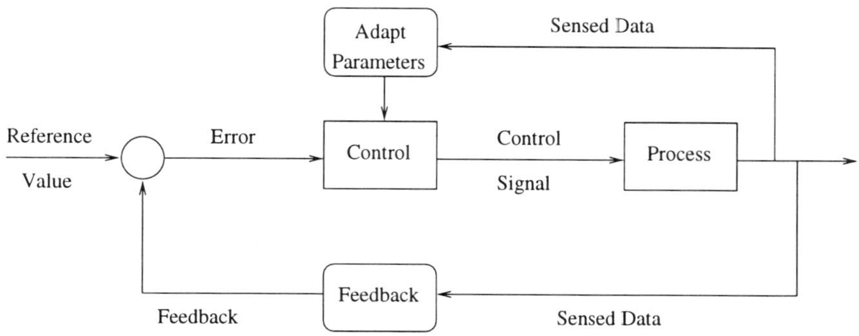

Fig. 4.4. Adaptive control

qualitatively; e.g. when $a = 0.2$ and $b = 10$, then there is a single solution (see Fig. 4.5). However, as the parameter a changes, there may be more than one solution (see Fig. 4.6). Finally, if systems are designed exploiting these features, then there may also be large jumps in solution values; Fig. 4.7 shows how with a slight increase in the value of a, the solution will jump from about 1.3 to about 8. Such qualitative properties can be exploited in designing a more robust controller. (See [36] for a detailed discussion of these issues in biological mechanisms.)

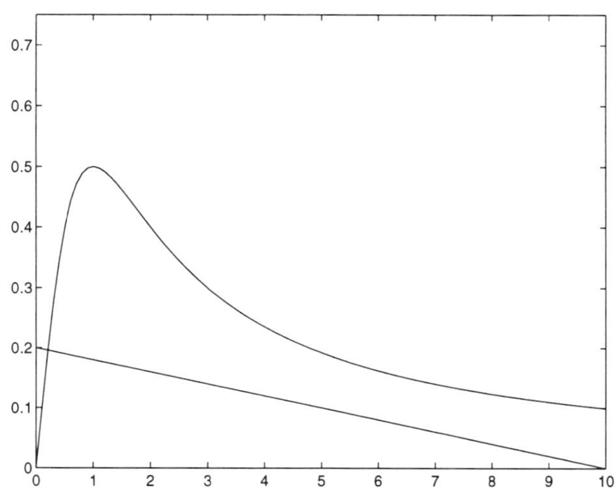

Fig. 4.5. Single solution (when $a = 0.2$ and $b = 10$)

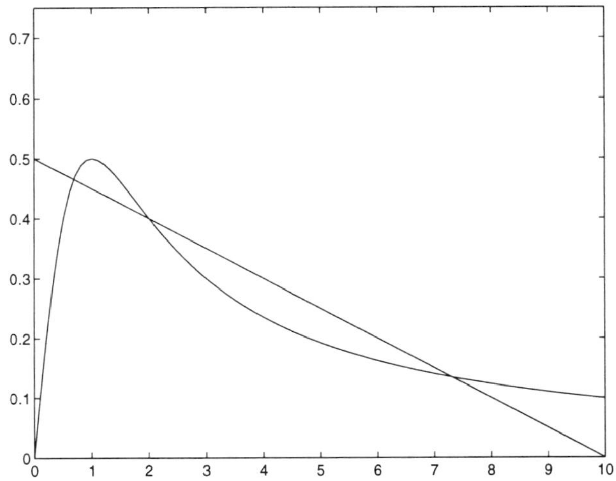

Fig. 4.6. Multiple solutions (when $a = 0.5$ and $b = 10$)

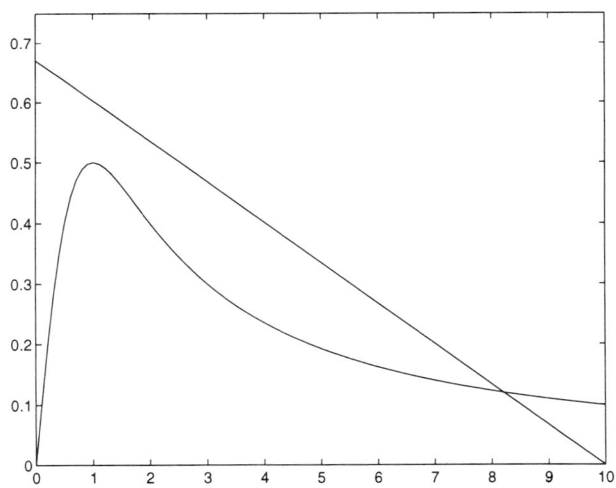

Fig. 4.7. A jump in solutions (when $a = 0.67$ and $b = 10$)

5. Conclusions

In this chapter we have presented an overview of current issues as well as some new directions in multisensor fusion and control. We have described an approach to the design and execution of wide area sensor networks. In addition, we have proposed new techniques for obtaining more robust multisensor fusion systems. However, one very important new area that we have not covered is the ability of multisensor fusion systems to learn during execution (see [43] for an approach to that). These are very exciting times, and we believe that major strides will be made in all these areas in the next few years.

References

[1] Abidi M A, Gonzalez R C 1993 *Data Fusion in Robotics and Machine Intelligence.* Academic Press, Boston, MA

[2] Allen P, Miller A T, Oh P Y, Leibowitz B S 1996 Integration of vision and force sensors for grasping. In: *Proc 1996 IEEE Int Conf Multisens Fusion Integr Intel Syst.* Washington, DC, pp 349–356

[3] Bajcsy R, Allen, P 1990 Multisensor integration. In: Shapiro S C (ed) *Encyclopedia of Artificial Intelligence.* Wiley, New York, pp 632–638

[4] Betgé-Brezetz S, Chatila R, Devy M, Fillatreau P, Nashashibi F 1996 Adaptive localization of an autonomous mobile robot in natural environments. In: *Proc 1994 IEEE Int Conf Multisens Fusion Integr Intel Syst.* Las Vegas, NV, pp 77–84

[5] Boissier O, Demazeau Y 1994 MAVI: A multi-agent system for visual integration. In: *Proc 1994 IEEE Int Conf Multisens Fusion Integr Intel Syst.* Las Vegas, NV, pp 731–738

[6] Brooks R R, Iyengar S S 1996 Maximizing multi-sensor system dependability. In: *Proc 1996 IEEE Int Conf Multisens Fusion Integr Intel Syst.* Washington, DC, pp 1–8

[7] Cauthorn, J 1997 Load profiling and aggregation using distributed measurements. In: *Proc 1997 DA/DSM DistribuTECH.* San Diego, CA

[8] Cooper S, Durrant-Whyte H 1996 A frequency response method for multisensor high-speed navigation systems. In: *Proc 1994 IEEE Int Conf Multisens Fusion Integr Intel Syst.* Las Vegas, NV, pp 1–8

[9] Dekhil M, Henderson T C 1996 Instrumented sensor systems. In: *Proc 1996 IEEE Int Conf Multisens Fusion Integr Intel Syst.* Washington, DC, pp 193–200

[10] Dekhil M, Henderson T C 1997 Instrumented sensor system architecture. Tech Rep UUCS-97-011, University of Utah

[11] Donald B R 1995 On information invariants in robotics. *Artif Intel.* 72:217–304

[12] Durrant-Whyte H 1988 *Integration, Coordination and Control of Multisensor Robot Systems.* Kluwer, Boston, MA

[13] Freund E, Rossmann J 1996 Intuitive control of a multi-robot system by means of projective virtual reality. In: *Proc 1996 IEEE Int Conf Multisens Fusion Integr Intel Syst.* Washington, DC, pp 273–280

[14] Garvey T D 1987 A survey of AI approaches to the integration of information. In: *Proc 1987 SPIE Conf Infrared Sensors Sens Fusion.* pp 68–82

[15] Gelb A 1974 *Applied Optimal Estimation.* MIT Press, Cambridge, MA

[16] Griss M L, Kessler R R 1996 Building object-oriented instrument kits. In: *Object Mag.*

[17] Hackett J K, Shah M 1990 Multisensor fusion: A perspective. In: *Proc 1990 IEEE Int Conf Robot Automat.* Cincinnati, OH, pp 1324–1330

[18] Hager G 1990 *Task-directed Sensor Fusion and Planning.* Kluwer, Boston, MA

[19] Henderson T C, Shilcrat E 1984 Logical sensor systems. *J Robot Syst.* 1:169–193

[20] Henderson T C, Weitz E, Hansen C, Mitiche A 1988 Multisensor knowledge systems: interpreting 3D structure. *Int J Robot Res.* 7(6):114–137

[21] Iyengar S S, Prasad L 1995 A general computational framework for distributed sensing and fault-tolerant sensor integration. *IEEE Trans Syst Man Cyber.* 25:643–650

[22] Iyengar S S, Sharma M B, Kashyap R L 1992 Information routing and reliability issues in distributed sensor networks. *IEEE Trans Sign Proc.* 40:3012–3021

[23] Joshi R, Sanderson A 1996 Multisensor fusion and model selection using a minimal representation size framework. In: *Proc 1996 IEEE Int Conf Multisens Fusion Integr Intel Syst.* Washington, DC, pp 25–32

[24] Kak A 1987 A production system environment for integrating knowledge with vision data. In: *Proc 1987 AAAI Work Spatial Reason Multisens Fusion.* St. Charles, IL, pp 1–12

[25] Kamberova G, Mandelbaum R, Mintz M 1996 Statistical decision theory for mobile robotics: Theory and application. In: *Proc 1996 IEEE Int Conf Multis Fusion Integr Intel Syst.* Washington, DC, pp 17–24

[26] Korona Z, Kokar M 1996 Model theory based fusion framework with application to multisensor target recognition. In: *Proc 1996 IEEE Int Conf Multisens Fusion Integr Intel Syst.* Washington, DC, pp 9–16

[27] Kristensen S, Christensen H I 1996 Decision-theoretic multisensor planning and integration for mobile robot navigation. In: *Proc 1996 IEEE Int Conf Multisens Fusion Integr Intel Syst.* Washington, DC, pp 517–524

[28] Luo R, Kay M G 1989 Multisensor integration and fusion in intelligent systems. *IEEE Trans Syst Man Cyber.* 19:901–931

[29] Mann R C 1987 Multisensor integration using concurrent computing. In: *Proc 1987 SPIE Conf Infrared Sensors Sens Fusion.* pp 83–90

[30] Matsuyama T 1994 Belief formation from observation and belief integration using virtual belief space in Dempster-Shafer probability model. In: *Proc 1994 IEEE Int Conf Multisens Fusion Integr Intel Syst.* Las Vegas, NV, pp 379–386

[31] Meystel A 1995 Multiresolutional semiosis. In: *Proc 1996 IEEE Int Symp Intel Contr.* Monterey, CA, pp 13–20

[32] Mitiche A, Aggarwal J K 1986 An overview of multisensor systems. *SPIE Opt Comp.* 2:96–98

[33] Mitiche A, Aggarwal J K 1986 Multisensor integration/fusion through image processing. *Opt Eng.* 25:380–386

[34] Mueller-Planitz C, Kessler R R 1997 Visual threads: The benefits of multithreading in visual programming languages. Tech Rep UUCS-97-012, University of Utah

[35] Murphy R 1996 Adaptive rule of combination for observations over time. In: *Proc 1996 IEEE Int Conf Multisens Fusion Integr Intel Syst.* Washington, DC, pp 125–131

[36] Murray J D 1993 *Mathematical Biology.* Springer-Verlag, Berlin, Germany

[37] Mutambara A G O, Durrant-Whyte H F 1994 Modular scalable robot control. In: *Proc 1994 IEEE Int Conf Multisens Fusion Integr Intel Syst.* Las Vegas, NV, pp 121–127

[38] Nakauchi Y, Mori Y 1996 Emergent behavior based sensor fusion for robot navigation system. In: *Proc 1996 IEEE Int Conf Multisens Fusion Integr Intel Syst.* Washington, DC, pp 525–532

[39] Rander P W, Narayanan P J, Kanade T 1996 Recovery of dynamic scene structure from multiple image sequences. In: *Proc 1996 IEEE Int Conf Multisens Fusion Integr Intel Syst.* Washington, DC, pp 305–312

[40] Rus D, Kabir A, Kotay K D, Soutter M 1996 Guiding distributed manipulation with mobile sensors. In: *Proc 1996 IEEE Int Conf Multisens Fusion Integr Intel Syst.* Washington, DC, pp 281–288

[41] Sakaguchi Y, Nakano K 1994 Haptic recognition system with sensory integration and attentional perception In: *Proc 1994 IEEE Int Conf Multisens Fusion Integr Intel Syst.* Las Vegas, NV, pp 288–296

[42] Takahashi E, Nishida K, Toda K, Yamaguchi,Y 1996 CODA: Real-time parallel machine for sensor fusion – Evaluation of processor architecture by simulation. In: *Proc 1996 IEEE Int Conf Multisens Fusion Integr Intel Syst.* Washington, DC, pp 836–844

[43] van Dam J W M, Krose B J A, Groen, F C A 1996 Adaptive sensor models. In: *Proc 1996 IEEE Int Conf Multisens Fusion Integr Intel Syst.* Washington, DC, pp 705–712

[44] Voyles R M, Morrow J D, Khosla P 1996 Including sensor bias in shape from motion calibration and sensor fusion. In: *Proc 1996 IEEE Int Conf Multisens Fusion Integr Intel Syst.* Washington, DC, pp 93–99

[45] Warrior, J 1997 Smart sensor networks of the future. *Sensors.* 2:40–45

[46] Weller G A, Groen F C A, Hertzberger L O 1990 A sensor processing model incorporating error detection and recovery. In: Henderson T (ed) *Traditional and Non-Traditional Robotic Sensors.* Springer-Verlag, Berlin, Germany, pp 351–363

Discrete Event Theory for the Monitoring and Control of Robotic Systems

Brenan J. McCarragher

Department of Engineering, Faculties, Australian National University, Australia

Discrete event systems are presented as a powerful framework for a large number of robot control tasks. Their advantage lies in the ability to abstract a complex problem to the essential elements needed for task level control. Discrete event control has proven to be successful in numerous robotic applications, including assembly, on-line training of robots, mobile navigation, control of perception capabilities, and human-robot shared control. This chapter presents a general description of the discrete event modelling and control synthesis for constrained motion systems. Additionally, methods for the effective monitoring of the process based on the detection and identification of discrete events are given. In each of these areas, open research questions are discussed. Advances in the discrete event modelling of robotic systems, especially event definitions based on a solid mathematical formulation, are essential for broadening the range of applications. Robust process monitoring techniques, as well as sensory fusion methods, are needed for successful practical implementations. Control synthesis methods which are truly hybrid, incorporating both the continuous and the discrete event control, would significantly advance the use of discrete event theory. Lastly, convergence proofs under practical assumptions relevant to robotics are needed.

1. Introduction and Motivation

Intuitively, a hybrid dynamic system consists of a discrete event decision-making system interacting with a continuous time system. A simple example is a climate control system in a typical home. The on-off nature of the thermostat is modelled as a discrete event system, whereas the furnace or air-conditioner is modelled as a continuous time system. Most research work in the area of hybrid systems is concerned with developing and proving control-theoretic ideas for specific classes of systems. Ostroff and Wonham [24] provide a powerful and general framework (TTM/RTTL) for modelling and analyzing real-time discrete event systems. Holloway and Krogh [10] use cyclic controlled marked graphs (CMG's) to model discrete event systems, allowing for the synthesis of state feedback logic. Brockett [7], Stiver and Antsaklis [26], and Gollu and Varaiya [9] have given more general formulations for the modelling and analysis of hybrid dynamical systems. Practical applications include mining [23], manufacturing [8], detailed robotic assembly [20], and even elevator dispatching [25]. These papers have tried to capture a wide range of system attributes in simple models to allow for tractable analysis and control optimization.

2. Discrete Event Modelling

2.1 Modelling using Constraints

We will consider a constrained motion system which involves the motion of a *workpiece* with possible constraints introduced by a fixed *environment*. Systems of this type are typical of assembly processes [20], and other constrained motion systems [7]. Hybrid dynamic modelling is particularly appropriate for assembly processes as assembly has natural discrete event dynamics due to the changes in the state of contact. This constrained motion framework can also be used to describe the mobile navigation problem where the environment (e.g. walls and obstacles) create constraints on the motion of the vehicle. A general initial structure as shown in Fig. 2.1 is proposed. The structure consists of three parts, which are the continuous time plant, the discrete event controller, and the interface.

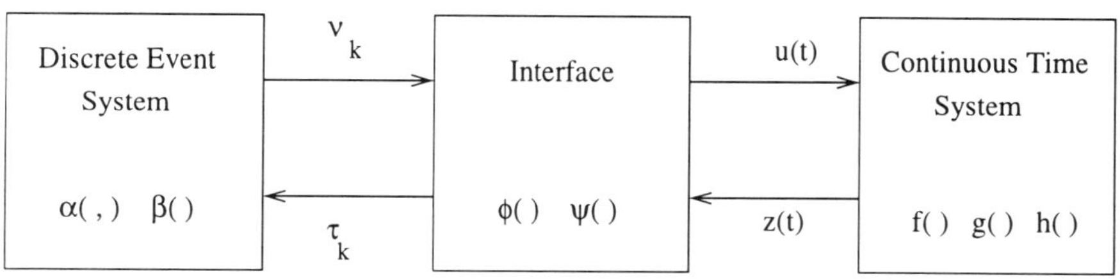

Fig. 2.1. Block diagram representation of hybrid dynamic system

2.1.1 The continuous-time system. Consider the general motion control of an object or workpiece. The equation of motion of the workpiece in free-space is given generally as

$$\dot{\mathbf{x}}(t) = f\left(\mathbf{x}(t), \mathbf{u}(t)\right) \tag{2.1}$$

where $\mathbf{x}(t)$ is the continuous time state vector, and $\mathbf{u}(t)$ is the input vector. As the workpiece interacts with the environment, the free-space dynamics of (2.1) become constrained by

$$\mathbf{0} = g_k\left(\mathbf{x}(t), \mathbf{u}(t)\right) \qquad t_k \leq t < t_{k+1} \tag{2.2}$$

where g_k represents the different constraint equations for the different discrete states, and t_k is the time that the kth discrete event occurs. Additionally, system is observed through the measurement of the state $\mathbf{z}(t)$ given by

$$\mathbf{z}(t) = h(\mathbf{x}(t)) \tag{2.3}$$

Since we have a hybrid system that interacts with the environment, the state may be easier to determine according to the measurement of both the force and the state. Hence, we have a state estimate $\dot{\hat{\mathbf{x}}}$ described by the following equation

$$\dot{\hat{\mathbf{x}}}(t) = \tilde{f}(\hat{\mathbf{x}}(t), F(t), \mathbf{z}(t)) \tag{2.4}$$

where $F(t)$ is the force measurement.

2.1.2 The discrete event system. The continuous time plant is controlled by a task-level, discrete event controller modelled as an automaton. The automaton is a quintuple, (S, E, C, α, β), where S is the finite set of states, E is the set of plant events, C is the set of controller events, $\alpha : S \times E \to S$ is the state transition function, and $\beta : S \to C$ is the output function. Each state in S, denoted γ_k, is defined to be a discrete state of constraint. Each event in C, denoted v_k, is generated by the discrete event controller, whereas each event in E, denoted τ_k, is generated by the conditions in the continuous plant. The dynamics of the discrete event controller are given generally by

$$\gamma_{k+1} = \alpha(\gamma_k, \hat{\tau}_k) \tag{2.5}$$

$$v_k = \beta(\gamma_k) \tag{2.6}$$

where $\gamma_k \subset S, \hat{\tau}_k$ is the estimate of $\tau_k \subset E$ and $v_k \subset C$. The index k specifies the order of the discrete states or events. The input and output signals of the discrete event controller are asynchronous sequences of events, rather than continuous time signals. The function α is functionally dependent on f and g as defined by the continuous-time system. Note that the state of the system is $\mathbf{x}$, whereas γ is the state variable of the discrete event system and is dependent on $\mathbf{x}$.

2.1.3 The interface. The interface is responsible for the communication between the plant and the controller, since these components cannot communicate directly due to the different types of signals being used. The interface consists of two maps ϕ and ψ. The first map ϕ converts each controller event into a plant input as follows

$$\mathbf{u}(t) = \phi(v_k) \qquad t_k \leq t < t_{k+1} \tag{2.7}$$

where v_k is the most recent controller event before time t. Initially, we will require that $\mathbf{u}(t)$ be a constant plant input for each controller event. Hence, the plant input is a piecewise constant signal which may change only when a controller event occurs. For synthesis purposes it is often easier to combine the control Eqs. (2.6) and (2.7) such that

$$\mathbf{u}(t) = \phi(\beta(\gamma_k)) \qquad t_k \leq t < t_{k+1} \tag{2.8}$$

The second map ψ converts the state space of the plant into the set of plant events.

$$\tau_k = \psi\left(\mathbf{x}(t)\right) \tag{2.9}$$

Note that Eq. (2.9) does not imply that τ_k changes continuously as $\mathbf{x}(t)$ changes. The state space of the plant is partitioned into contiguous regions. The function ψ generates a new plant event only when the state first enters one of these regions. The function ψ is usually well defined. However, the state variable $\mathbf{x}(\mathbf{t})$ needs to be estimated according to (2.4). Thus, we have an estimator equation for the determination of the discrete plant events.

$$\hat{\tau}_k = \psi\left(\hat{\mathbf{x}}(t)\right) \tag{2.10}$$

Note that $\tilde{f}$ of Eq. (2.4) does not need to be close to the actual function f, provided that $\hat{\tau}_k$ is a good estimate of τ_k.

2.2 An Assembly Example

It is important to note that the DES modelling of assembly is motivated by the key discrete event features of the process itself. The first key element is that, for polygonal models of the workpiece and the environment, the states of contact, or constraints, between the two parts are discrete. In modelling the manipulation process as a DES, we define the state of contact to be the discrete state vector. The second key element is that the states of contact (or constraints) change at discrete, and often unknown, instances. It is an abrupt change where the dynamics change instantly. Moreover, due to the uncertainty of location and speed, the instant when the transition occurs is often unknown. A transition that results in a change of state, either a loss of contact or a gain of contact, is defined as a discrete event. The discrete events are key because, at these points the geometric constraints on the workpiece dynamics change, requiring a change in both the trajectory and the controller. DES modelling makes transitions one of the central control features of the model.

Three dimensional polygonal objects can be fully described using three components: surfaces, edges and vertices, as shown in Fig. 2.2. Each of these components is labeled: surfaces with upper case letters, edges with numbers, and vertices with lower case letters. The labeling is done for both the workpiece and the environment. Thus, any contact state can be defined by listing the pairs that are in contact. For example, the surface-edge contact shown in Fig. 2.2-(a) is defined by the pair $(A-16)$. The surface-surface contact shown in Fig. 2.2-(b) is defined by the pair $(E-K)$. Multiple points of contact are given by listing all relevant pairs. For example, $(A-16)(10-J)$ describes the contact state shown in Fig. 2.2-(c).

By straight enumeration there are six types of contact pairs. These are surface-surface, surface-edge, surface-vertex, edge-edge, edge-vertex and vertex-vertex. However, of these six only two are task primitives [17]. For three dimensional polygonal objects, the two task primitive contact types are the **surface-vertex** and the **edge-edge** contact types. The other four types

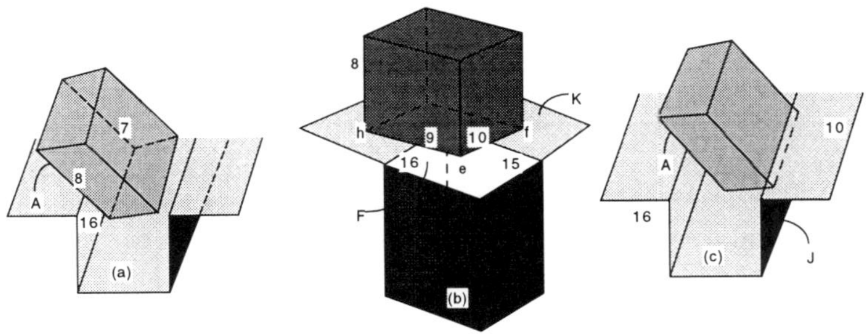

Fig. 2.2. Constraint states. (a) $(A - 16)$ (b) $(E - K)$ (c) $(A - 16)(10 - J)$

of contact–surface-surface, surface-edge, edge-vertex, and vertex-vertex–can be comprised of the task primitive contact types.

We formalize the discrete event model of assembly using Petri nets. Petri nets are a compact mathematical way of describing the geometric constraints and the admissible transitions for an assembly task, highlighting the events. Moreover, Petri nets are a useful method for describing the indeterministic nature of robotic assembly, by incorporating transitions that are possible given the uncertainties, unknowns and errors in the system. The ability to address these unknowns is one of the primary strengths of the Petri net modelling method [20, 21].

A significant advantage to discrete event modelling of assembly using task primitive contacts is that redundant information is not included. The collection of contact states contain redundant information, especially regarding the mathematical constraints on the motion of the workpiece. The discrete event description using Petri nets eliminates this redundant information by recording only the constraint information that is necessary to describe the motion; that is, recording only the constraint information of the primitive contact pairs. Thus, the constrained motion of the workpiece in all contact states can be obtained through combinations of the mathematical constraints of the primitive contact pairs.

2.3 Research Challenges

2.3.1 Rigorous event definition. One of the most significant challenges in the use of discrete event systems theory in robotics is a rigourous definition of discrete states and discrete events. It is not enough to simply place a discrete event framework over an existing control system. Rather, there must be some inherent discrete event dynamics for the DE framework to be effective. Indeed, these underlying discrete dynamics needs to be the foundation for the discrete state and discrete event definitions. A significant research challenge is the effective discrete event modelling of robotic systems. In support of this challenge is the development of automated computer modelling tools (e.g. assembly Petri nets, automata).

2.3.2 Application areas. In conjunction with rigourous event modelling, there is a challenge to expand the robotic application areas that would benefit from a discrete event systems theory. The example of assembly presented a good application area because assembly is inherently a constrained motion system. The definitions of discrete state and discrete event are very natural to the problem.

Other application areas would benefit from a formal discrete event approach. In particular there is significant potential for advanced control within mobile navigation. The problem of mobile navigation is immense. Different strategies are required for static, dynamic, partially known and unknown environments. Attempts have been made to use a DE approach to the problem [16]. However, this work suffered from an ad hoc definition of discrete state and discrete event based on the desired motions of the robots. To be effective, the mobile navigation model needs to be rigourously based on the discrete event dynamics of the system. For example, a constraint to avoid collision with a wall needs to include the position of the wall and the state (position and velocity) of the vehicle.

Another area that looks promising is human-robot shared control [1]. In this field, discrete event theory allows for a clear definition of interactions between the two controllers, easing the complications associated with the control interaction. The DE approach facilitates interactions on both continuous and discrete levels. The method also allows for 'control guides' that aid the human operator in the execution of the task. There is a strong possibility that the DE approach for shared control provides a global framework that can incorporate other shared control methodologies.

2.3.3 Hierarchical integration in production systems. A significant advantage of discrete event systems theory in manufacturing systems is the ability to incorporate many levels of control operation in a consistent, hierarchical fashion. Using the DE framework, all levels of the manufacturing system can be controlled using a common model, representation and language. For example, an entire factory operation may be represented with a discrete event model, and at the same time, the operation of an individual workstation in the factory is controlled using DE theory. The communication between the two systems can be quite seamless.

The advantages for robotic control are just as significant. Hierarchical Discrete Event control allows for robotic work cells and robots themselves to be included in the overall production control systems. Such integration can be applied to mining and other production systems, as well as manufacturing. For example, the lowest level of the hierarchy would be the detailed control of the assembly process. The next level up could be the control of the assembly work-cell, including a discrete event model of transport using a DE model for mobile navigation of the AGVs. The third level would connect all the assembly work-cells to control the entire assembly process. Lastly, a discrete event model could include supply, production, warehousing and transport, treating the assembly DE model as a state or transition in the highest level model.

The clear advantage for robotics in this scenaric is its direct inclusion into the overall production system. Using DE theory, robotic control is demystified and simplified, important goals for the broader acceptance and usage of robots. Also, the DE approach incorporates robots as part of the complete automation system. In this way, robotic researchers and developers are forced to examine the entire production and automation problem, rather than just the robot control in isolation.

3. Discrete Event Control Synthesis

3.1 Controller Constraints

The control commands are determined by first establishing a *desired event* for each state. The desired event is selected to move to a state "closer" to the target state, that is, to move towards completion. The desired events may be determined manually or automatically, depending upon the application. For any given state, we use the desired event and geometric considerations of the constraints to establish three conditions on the command to be executed.

3.1.1 Maintaining condition. The motion of the system described by (2.1) is constrained by (2.2). The first possible task of the controller is to ensure that the control commands satisfy this geometric constraint. To derive admissible velocities that satisfy the geometric constraint, we can differentiate (2.2) to give

$$\frac{\partial}{\partial \mathbf{x}} \left[g_j \left(\mathbf{x} \right) \right] \frac{d}{dt} \mathbf{x}(t) = \mathbf{0} \qquad t_k \le t < t_{k+1} \tag{3.1}$$

where g_j is the constraint function for this contact. This can be rewritten as

$$\mathbf{a}_j^T \dot{\mathbf{x}}(t) = \mathbf{0} \qquad t_k \le t < t_{k+1} \tag{3.2}$$

where $\mathbf{a}_j = \frac{\partial}{\partial \mathbf{x}} g_j(\mathbf{x})$ is a column vector with length equal to the number of degrees of freedom. Equation (3.2) is our maintaining condition in that it must be satisfied to maintain the contact or geometric constraint. When g_j is a distance measure, Eq. (3.2) becomes a requirement that the distance between the points of contact remains zero (i.e. the points remain in contact).

3.1.2 Enabling condition. In addition to determining motion that maintains a constraint, it is desired to determine the motion such that the workpiece encounters the next discrete state γ_{k+1}. Since the system is not in γ_{k+1}, the following must be true

$$g_j(\mathbf{x}(t)) = K \qquad t_k \leq t < t_{k+1} \tag{3.3}$$

where, without loss of generality, K is a positive constant. In order to direct the system such that $K \to 0$, we require the time derivative to be negative.

$$\mathbf{a}_j^T \dot{\mathbf{x}}(t) < \mathbf{0} \qquad t_k \leq t < t_{k+1} \tag{3.4}$$

Equation (3.4) is our enabling condition. It is a necessary condition for discrete event τ_{k+1} to occur. When g_j is a distance measure, Eq. (3.4) becomes a requirement that the distance decreases, that is, the intended points of contact move closer together.

3.1.3 Disabling condition. The third condition, the disabling condition, is derived directly from the enabling condition. Since (3.4) is a necessary condition for a discrete event to occur, a sufficient condition for a discrete event not to occur is obtained by changing the direction of the inequality.

$$\mathbf{a}_j^T \dot{\mathbf{x}}(t) \geq \mathbf{0} \qquad t_k \leq t < t_{k+1} \tag{3.5}$$

where j indicates the discrete states (constraint equations) that are not desired to occur. Essentially, this disabling condition prevents K from decreasing in magnitude. When g_j is a distance measure, Eq. (3.5) becomes a requirement that the distance between the possible points of contact does not decrease (i.e. the points stay apart).

3.2 Command Synthesis

The desired event determines which of the above conditions should be applied for each possible constraint. The maintaining condition (3.2) is used when it is desired to maintain a constraint. Note, when it is desired to immediately violate the current constraint by breaking the contact, the maintaining condition is not used. The enabling condition (3.4) is used to activate a currently inactive constraint. The disabling condition (3.5) is used to prevent unwanted constraints from becoming active. From the desired event, we have a set of constraints on the control command. The control command is now determined by satisfying this set of constraints. Any method for satisfying the set of constraints will yield an acceptable discrete event velocity command. One method [5], which uses a search technique to maximise the minimum distance to each constraint for maximum robustness, is suggested.

Potential fields can also be used to generate continuous velocity commands [2] because they provide a straightforward method which can deal with difficult environments without a complex set of path planning rules.

The fields can be modified by changing one or two variables which makes them attractive for online modification. The potential fields can also be used to generate barriers which are useful in restricting input. Potential fields can be divided into two main groups, attractive and repulsive potentials. The attractive potentials are used for maintaining and enabling constraints, whereas the repulsive potentials are used for disabling constraints. Attractive potentials can be represented by quadratic and conical wells [14, 15]. This type of well is centrally attractive at any distance and is utilised in order for the robot to reach a target position, the center of the well. Repulsive potentials are also important in order to repel the manipulator from a constraint or a boundary which is not to be crossed. These repulsive potentials can also be used to constrain commanded motion. The continuous velocity for the robot manipulator is generated by calculating the derivative of the composite potential field.

3.3 Event-level Adaptive Control

Discrete event control offers considerable advantages for constrained manipulation including excellent error-recovery characteristics. However, despite determining a velocity command which satisfies the constraint equations of Sect. 3.1 errors can still occur due to model inaccuracies, tracking control errors, or other unknowns. Unfortunately, the errors will result in a sub-optimal trajectory. In these situations, it is desired to have the system adjust to the new information and adapt the desired velocity commands. The ability to adapt is particularly important in an industrial setting where new products are frequently introduced and the production line needs to be "tuned" to the new tasks.

An effective means of task-level adaptation is to adjust the model so that the event conditions more accurately reflect the actual system [18, 22]. Consider the adaptation of a maintaining condition. Here, the estimate of the constraint vector $\hat{\mathbf{a}}$ and the velocity vector $\dot{\mathbf{x}}$ are orthogonal. Yet, the velocity vector is not orthogonal to the actual constraint vector $\mathbf{a}$, indicating the need for adaptation. By adding a portion of the velocity vector to the estimated constraint vector, the difference between the estimated and the actual constraint vectors decreases. Hence, the following adaptation law is proposed

$$\hat{\mathbf{a}}_{b+1} = \hat{\mathbf{a}}_b - \Delta\lambda\dot{\mathbf{x}}_b \tag{3.6}$$

where $\Delta = -sgn(\mathbf{a}^T\dot{\mathbf{x}})$ is a switching function determined by the error condition to be adapted; $\lambda > 0$ is the adaptation rate; and $\dot{\mathbf{x}}_b$ is the velocity command vector for the bth trial. The adaptation law is hybrid, based on the direction of the performance gradient and an estimate of its magnitude. The direction comes from a recognition of the constraint which was violated, and the estimate of the magnitude comes from the continuous velocity vector.

Given this adaptation law, two issues arise. The first is demonstrating the convergence of the estimated model constraint to the actual parameters. The second is the selection of λ such that the adaptation remains stable. Both of these issues can be answered using Lyapunov theory. For the complete proof of Lyapunov stability, the reader is referred to [22]. The result of that proof is that stability, and hence convergence to zero modelling error, is guaranteed if the following condition on the adaptation rate is met.

$$\lambda < \frac{2\Delta\tilde{\mathbf{a}}_b^T\dot{\mathbf{x}}_b}{||\dot{\mathbf{x}}_b||^2} \tag{3.7}$$

Examination of how to satisfy Eq. (3.7) for each of the discrete event conditions yields the following requirements. For simplicity and to highlight the adaptation equations, we will assume that the velocity vector has been normalised to $||\dot{\mathbf{x}}_b|| = 1$. This assumption has little effect as only the direction of the velocity vector is important.

For the maintaining condition, equation (3.7) is satisfied provided

$$\lambda < -2\Delta\mathbf{a}^T\dot{\mathbf{x}}_b \tag{3.8}$$

and, for the enabling and disabling conditions, equation (3.7) is satisfied if

$$\lambda < 2\Delta\hat{\mathbf{a}}^T\dot{\mathbf{x}}_b \tag{3.9}$$

The adaptation law has been tested by simulation and experiments [18, 22]. In many experiments, the system was able to adapt the discrete event command based on the occurrence of events. Indeed, the adaptive, event-level control has been used to train assembly lines in an industrial setting [18]. This adaptation law allows the adjustment of the rate of convergence and error recovery properties but the proper selection of λ is not guaranteed.

3.4 Research Challenges

3.4.1 Guaranteed convergence / stability. The primary duty of the discrete event control system is to drive the system to converge to the final desired discrete state, denoted γ_f. The papers by Astuti and McCarragher [3, 4] argue that most work in the literature on the convergence of discrete event systems is not applicable to robotic systems because these convergence proofs fail to recognise two important issues in practical implementation. The first important issue is that robotic systems have tracking errors. As such, an unrealistically large control effort is required to enable and disable events accurately. Accurate event control is a prerequisite for most of the convergence proofs. Secondly, the literature tends to assume that all events are perfectly recognisable. Unfortunately, in robotics, event detection is a difficult and error-prone process.

In response to these important practical considerations, [3] proposes the concept of *p-convergence* for discrete event systems. By identifying the probability of event occurring (rather than a guarantee), p-convergence is developed. p-convergence is a generalisation of a finite-time convergence proof given in [6], yet p-convergence is less restrictive. The finite-time conditions guaranteed convergence within a finite number of events, whereas the conditions for p-convergence allow the number of events to tend to infinity for guaranteed convergence.

The concept of p-convergence is a more realistic design goal for robotic discrete event systems due to the reasonable modelling and control effort needed, and due to the allowance for tracking and sensing errors. A trade-off is achieved between guaranteed convergence and the amount of control effort. Additionally, as a design objective for robotic discrete event systems, p-convergence lends itself to the formulation of an optimal control problem, finding a controller which maximizes the probability of entering the invariant set of discrete states while minimizing the number of events [4].

The standard stability problem seeks to find a set of control commands given by Eq. (2.8) that will cause the system to asymptotically converge to the desired final discrete state γ_f. Standard is used to imply that there is full knowledge of the system. That is, equations (2.1), (2.2), (2.3), (2.5), (2.6), (2.7) and (2.9) are known exactly, and the estimation equations (2.4) and (2.10) are not needed since these variables are known exactly. The challenge becomes increasingly difficult when tracking errors, parametric modelling errors, sensing noise or sensing delay are considered.

3.4.2 Hybrid synthesis. To date, discrete event control synthesis techniques used in robotics have been based primarily on the discrete state alone, as per Eq. 2.6. Previous research has shown that control based solely on discrete events, while effective in simplifying complex control problems, is limited in the robotics context. Since robotics is a hybrid dynamic problem, rather than than a pure discrete event system, control synthesis based on both discrete and continuous state vectors is expected to increase advanced control operations. The adaptive control problem of Sect. 3.3 highlights the benefits of using both the discrete and continuous state vectors for control synthesis.

Recently, there has been an emphasis on broadening the information base for discrete event decisions to explicitly include the continuous state vector. As such, the research challenge is to synthesize a discrete control command as follows

$$v_k = \beta(\gamma_k, \mathbf{x}(t_k)) \tag{3.10}$$

where $\mathbf{x}(t_k)$ is the continuous state vector at the time of event k. This hybrid synthesis formulation also has implications for the interface, which now needs to pass both the discrete state and continuous state to the discrete event controller.

An alternative hybrid synthesis formulation can be derived from a modification of Eq. 2.7. This interface equation would need to explicitly include the continuous state vector.

$$\mathbf{u}(t) = \phi(v_k, \mathbf{x}(t_k)) \qquad t_k \leq t < t_{k+1} \tag{3.11}$$

A hybrid synthesis formulation further complicates the stability and convergence problem. Now, stability proofs also must be hybrid. Unfortunately, few mathematical tools exist for the analysis of hybrid systems. Indeed, the development of mathematical tools for the analysis of hybrid systems are strongly needed, and a very good research challenge.

4. Process Monitoring

4.1 Monitoring Techniques

Process monitoring is the task of determining the current status of the robotic process, and is among the key techniques for improving reliability and preventing failures. Traditionally, it is attempted to determine the exact state vector of the system. Discrete event modelling, however, reduces the problem to determining the discrete state vector, since the constrained dynamics and control commands can then be determined. Undoubtedly, the specifics of process monitoring techniques will depend heavily on the specifics of application area. To date, most of the discrete event monitoring work in robotics has been in the area of event detection for constrained manipulation and assembly. Additionally, some work has been done in event recognition for mobile navigation.

We currently use several methods for event recognition in assembly. First, we use a qualitative processing approach to analyse the force signal for the detection and identification of discrete state transitions. There are significant advantages to the qualitative approach to process monitoring. First, qualitative monitoring is faster than quantitative since detailed calculations are not necessary. The advanced speed results from the detection of dynamic effects as opposed to a quasi-static method which waits for force transients to die out. Also, a quantitative method would require estimates of several quantities such as friction that are not easily available. Lastly, the qualitative method is less susceptible to noise in the system since exact numbers are not being used. The interested reader is referred to [19] for a comprehensive derivation of the qualitative process monitor.

The second method for event recognition uses a Hidden Markov Model (HMM) for each transition. HMMs use quantitative data, but do not have the problems of unreliable estimates. Instead, HMMs are trained on a set of samples, from which the stochastic 'signature' of the transition signal is determined. The advantage of a HMM approach is that it is more reliable than the qualitative method due to the use of more, quantitative information. On the other hand, HMMs have a significantly longer process time, which tends to slow the assembly process. The interested reader is referred to [13] for a full explanation of the HMM transition recognition method.

Third, a method for combining dynamic force and static position measurements for the monitoring of assembly has been developed [12]. A multilayer perceptron (MLP) network is used as a classifier where the individual network outputs correspond to contact state transitions occurring during the assembly process. When a contact state transition occurs, the MLP output with the largest value is chosen. The recognised contact state is sent to a discrete event controller which guides the workpiece through a series of contact states to the final desired configuration. The MLP has been successfully implemented with high recognition rates. One advantage of the proposed MLP method compared to other existing solutions for recognition of contact state transitions is that it models both dynamic and static behaviour. The dynamic force measurements depend on a number of system parameters, such as workpiece and environment stiffnesses, sensor noise and dynamics and the individual joint PID gains of the robot manipulator. These factors may vary from one contact state to another and hence help to improve the performance of the discrete event recognition. The position measurements, on the other hand, contain mostly static information during the short time from an event occurs until it is recognised by the monitor.

4.2 Control of Sensory Perception

Reliable sensing is essential for successful control of plants in uncertain environments. Traditionally, control systems receive measurements from a fixed sensing architecture where all the sensors are used all the time. Hence, the bandwidth of the overall control structure is limited by the slowest sensor. We present a new technique for the real-time control of sensory perception. Typically, only a few sensors are needed to verify nominal operation. When an anomaly develops, additional sensors are utilised. The benefits of the proposed method are an increased reliability compared to individual sensors while the bandwidth is kept high.

The control of sensory perception is well suited to the hybrid dynamic framework, Fig. 2.1. The process monitors provide feedback to the discrete event controller only when discrete events occur. Hence, processing time is available between events for use by the sensory perception controller. A sensory perception controller (SPC) has been implemented for the discrete event control of a robotic assembly task. The three process monitoring techniques are available to the sensory perception controller. The method used for the dynamic sensory perception is based on stochastic dynamic programming and is described in detail in [11].

The method starts with an initial confidence level of zero and all monitors enabled. Then the sensory perception consists of two parts. First, an iterative dynamic programming (DP) algorithm evaluates all possible orderings of enabled process monitors by calculating the DP value function V. The dynamic programming model is formulated as an optimal stopping problem. At each iteration two actions are evaluated; a_1 – terminate the sensory perception, or a_2 – consult another process monitor. Second, the (SPC) selects the ordering of enabled process monitors with the highest V. If the optimal action for this ordering is a_2, then the first monitor in the ordering is consulted. The confidence level output from the monitor is recorded. Next the monitor is disabled and the sensory perception problem is repeated with the new initial confidence level. The SPC terminates when the optimal action is a_1 or all monitors are disabled. The final recognised discrete event is sent to the discrete event controller.

Table 4.1. Evaluation of different process monitoring techniques and the sensory perception controller

Monitor	Rate	CPU Time
Position	79%	0.10
Template Matching	85%	0.08
Hidden Markov Models	87%	0.87
SPC	97%	0.38

The performance of the SPC was evaluated from a sample set of 100 discrete events and compared with the individual process monitoring techniques. The results are given in Tab. 4.1. The table shows the average successful recognition rates and the costs of the individual process monitors. With individual average recognition rates of 79%, 85% and 87% the average recognition rate of the sensor selection controller is as high as 97% which is better than any individual process monitor. The CPU time spent by the SPC depends on the number of monitors used for each event. The average CPU time for the sample set of 100 discrete events was found to be 0.38 seconds. These results clearly show the benefits of fusing several sensing techniques for the process monitoring of robotic assembly.

The main advantage of the sensory perception control structure is an improved event recognition rate compared to individual event monitors while the total cost is kept low. For cases where it is too expensive to use all the event monitors simultaneously, the maximum total cost can be limited and hence the proposed solution is well suited for use in real-time control systems with bandwidth requirements.

4.3 Research Challenges

4.3.1 Event and state monitoring. The methods presented here have concentrated on event monitoring for robotic systems, primarily for constrained manipulation systems. A fair amount of literature exists concentrating on state monitoring (see [13]). However, very little work has been done to effectively combine the two monitoring philosophies in a coherent and rigourous way. Event monitoring would have the advantage of detecting dynamic events as well as limiting the range of possible discrete state vectors. The state monitoring would add robustness and certainty to the decisions made by the event monitor. Advanced control methods, such as the hybrid control synthesis and the adaptive event control, would most likely require both event and state monitoring to be effective. The real challenge, however, is to develop a synergistic method of process monitoring. It is not enough to take two existing methods and apply them to the same problem. Rather, the event and state monitors must derive some advantage from the existence of the other method. For example, the event monitor must know that an event is likely due to the feedback of the state monitor. Additionally, the state monitor should be more accurate due to the information gathered by the event monitor.

4.3.2 Sensory fusion. The benefits of sensory fusion were demonstrated in Tab. 4.1 of Sect. 4.2. In that example, the sensory fusion method had a higher recognition rate than any individual monitor, and also had a very low average cost. Despite its successes, many challenges still exist in the area of sensory fusion for discrete event systems in robotics. First, in the method presented, all process monitors gave events as output. There was no recognition of the different modes of operation of the sensors. That is, the SPC would make no distinction between a vision system and a force sensor. Yet, the type of information available from each sensor is quite different and should be treated differently. Clearly, a force sensor is of little use unless there is a contact situation.

The difference in sensor mode could be accounted for using a fusion method with a discrete state dependency. The ability to alter the sensory fusion algorithm according to the discrete process state is another clear advantage to the discrete event approach to robotics. For example, a given discrete state may indicate no-contact. In this state, the fusion algorithm should not consider the data from the force sensor. In another state where vision is occluded, force sensing would be imperative. Indeed, the sensory control could be expanded even to include active sensing based on the discrete state of the robotic system. The ability of the sensing system to react to the discrete state of the system is a very important advance in robotics research and implementation, yet the problem is often overlooked.

Acknowledgement. This work has been supported in part by the Australian Research Council, Large Grants Program.

References

[1] Aigner P, McCarragher B 1996 Human integration into control systems: Discrete event theory and experiments. In: *Proc 2nd World Automat Congr.* Montpellier, France

[2] Aigner P, McCarragher B 1997 Human integration into robot control utilising potential fields. In: *Proc 1997 IEEE Int Conf Robot Automat.* Albuquerque, NM

[3] Astuti P, McCarragher B 1996 The stability of discrete event systems using Markov processes. *Int J Contr.* 64:391-408
[4] Astuti P, McCarragher B 1997 Controller synthesis of manipulation hybrid dynamic systems. *Int J Contr.* 66
[5] Best M J, Ritter K 1985 *Linear Programming: Active Set Analysis and Computer Programs.* Prentice-Hall, Englewood Cliffs, NJ
[6] Brave Y, Heymann M 1990 Stabilization of discrete-event processes. *Int J Contr.* 51:1101–1117
[7] Brockett R 1993 Hybrid models for motion control systems. In: Trentelman H L, Willems J C (eds) *Essays on Control: Perspectives in the Theory and its Applications.* Birkhäuser, Boston, MA, pp 29–53
[8] Gershwin S 1989 Hierarchical flow control: A framework for scheduling and planning discrete events in manufacturing systems. *Proc IEEE*
[9] Gollu A, Varaiya P 1989 Hybrid dynamical systems. In: *Proc 28th IEEE Conf Decision Contr.* Tampa, FL, pp 2708–2712
[10] Holloway L E, Krogh B H 1990 Synthesis of feedback control logic for a class of controlled Petri nets. *IEEE Trans Automat Contr.* 35
[11] Hovland G E, McCarragher B J 1996 Control of sensory perception using stochastic dynamic programming. In: *Proc 1st Australian Data Fusion Symp.* Adelaide, Australia
[12] Hovland G E, McCarragher B J 1997 Combining force and position measurements for the monitoring of robotic assembly. *IEEE/RSJ Conf Intel Robot Syst.* Grenoble, France
[13] Hovland G E, McCarragher B J 1998 Hidden Markov models as a process monitor in robotic assembly. *Int J Robot Res.* to appear
[14] Khatib O 1986 Real-time obstacle avoidance for manipulators and mobile robots. *Int J Robot Res.* 5(1):90–98
[15] Khosla P K, Volpe R 1988 Superquadratic artificial potentials for obstacle avoidance and approach. In: *Proc 1988 IEEE Int Conf Robot Automat.* Philadelphia, PA, pp 1778–1784
[16] Košecká J, Bajcsy R 1995 Discrete event systems for autonomous mobile agents. *Robot Autonom Syst.* 12:187–198
[17] McCarragher B 1996 Task primitives for the discrete event modelling and control of 6 DOF assembly tasks. *IEEE Trans Robot Automat.* 12
[18] McCarragher B 1997 Adaptive discrete event control for assembly: Theory and industrial implementation. *Robot Autonom Syst.* 14
[19] McCarragher B J, Asada H 1993 Qualitative template matching using dynamic process models for state transition recognition of robotic assembly. *ASME J Dyn Syst Meas Contr.* 115:261–275
[20] McCarragher B, Asada H 1995 The discrete event modelling and trajectory planning of robotic assembly tasks. *ASME J Dyn Syst Meas Contr.* 117
[21] McCarragher B, Asada H 1995 The discrete event control of robotic assembly tasks. *ASME J Dyn Syst Meas Contr.* 117
[22] McCarragher B, Austin D 1998 Model adaptive discrete event control for constrained motion systems. *IEEE Trans Automat Contr.* to appear
[23] McCarragher B, Read D 1997 Perturbation analysis of a discrete event model of a mining operation. *1997 World Manufactur Congr.* Aukland, New Zealand
[24] Ostroff J S, Wonham W M 1990 A framework for real-time discrete event control. *IEEE Trans Automat Contr.* 35(4)
[25] Pepyne D L, Cassandras C 1996 Optimal dispatching control for elevator systems during peak traffic. In: *Proc 35th IEEE Conf Decision Contr.* Kobe, Japan, pp 3837–3842

[26] Stiver J, Antsaklis P 1992 Modelling and analysis of hybrid control systems. In: *Proc 31st IEEE Conf Decision Contr.* Tucson, AZ, pp 3748–3751

Scheduling of Flexible Manufacturing Systems

Peter B. Luh

Department of Electrical and Systems Engineering, University of Connecticut, USA

This chapter presents the state-of-art formulations and solution methodologies for the scheduling of flexible manufacturing systems. A case study is drawn from a flexible manufacturing system for the apparel industry. A number of important issues are then identified, and new promising research approaches indicated.

1. Introduction

Manufacturing converts materials into useful objects, and is one of the few ways that wealth is created. To meet the demand for the targeted products, manufacturing facilities are often organized in specific layouts selected from a spectrum of possibilities. Examples include "continuous flow" for a very standard and high volume product (such as sugar), "line flow" for high volumes of several major products (such as air conditioners), " job shop" for low volumes of many products (such as components of prototype jet engines), and "project" for one-of-a-kind items (such as a custom designed house).

Among the spectrum of possibilities, mid-volume, mid-variety manufacturing is among the least developed ones. Driven by global competitiveness, however, the market is gradually shifting to semi-customized products in order to meet the increasingly diverse and rapidly changing demand at low cost. Highly flexible systems are therefore required for a manufacturer to survive and to thrive in this mid-volume, mid-variety paradigm.

Building on the advancements in numerically controlled (NC) machines, the group technology methods, and information technology, flexible manufacturing systems (FMSs) were created to meet the new challenges. An FMS is a set of automatic machine tools or fabrication equipment linked together by an automatic material handling system and controlled by a computer system for the flexible fabrication of parts or assembling of products that fall within predetermined families. A part is fixed on a *fixture* which is in turn mounted on a *pallet*, and is delivered to a machine for processing. The required tools for the operation are stored in the machine's *tool magazine*, and are automatically installed as needed. Once the operation is finished, the part is removed from the machine, and automatically transferred to the next machine for processing with minimum human intervention. Under adequate computer control, an FMS should be able to switch from one part type to a completely different part type, and should be able to respond to unforeseen and unpredictable disturbances such as urgent orders, machine breakdowns, or unavailability of raw materials. For a general description of FMSs, please see [29] and [25].

1.1 Classification of FMS

From system operation's point of view, FMSs can be classified into two categories according to [34]: "flexible flow systems" and "general flexible manufacturing systems" (called the "general flexible machining systems" in [34]). Flexible flow systems are designed to produce a few "similar" part types in relatively large volumes. Consequently, machine layouts are similar to those of conventional flow lines, and most part types follow the same *serial* production stages. General flexible manufacturing systems are designed to manufacture a wide variety of part types simultaneously. They are thus *similar* to conventional job shops, and permit a random routing of part types. These systems could either be dedicated to produce a fixed set of part types with required ratios, or be of general purpose to produce a wide variety of part types within a family to satisfy time varying demand.

1.2 Key Issues in Operating an FMS

Effective planning and operation of an FMS's production is a very important task. The problem, however, has been recognized to be extremely difficult because of the following factors:

- the complicated processing environment, including the simultaneous consideration of machines, pallets, fixtures, tools, finite buffers and transportation with the possibility of deadlocks; certain multi-axle machines may process multiple parts at the same time and require the synchronization of parts, etc.;
- combinatorial nature of integer optimization;
- the large sizes of real systems; and
- the need for fast and almost real time responses.

To overcome the above difficulties, decisions in practice are often made top down in a hierarchical order. The following five problems were identified in [41]:

- Part type selection. From the set of part types to be produced, select a subset for simultaneous processing over some period of time.
- Machine grouping. Partition the machines into groups so that the machines belonging to a particular group are identically tooled and can perform the same set of operations.
- Production ratio setting. Determine the ratios at which the part types selected above are to be produced.
- Resource allocation. Allocate the limited number of pallets and fixtures among the selected part types.

- Part loading. Allocate the operations and required tools among the machine groups.

The above represents a hierarchical framework, where decisions are made in a top-down fashion. Many times, however, the lines between adjacent layers may not be clearly drawn, and integrated consideration of several layers, if possible, may turn out to be more effective.

1.3 Scope of This Chapter

This chapter presents the state-of-art formulations and solution methodologies for the scheduling of non-dedicated general flexible manufacturing systems (corresponds to Resource allocation and Part loading in the above hierarchical framework). The issues may not seem to be control related on the surface. The readers will find out, however, that dynamic programming, Lagrangian relaxation, and nonlinear programming, which are basic tools for the control community, play a key role in developing the latest generation of solution methodology. A case study presented in Sect. 4. is drawn from a flexible manufacturing system for the apparel industry. A number of important problems for the control community are then identified, and new promising research approaches indicated.

2. Problem Formulation

Since a non-dedicated general flexible manufacturing system is similar to a conventional job shop, we shall first present the formulation of a job shop scheduling problem. The presentation will be intuitive, without using equations. For mathematically oriented readers, please refer to [27] for details. Differences between FMS vs. job shop scheduling will then be highlighted, and a mathematical formulation of an FMS for apparel production will be presented in Sect. 4.

2.1 Formulation of a Job Shop Scheduling Problem

As mentioned earlier, job shop is a typical layout for low volume high variety manufacturing. In a job shop, machines of similar functionality are grouped together in one area, such as all grinding machines in one area, all milling machines in another. A part then travels from area to area according to the established sequence of operations. Typical examples include machine shops, tool and dye shops, many plastic molding operations, and hospitals. Job shop scheduling problems take many different forms, include these presented in [33, 2], and [32]. The formulations differ in the selection of variables, the modeling of constraints, and the choosing of objective function to be optimized. The following job shop formulation is based on [27].

Suppose that machines in a job shop belong to several machine types, and each machine type has one or multiple identical machines. Each part to be processed has to go through a sequence of operations, and each operation requires a machine of a particular type to process for a specific amount of time. The scheduling is subject to the following constraints:

- Machine Capacity Constraints. The number of parts being processed on a machine type may not exceed the number of machines available at any time.
- Operation Precedence. An operation cannot be started until its preceding operation has been completed.
- Processing Time Requirements. Each operation must be assigned the required amount of time for processing on the specified machine type.

By appropriately selecting decision variables, all the above can be mathematically represented as linear equality or inequality constraints.

Many objective functions have been presented. The most common one is the "makespan," i.e. the span of time from the beginning of the first operation to the completion of all the parts to be scheduled. Without due dates in mind, schedules generated using this criterion many times behave poorly in terms of on-time delivery performance. Other objective functions include flow time (sum of all part completion times), lateness and tardiness related penalties, throughput, and/or machine utilization. Lateness of a part is defined as the part's completion time minus its due date, and is positive when the part is completed late and negative when it is completed early. Tardiness is similar to lateness except that its value is set to zero when the part is completed early. The dominant goal of manufacturing scheduling, nevertheless, is generally recognized as on-time delivery of parts with low work-in-process inventory. This can be modeled as a weighted sum of tardiness and earliness penalties, where earliness of a part is the amount that the part's release time (beginning time of the first operation) leads an *desired release time*.

The overall problem is thus to minimize the weighted tardiness and earliness penalties subject to the above mentioned constraints by selecting appropriate operation beginning times. Since all the constraints are linear and the objective function is part-wise additive, the model is "separable." This is essential for *Lagrangian relaxation* to be effective as will be explained in Sect. 3.

2.2 Differences between FMS and Job Shop Scheduling

The major differences between FMSs and job shops from the scheduling's viewpoint include [22, 34]:

- Alternative routing. Because of flexible hardware and software, FMSs generally enjoy high routing flexibility. Although flexibility does exist for job shops, it is usually not fully exploited especially during on-line operations because of the lack of timely information and real-time decision capability.

- Buffer limitation. Buffers within an FMS are generally of finite capacity. Finite buffers can lead to blocking and starving of machines, and deadlock of the system. Buffer capacities in job shops are generally considered as infinite except in very few selected cases.
- Transportation capacity and time. The issue of transportation exists for both job shops and FMSs, however, it is mostly ignored in the job shop literature. The issue becomes acute for FMSs because of limited work-in-process inventory caused by finite capacity buffers; tight coupling among the machines; and automated but limited transportation capability (e.g. having a finite number of automated guided vehicles or tow-line carts). Transportation itself is a complicated problem, and the coupling of transportation with part processing makes the problem even more complicated. For example, one may not be able to pre-determine the time required to transport a part from one machine to another since this time depends on the locations of automated guided vehicles at the instant when the service is requested.
- Reduction of the number and time required for setups. In FMS process planning, one generally attempts to assign as many operations to a single setup as possible. This reduces the number of setups required for manufacturing parts using an FMS. Furthermore, the programmability of instructions for FMS operations reduces setup times between consecutive operations.
- Deterministic but longer processing times. Since FMS operations are computer controlled, processing times are generally highly predictable. Because of the different process planning practice as mentioned in (4.4) and the attempt to have multiple identical parts clamped onto a single fixture, the processing time per machine load is generally much longer than it would be in an equivalent classical machining center.
- Pallet, fixture, tool, and tool magazine limitations. The numbers of pallets and fixtures are limited. Part processing also requires various tools which are stored in the machine's tool magazine. The number of tools available and the capacity of the tool magazine are also limited.

Some of the above features (e.g. alternative routings, finite capacity buffers, pallets and fixtures) have been addressed in the "extended" job shop literature and handled mathematically as mentioned above. Others (e.g. transportation time) have not been explicitly considered at a detailed level except in selected simulation studies.

3. Solution Methodology

Similar to the previous section, approaches for job shop scheduling will first be reviewed. Methods relevant to the scheduling of non-dedicated general flexible manufacturing systems will then be summarized. For a detailed presentation of general scheduling approaches, please see [33, 2, 36]. For a review of FMS scheduling approaches, including those for flexible flow systems and dedicated flexible manufacturing systems, please see [34].

3.1 Approaches for Job Shop Scheduling

Given the economic and logistical importance of the scheduling problem, many of the early efforts centered on obtaining optimal schedules. Two prominent optimization methods are the branch and bound method [11]) and dynamic programming, e.g. [33]. However, it was discovered that job shop scheduling is among the hardest combinatorial optimization problems and is NP-complete [12]), and the generation of optimal schedules often requires excessive computation time regardless the methodology.

In practice, material planning systems (e.g. MRP or MRP II) are often used for high level production planning. Since resource capacities are ignored in these systems, resulting schedules are generally infeasible. Simple heuristics or dispatching rules are often used to determine schedules at the local (resource or machine) level. Many heuristic methods have been presented based on due dates, criticality of operations, operation processing times, and machine utilization, e.g. [4]. Many artificial intelligence (AI) approaches also use heuristics for scheduling, e.g. [22]. These heuristic-based approaches usually generate feasible schedules fast, but it is difficult to evaluate the quality of the schedules. Also, most heuristics do not provide for iterative improvement of the schedules.

Attempts to bridge the gap between heuristic and optimization approaches have also been undertaken [1, 27, 44]. In [1], for example, a heuristic for job shop scheduling was developed based upon optimally solving single machine sequencing problems. A criterion for measuring machine busyness was developed, and the job sequence for the busiest machine (the bottleneck) was first developed. The job sequence for the next busiest machine was then determined, and the solution was fed back into the previously solved machine problem by a “local re-optimization.”

Lagrangian relaxation has recently been developed to be an effective and practical method for job shop scheduling [27, 10, 7, 19, 46]. Similar to the pricing concept of a market economy, the method replaces “hard” machine capacity constraints by using “soft” prices (i.e. Lagrange multipliers, or “shadow prices” in the economic literature). A part has to pay a certain “price” to use a machine at each time unit. Given the values of multipliers, the overall problem can be decomposed into smaller subproblems, i.e. one for each part to find the optimal beginning times of its operations. These subproblems are much easier to solve as compared to the original problem, and solutions can be efficiently obtained by using *dynamic programming*. These prices are then iteratively adjusted based on the degree of constraint violations following again the market economy concept (i.e. increase the prices for over-utilized time units and reduce the prices for under-utilized time units). At the termination of such price updating iterations, a few constraints may still be violated. Simple heuristics are then applied to adjust subproblem solutions to form a feasible schedule satisfying all constraints.

The resolution of the original problem is thus through a two-level approach, where *low level* consists of solving individual subproblems. Coordination of subproblem solutions is done through the iterative updating of Lagrange multipliers at the *high level* to ensure near optimality of the overall solution. In the optimization terminology, the "dual function" is maximized. Furthermore, the "dual cost" is a lower bound on the optimal cost, and enables quantitative measurement of solution quality. Methods for maximizing the dual functions include the commonly used subgradient methods, the reduced complexity bundle method [43], and the recently developed "interleaved subgradient method" [19] and "surrogate subgradient method" [47].

The basic job shop scheduling model and the corresponding Lagrangian relaxation method have been extended to consider machining centers with pallets and tool magazines [45]; finite capacity buffers, alternative routings, and machines with setup requirements [26]; and batch processing facilities such as heat treat ovens that can process multiple parts at the same time [28]. In all these cases, the separability of the formulation is preserved, and Lagrangian relaxation has been successfully applied to obtain near-optimal solutions in a computationally efficient manner.

3.2 Methods for FMS Scheduling

Most approaches for non-dedicated general flexible manufacturing systems mimic job shop scheduling methods with additional consideration of FMS features. In view of problem complexity, most approaches are based on heuristics, and shall not be elaborated here. Examples include [38, 40, 35, 37, 18]. Optimization-based approaches are limited, and include dynamic programming [5]. The direct application of dynamic programming, however, suffers from the *curse of dimensionality*. For a detailed review of FMS scheduling approaches, please see [34, 13]. Other references include [21, 17, 30, 9].

4. A Case Study of the Apparel Production

The section highlights the formulation and resolution of the apparel production scheduling problem presented in [42]. The problem was motivated by the issues faced by Cannondale Corporation, a manufacturer of clothing for bicyclists (Cannondale also manufactures bicycles, separate from apparel). The FMS have 50 to 100 stations and over 100 machines, and one week of production consists of about 10 to 40 lots, where lot sizes typically range from 100 to 1000. A complicating factor is that the time to process a production lot depends on the quantity of machines allocated to the cell in which the lot will be processed, and these scheduling and resource allocation decisions are highly coupled. An accurate and low-order model that integrates scheduling and resource allocation was developed, and the model is solved using Lagrangian relaxation. The combination of an accurate low-order model, Lagrangian relaxation, and the reduced complexity bundle method generates good solutions in a reasonable amount of time using data from Cannondale's factory, and shows the approach is practical.

4.1 Description of the FMS for Apparel Production

The backbone of the FMS under consideration is a circular, unidirectional, automated material handling system. The material handling system has a fixed number of stations to host sewing machines. Since there are more sewing machines than stations, machines are on carts to facilitate quick movement in and out of stations to meet the demand. In this environment, a *cell* consists of a set of dedicated stations and machines where the FMS is programmed to route garments between stations in the cell. Furthermore, the cell exists only long enough to process its production lot, meaning that once the lot is complete, the stations and machines used by the cell are released for use by future cells. The setup time of a cell is the time to move machines (if necessary), and the time to change threads on machines. For simplicity of presentation, they are ignored here.

During processing, a garment and associated raw material are attached to a hanger and hangers are guided along a conveyor, which is a suspended track. To process a production lot in its cell, a specified sequence of operations must be performed on each garment. Each operation has a set of eligible machine type(s), and requires a specified amount of time per garment. A machine assigned to an operation is set up and dedicated to that operation during the processing of the lot. Although a garment requires one machine to perform each operation, garments of the same lot can be processed in parallel by assigning multiple machines to that operation. When a garment arrives at a station, the hanger is automatically stored in the buffer. After the garments previously in the buffer are processed, the operator removes the garment and required raw material from the hanger, performs the operation, and puts the garment back on the hanger. The material handling system is programmed to route garments according to the sequence of operations and the stations containing the machine(s) assigned to the operations. If the buffer(s) for the next operation are not full, the material handling system takes the hanger and routes it to one of the machines assigned to perform the next operation. The machine chosen is the one with the fewest garments in its buffer. If the buffer(s) for the next operation are full, the current operation is blocked and cannot continue until an opening occurs.

4.2 Mathematical Problem Formulation

The formulation is an extension of the job shop scheduling formulation presented above [27]. The due date d_ℓ, tardiness weight w_ℓ, earliness weight β_ℓ, and lot size of lot ℓ, $\ell = 1, 2, .., L$, are assumed given. The decision variables are the beginning time b_ℓ to set up and process each lot ℓ, and the number of machines $m_{\ell jh}$ of type h to allocate to operation j of lot ℓ. The beginning time is an integer variable from 1 to K, where the scheduling horizon is divided into K equal time intervals.

4.2.1 Objective function. The objective function to be minimized is the weighted sum of production lot tardiness and lot release earliness to meet customer demand and to reduce work-in-process inventory, i.e.

$$\min_{\{b_\ell\},\{m_{\ell jh}\}} J(b_\ell, m_{\ell jh}), \quad \text{where } J(b_\ell, m_{\ell jh}) = \sum_{\ell=1}^{L} (w_\ell T_\ell + \beta_\ell E_\ell), \tag{4.1}$$

where T_ℓ is the tardiness of lot ℓ defined as $T = max[0, c_\ell - d_\ell]$, c_ℓ is the completion time of lot ℓ, E_ℓ is the release earliness defined as $E_\ell = max[0, s_\ell - b_\ell]$, s_ℓ is the desired start date of lot ℓ. The beginning and completion times are related according to:

$$c_\ell = b_\ell + u_\ell + t_\ell - 1, \tag{4.2}$$

where the beginning time b_ℓ is defined to be the beginning of a time period, the completion time c_ℓ is defined to be at the end of a time period, u_ℓ the setup time for lot ℓ, and t_ℓ the time to process all the garments of lot ℓ. The time t_ℓ to process lot ℓ is the duration from processing the first garment of the lot to the finish of the last garment in the lot. This time depends on the number of machines assigned to the lot.

4.2.2 Lot processing time. An accurate approximation to t_ℓ can be derived by observing that for a given allocation of machines, each lot has a bottleneck operation. Therefore t_ℓ is the sum of three parts:

1. t_ℓ^1: the time to initially feed the bottleneck operation of lot ℓ,
2. t_ℓ^2: the time for the bottleneck to process all the garments in the lot once the bottleneck is fed, and
3. t_ℓ^3: the time for the garments to finish operations downstream of the bottleneck once all garments have been processed at the bottleneck.

For lot ℓ, the maximum rate of the bottleneck operation is

$$r_\ell = \min_j [m_{\ell j}/t_{\ell j}], \tag{4.3}$$

where $t_{\ell j}$ is the per-garment processing time for operation j of lot ℓ ($t_{\ell j}$ is assumed to be independent of the eligible machine type selected or the operator assigned as is essentially the case for Cannondale), and $m_{\ell j}$ is the total number of machines assigned to perform operation j. The equation for $m_{\ell j}$ is

$$m_{\ell j} = \sum_{h \in \mathcal{H}_{\ell j}} m_{\ell j h} \geq 1, \tag{4.4}$$

where $\mathcal{H}_{\ell j}$ is the set of machine types that can perform operation j of lot ℓ.

The time to process all the garments on the bottleneck is the lot size divided by the rate of the bottleneck, or

$$t_\ell^2 = N_\ell / r_\ell, \tag{4.5}$$

where N_ℓ is the lot size. Equation (4.5) assumes that the bottleneck machines operate continuously during the time t_ℓ^2, and keeping these machines busy during that time can be achieved by following a flowline release policy such as kanban or drum-buffer-rope [14]. The other components of lot processing time t_ℓ^1 and t_ℓ^3 are negligible in the case of Cannondale.

Although the expression for t_ℓ is an *approximation*, its accuracy was deemed satisfactory by simulating real cell configurations using SimFactory, a commercial software package. Also, constraints (4.3) through (4.5) clearly show the dependence of t_ℓ on $m_{\ell j h}$, and thus the scheduling and capacity allocation decisions are interdependent.

4.2.3 Machine and station availability. The availabilities of machines and stations are limited. Also, machines and stations assigned to lot ℓ are unavailable to other lots during the time from b_ℓ to c_ℓ. The total number of machines of type h, $h = 1, 2, ..., H$, being used in each time period k must be less than or equal to M_{kh}, the total number of machines of type h available at that time, or

$$\sum_{\ell=1}^{L} \sum_{j=1}^{J_\ell} \delta_{\ell k} m_{\ell j h} \leq M_{kh}, \qquad k = 1, ..., K; \quad h = 1, ..., H, \tag{4.6}$$

where $\delta_{\ell k}$ is one if lot ℓ is active at time k (or, equivalently, if $b_\ell \leq k \leq c_\ell$) and zero otherwise. In (4.7), the value of $m_{\ell j h}$ for $h \notin \mathcal{H}_{\ell j}$ is implied to be zero.

Finally, the station availability constraints require that the total number of allocated machines of all types must be less than or equal to S_k, the number of stations available at each time k (each assigned machine requires one station), or

$$\sum_{h=1}^{H} \sum_{\ell=1}^{L} \sum_{j=1}^{J_\ell} \delta_{\ell k} m_{\ell j h} \leq S_k, \qquad k = 1, ..., K. \tag{4.7}$$

Again, $m_{\ell jh}$ is zero for $h \notin \mathcal{H}_{\ell j}$.

The optimization problem is to minimize (4.1) subject to constraints (4.2)–(4.7). The objective function and the system-wide constraints (4.6) and (4.7) consist of sums of functions of individual lots. This characteristic will be exploited by decomposing the problem into lot subproblems using Lagrangian relaxation.

4.3 Solution Methodology

As presented in Sect. 3. our scheduling goal is to obtain good solutions in a reasonable amount of time and with quantifiable quality by using the Lagrangian relaxation method.

4.3.1 Lagrangian relaxation. The method of Lagrangian relaxation forms a dual function by relaxing machine and station capacity constraints with Lagrange multipliers $\{\pi_{kh}\}$ and $\{\tau_k\}$, respectively, and consequently decouples the lots. The relaxed problem is

$$\begin{aligned} J_{LR}(\pi_{kh}, \tau_k) \quad \equiv \quad & \min_{\{b_\ell\},\{m_{\ell jh}\}} \{\sum_{\ell=1}^{L}(w_\ell T_\ell + \beta_\ell E_\ell) + \\ & \sum_{k=1}^{K}\sum_{h=1}^{H}\pi_{kh}[\sum_{\ell=1}^{L}\sum_{j=1}^{J_\ell}\delta_{\ell k}m_{\ell jh} + s_{kh} - M_{kh}] + \\ & \sum_{k=1}^{K}\tau_k[\sum_{h=1}^{H}\sum_{\ell=1}^{L}\sum_{j=1}^{J_\ell}\delta_{\ell k}m_{\ell jh} + s_k - S_k]\}, \end{aligned} \tag{4.8}$$

subject to constraints (4.2)–(4.5). In the above, s_{kh} and s_k are slack variables for machine and station capacity constraints, respectively.

The above minimization can be decomposed into individual lot subproblems:

$$R_\ell^1 = \min_{b_\ell,\{m_{\ell jh}\}} \{w_\ell T_\ell + \beta_\ell E_\ell + \sum_{h=1}^{H}\sum_{j=1}^{J_\ell} m_{\ell jh}[\sum_{k=b_\ell}^{c_\ell}(\pi_{kh} + \tau_k)]\}, \tag{4.9}$$

subject to (4.2)–(4.5).

Let R_ℓ^1 denote the optimal cost of the lot ℓ subproblem. Then the Lagrangian dual problem is [31, 3]

$$\max_{\{\pi_{kh}\},\{\tau_k\}} J_{LR}(\pi_{kh}, \tau_k), \tag{4.10}$$

where now

$$J_{LR}(\pi_{kh}, \tau_k) = -\sum_{k=1}^{K}\sum_{h=1}^{H} M_{kh}\pi_{kh} - \sum_{k=1}^{K} S_k\tau_k + \sum_{\ell=1}^{L} R_\ell^1 + \sum_{k=1}^{K}\sum_{h=1}^{H} R_{kh}^2 + \sum_{k=1}^{K} R_k^3. \tag{4.11}$$

In the above, R_{kh}^2 and R_k^3 are the optimal costs of the slack subproblems:

$$R_{kh}^2 = \min_{0 \le s_{kh} \le M_{kh}} \{\pi_{kh} s_{kh}\}, \tag{4.12}$$

$$R_k^3 = \min_{0 \le s_k \le S_k} \{\tau_k s_k\}. \tag{4.13}$$

4.3.2 Solving the subproblems. The number of possible cell configurations (i.e. allocation of machines to operations for a given lot) is enormous. However, several different configurations can produce the same lot processing time. For a given lot processing time, the cell configuration with the fewest number of machines producing that processing time is called a "minimum-machine configuration." To simplify subproblem solving but without loss of generality, only minimum-machine configurations are considered. The minimum-machine configurations for each lot and associated lot processing times are stored in the so called "cell configuration table." The method of solving the lot subproblems in (4.9) is then to enumerate all combinations of lot beginning time and lot processing time, where the lot processing time determines $\{m_{\ell j}\}$ via the cell configuration table.

The worst-case number of enumerations is K^2, where all K lot beginning times and all K lot processing times are enumerated. For a given b_ℓ and t_ℓ (therefore $\{m_{\ell j}\}$), the values of $m_{\ell jh}$ are determined so as to minimize the subproblem cost in (4.9), and this minimization can be easily carried out. The slack subproblem for s_{kh} in (4.12) is easily solved by checking the sign of π_{kh}; the subproblem for s_k in (4.13) is solved similarly.

4.3.3 Solving the dual problem. This dual problem is polyhedral concave, and the subgradient method [39] is commonly used to solve this type of problems. The method, however, may suffer from slow convergence as multipliers zigzag across ridges of the hyper-surface of the dual function.

The bundle method [15] overcomes the slow convergence of the subgradient method by accumulating subgradients of points in a neighborhood of the current iterate and store them in a "bundle." The method then finds an ϵ-ascend direction by solving a quadratic programming problem with complexity $O(b^3)$, where b is the number of elements in the bundle. The bundle method can also detect if ϵ optimal solution is reached. It, however, the method becomes very computation intensive as the bundle size increases.

The recently developed Reduced Complexity Bundle Method (RCBM) [43] reduces the complexity of $O(b^3)$ to $O(b^2)$ by performing a projection of a bundle element onto a linear subspace instead of solving a quadratic programming problem. The RCBM is used to update the multipliers.

4.3.4 Obtaining a feasible schedule. Because of the relaxation of constraints for an integer optimization problem, subproblem solutions generally produce an infeasible schedule, i.e. machine and/or station capacity constraints are violated. Other constraints are always satisfied since they were carried through to the subproblem level. A heuristic *list-scheduling procedure*, similar to that presented in [27], is used to adjust subproblem solutions to form a feasible schedule.

4.4 Numerical Results

The numerical results of six examples based on Cannondale's system are presented below. There are 9 machines types, 105 total machines, and 60 stations. All machines and stations are available throughout the time horizon of $K = 50$, where each time slot k represents one hour. The number of Lagrange multipliers is $HK + K = (9)(50) + 50 = 500$. Each of the six examples represents one week of work. The lots have various due dates and weights. RCBM was terminated when either of the following two conditions was met: C1) an ϵ-optimal point was detected, or equivalently, when an optimal dual point lies within the ball of radius δ centered at the current iterate, or C2) the duality gap is less than or equal to 1%. Table 4.1 summarizes each example.

Table 4.2 shows the numerical results. In Tab. 4.2, the duality gap equals (primal cost − dual cost) / (dual cost). In all the examples here, the stopping condition C1 was met. The dual costs reported in Tab. 4.2 are therefore ϵ-optimal, but not necessarily optimal. The number of function evaluations shows the number of times the dual cost was evaluated (i.e. the number of times the subproblems were solved). The CPU time is on the same 60 MHz personal computer.

Table 4.1. Description of examples

Ex.	No. of lots	No. of garments	Ave. No. of operations/lot	No. of primal decision variables
2	10	7046	6.4	74
3	20	4486	6.9	157
4	25	7649	6.6	189
5	30	7798	6.6	228
6	35	7970	6.5	264
7	40	8908	6.7	309

The duality gap results in Tab. 4.2 show that the primal schedules are 16% to 29% from optimal. Although the dual cost provides a lower bound to the optimal primal cost, the optimal dual cost is not necessarily a tight lower bound. Despite the larger-than-desired duality gaps, the method still produces near-optimal schedules, and does this in less than 3.5 CPU minutes

Table 4.2. Numerical results

Ex.	Primal cost	Dual cost	Duality gap	No. RCBM iterations	No. func. evaluations	CPU time (min:sec)
1	221	191	16%	37	727	1:25
2	200	155	29%	37	543	1:29
3	341	283	20%	37	679	2:44
4	430	369	17%	30	621	3:01
5	380	316	20%	45	741	3:28
6	524	411	28%	33	450	2:26

on a personal computer. If desired, a tighter bound could be obtained by using a branch and bound enumeration procedure, where the primal and dual costs in Tab. 4.2 are used as initial bounds. A branch and bound method might not be practical due to the potentially large CPU time.

5. New Promising Research Approaches

Because of problem complexity, most approaches for FMS scheduling are based on heuristic rules. New opportunities, however, are emerging in view of the advancements in computer technology, and progress in system theory and mathematical optimization.

One potentially beneficial improvement is to on-line update the multipliers. Although dynamic programming was not used in the case study of Sect. 4. as a result of model simplification, it is generally needed to solve subproblems when operation precedence constraints are involved. It is known that dynamic programming provide "closed-loop" solutions that can react to system disturbances. Within the Lagrangian relaxation framework, however, dynamic programming are solved for a fixed set of multipliers. These multipliers are iteratively updated during the solution process, but are fixed at the termination of the algorithm. They thus are "open-loop" in nature, and cannot react to disturbances without being further updated. Consequently, the overall solution is "semi closed-loop." If the multipliers can be continuously updated using the latest information, closed-loop control can be achieved.

In addition, future uncertainties can be proactively considered by using stochastic dynamic programming in place of standard dynamic programming, e.g. [8] to improve system performance. Within this stochastic framework, "ordinal optimization" [16, 6] turns out to be valuable to perform short simulation runs so as to select a good dual solution to feed the heuristics [24]. This is because a good dual solution may not correspond to a good feasible solution in view of the heuristic nature of how feasible schedules are constructed. One therefore has to try out several candidate dual solutions with high dual costs to find which one generates a good feasible schedule. In the stochastic setting, each dual solution is in fact a policy, indicating what to do under which circumstance. The tryout of a single dual solution thus involves simulation, and is a very time consuming task. Ordinal optimization can be used to perform short simulation runs on selected candidate dual solutions to determine their "order" or "ranking." A winner of the short tryout is then the dual solution to be selected to feed to the heuristics, and rigorous simulation runs can then be performed to obtain performance statistics.

Further improvement of the high level algorithm is needed to handle larger or more complicated cases.

The investigation of heuristics based on the theory of stochastic processes to understand their properties (e.g. stability and performance bounds) beyond performing brute force simulation is an exciting area, and exemplary work include [20, 13].

Deadlock has been mostly ignored in the scheduling literature. The combination of Petri net and scheduling seems promising in deadlock prevention and resolution. Selected work includes [23].

Another challenge is to simultaneously consider machines and material handling. As mentioned in Sect. 2.2, the material handling system itself is very sophisticated, and its interaction with machines further complicated the modeling and resolution process. Limited research includes [5, 37].

Acknowledgement. This work was supported in part by the National Science Foundation under DMI-9500037, and the Advanced Technology Center for Precision Manufacturing, University of Connecticut.

References

[1] Adams J, Balas E, Zawack D 1988 The shifting bottleneck procedure for job shop scheduling. *Manag Science.* 34:391–401
[2] Baker K 1995 Elements of sequencing and scheduling. Dartmouth College, Hanover, NH
[3] Bertsekas D P 1995 *Nonlinear Programming.* Athena Scientific, Belmont, MA
[4] Blackstone J H, Phillips D T, Hogg G L 1982 A state-of-the-art survey of dispatching rules for manufacturing job shop operations. *Int J Prod Res.* 20:27–45
[5] Blazewicz J, Eiselt H A, Finke G, Laporte G, Weglarz J 1991 Scheduling tasks and vehicles in a flexible manufacturing system. *Int J Flex Manufactur Syst.* 4:5–16
[6] Chen C H 1995 An effective approach to smartly allocate computing budget for discrete event simulation. In: *Proc 34th IEEE Conf Decision Control.* New Orleans, LA, pp 2598–2605
[7] Chen H, Chu C, Proth J M 1995 A more efficient Lagrangian relaxation approach to job-shop scheduling problems. In: *Proc 1995 IEEE Int Conf Robot Automat.* Nagoya, Japan, pp 496–501
[8] Chen D, Luh P B, Thakur L S 1997 Modeling uncertainty in job shop scheduling. In: *Proc 1st Int Conf Operat Quantitative Manag.* Jaipur, India, pp 490–497

[9] Choi J, Hitomi K 1994 A method of flexible scheduling for flexible manufacturing systems. *Int J Prod Econ.* 33:247–255.

[10] Czerwinski C, Luh P B 1994 Scheduling parts with bills of materials using an improved Lagrangian relaxation technique. *IEEE Trans Robot Automat.* 10:99–111

[11] Fisher M L 1973 Optimal solution of scheduling problems using Lagrange multipliers, part I. *Operat Res.* 21:1114–1127

[12] Garey M R, Johnson D S 1979 *Computers and Intractability.* Freeman, San Francisco, CA

[13] Gershwin S B 1994 *Manufacturing Systems Engineering.* Prentice-Hall, Englewood Cliffs, NJ

[14] Goldratt E M, Fox R E 1986 *The Race.* North River Press, New York

[15] Hiriart-Urruty J-B, Lemarechal C 1993 *Convex Analysis and Minimization Algorithms I and II.* Springer-Verlag, Berlin

[16] Ho Y C, Sreenivas R S 1992 Ordinal optimization of DEDS. In: *Discrete Event Dynamic Systems: Theory and Applications 2.* Kluwer, Boston, MA, pp 61–88

[17] Inman R R, Jones P C 1993 Decomposition for scheduling flexible manufacturing systems. *Operat Res.* 41:608–617

[18] Ishii N, Talavage J J 1994 A mixed dispatching rule approach in FMS scheduling. *Int J Flex Manufactur Syst.* 6

[19] Kaskavelis C A, Caramanis M C 1997 Efficient Lagrangian relaxation algorithms for real-life-size job-shop scheduling problems. Working Paper, Department of Manufacturing Engineering, Boston University, personal communications

[20] Kumar P R, Seidman T I 1990 Dynamic instabilities and stabilization methods in distributed real-time scheduling of manufacturing systems. *IEEE Trans Automat Contr.* 35:289–298

[21] Kusiak A 1989 Aggregate scheduling in a flexible machining and assembly system. *IEEE Trans Robot Automat.* 5:451–459

[22] Kusiak A 1990 *Intelligent Manufacturing Systems.* Prentice-Hall, Englewood Cliffs, NJ

[23] Lee D Y, DiCesare F 1994 Scheduling flexible manufacturing systems using Petri nets and heuristic search. *IEEE Trans Robot Automat.* 10:123–132

[24] Liu F, Luh P B, Moser B 1997 Scheduling of design projects with resource constraints and uncertain number of design iterations. In: *Proc IEEE/ASME Int Conf Advanc Intel Mechatron.* Tokyo, Japan

[25] Luggen, W W 1991 *Flexible Manufacturing Cells and Systems.* Prentice-Hall, Englewood Cliffs, NJ

[26] Luh, P B, Gou L, Zhang Y, Nagahora T, Tsuji M, Yoneda M, Hasegawa T, Kyoya Y, Kano T 1997 Job shop scheduling with group-dependent setups, finite buffers, and long time horizon. In: *Mathematics of Industrial Systems.* Annals of Operations Research, to appear

[27] Luh P B, Hoitomt D J 1993 Scheduling of manufacturing systems using the Lagrangian relaxation technique. *IEEE Trans Automat Contr.* 38:1066–1079

[28] Luh P B, Wang J H, Wang J L, Tomastik R N 1997 Near optimal scheduling of manufacturing systems with presence of batch machines and setup requirements. *CIRP Annals.* 46:397–402

[29] Maleki, R A 1991 *Flexible Manufacturing Systems.* Prentice-Hall, Englewood Cliffs, NJ

[30] Nascimento M A 1993 Giffler and Thompson's algorithm for job shop scheduling is still good for flexible manufacturing systems. *J Operat Res Soc.* 44:521–524

[31] Nemhauser G, Wolsey L 1988 *Integer and Combinatorial Optimization.* Wiley, New York

[32] Ovacik I M, Uzsoy R 1997 *Decomposition Methods for Complex Factory Scheduling Problems.* Kluwer, Boston, MA

[33] Pinedo M 1995 *Scheduling – Theory, Algorithms and Systems.* Prentice-Hall, Englewood Cliffs, NJ

[34] Rachamadugu R, Stecke K E 1994 Classification and review of FMS scheduling procedures. *Prod Plan Contr.* 5(1):2–20

[35] Raman N, Talbot F B, Rachamadugu R V 1989 Due date based scheduling in a general flexible manufacturing system. *J Operat Manag.* 8:115–132

[36] Rodammer F A, White K P 1988 A recent survey of production scheduling. *IEEE Trans Syst Man Cyber.* 18:841–851

[37] Sabuncuoglu I, Hommertzheim D L 1993 Experimental investigation of an FMS due-date scheduling problem: Evaluation of machine and AGV scheduling rules. *Int J Flex Manufactur Syst.* 5

[38] Shanker K, Tzen Y J 1985 A loading and dispatching problem in a random FMS. *Int J Prod Res.* 23:579–595

[39] Shor N Z 1985 *Minimization Methods for Non-Differentiable Functions.* Springer-Verlag, Hiedelberg, Germany

[40] Slomp J, Gaalman G J C, Nawijn W M 1988 Quasi on-line scheduling procedures for flexible manufacturing systems. *Int J Prod Res.* 26:585–598

[41] Stecke K E 1983 Formulation and solution of nonlinear integer production planning problems for flexible manufacturing systems. *Manag Science.* 29:273–288

[42] Tomastik R N, Luh P B, Liu G 1996 Scheduling flexible manufacturing systems for apparel production. *IEEE Trans Robot Automat.* 12:789–799

[43] Tomastik R N, Luh P B, Zhang D Y 1996 A reduced-complexity bundle method for maximizing concave nonsmooth functions. In: *Proc 35th IEEE Conf Decision Contr.* Kobe, Japan, pp 2114–2119

[44] Ventura J A, Weng M X 1995 Minimizing single-machine completion time variance. *Manag Science.* 41:1448–1455

[45] Wang J, Luh P B 1996 Scheduling of a machining center. *Math Comp Model.* 24(11/12):203–214

[46] Wang J, Luh P B, Zhao X, Wang J 1997 An optimization-based algorithm for job shop scheduling. *SADHANA.* Journal of Indian Academy of Sciences, 22:241–256

[47] Zhao X, Luh P B, Wang J 1997 The surrogate gradient algorithm for Lagrangian relaxation method. *36th IEEE Conf Decision Contr.* San Diego, CA

Task Synchronization via Integration of Sensing, Planning, and Control in a Manufacturing Work-cell

Tzyh-Jong Tarn[1], Mumin Song[1], and Ning Xi[2]

[1] Department of Systems Science and Mathematics, Washington University, USA
[2] Department of Electrical Engineering, Michigan State University, USA

This chapter presents a novel approach for task synchronization of a manufacturing work-cell. It provides an analytical method for solving the challenging problem in intelligent control, i.e. the integration of low level sensor data and simple control mechanisms with high level perception and behaviour. The proposed Max-Plus Algebra model combining with event-based planning and control provides a mechanism to efficiently integrate sensing, planning and real time execution. It also enables a planning and control system to deal with the tasks involving both discrete and continuous actions. Therefore, task scheduling, which usually deals with discrete type of events, as well as action planning, which usually deals with continuous events, can be treated systematically in a unified framework. More important, the unique feature of this approach is that interactions between discrete and continuous events can be considered in the same framework. As a result, the efficiency and reliability of the task schedule and action plan can increase significantly. A typical robotic manufacturing work-cell is used to illustrate the proposed approach. The experimental results clearly demonstrate the advantages of the proposed approach.

1. Introduction

The tasks in a robotic system involve multiple segments of actions, such as moving a robot, making contact or picking up a part. All segments are connected or dependent upon each other logically and temporally. Task planning for such systems involves two issues: determining the sequence of actions, called task scheduling, and planning the actions them-self, called action planning. Therefore, the problem of designing a robotic system amounts to solving a three level problem: task scheduling, action planning and control, as shown in Fig. 1.1. In the task scheduling level, only discrete events are considered. The result of task scheduling is a sequence of logical commands. Various methods have been proposed for this type of task scheduling, including operation research type of approaches [11], heuristic approaches [9, 7, 15, 16], AND/OR graph approach [6], Petri Net approach [8], as well as the recently developed discrete event system approach [3]. Various methods have also been proposed to solve the problem of robotic action planning [10, 17, 19].

However, task scheduling, action planning and control have been treated as separate problems. The basic reason for this kind of approach is that there does not exist a model or framework which could describe both the scheduling and planning levels of the system. Furthermore, no efficient and simple method could be found to analyze and design systems involving both discrete and continuous events. However, in order to increase the efficiency, reliability and safety of robotic systems, the consideration of task scheduling and action planning in a unified framework could be an important step. For instance, an execution failure of action control could cause down time for the entire system. However, if the task schedule can adapt to this kind of unexpected event, the failure can be automatically corrected so that a local disturbance will not become a global one. Obviously, this requires an interaction between the different levels of design.

The challenge is to develop a mechanism for integration of high level system behaviour and perspective with low level system control and sensing to achieve an intelligent task scheduling, action planning and control. The major difficulty in developing a method for modeling, analysis and design of integrated schedule, plan and control of robotic systems is that such systems involve both discrete and continuous events. These are so called hybrid systems [4, 5].

For several years, considerable effort has been made to investigate hybrid systems. A three layer hierarchical model of controller and planner was introduced [14] by adding a high level monitoring layer to a basic system in order to deal with discrete decisions. Recently, several new methods have been proposed for designing a hybrid system. Nerode et al. [13] present a Computer-Aided Control Engineering environment which support automatic generation of automata that simultaneously comply with discrete and continuous dynamics. Bencze et al. [2] design a Real-time/Boolean Translator to interface between decision-making logic and manipulator controller. McCarragher et al. [12] applied hybrid system structure to formulate transitions between constrained motions for a peg-in-hole task in a robotic manufacturing system. However, these methods are either heuristics or one-of-a-kind designs.

This chapter presents a novel analytical method for modeling and design of hybrid system. First, a Max-Plus Algebra model of a manufacturing work-cell will be introduced. The relationship between discrete events and continuous events involved in the system will be described by the Max-Plus Algebra dynamic model. Combining the Max-Plus Algebra model with the event-based planning and control scheme, and incorporating a multi-sensor data fusion scheme, an integrated sensing, planning and control scheme is obtained. Finally, the experimental results for the robotic operation in the manufacturing work-cell clearly demonstrate the advantages of the method.

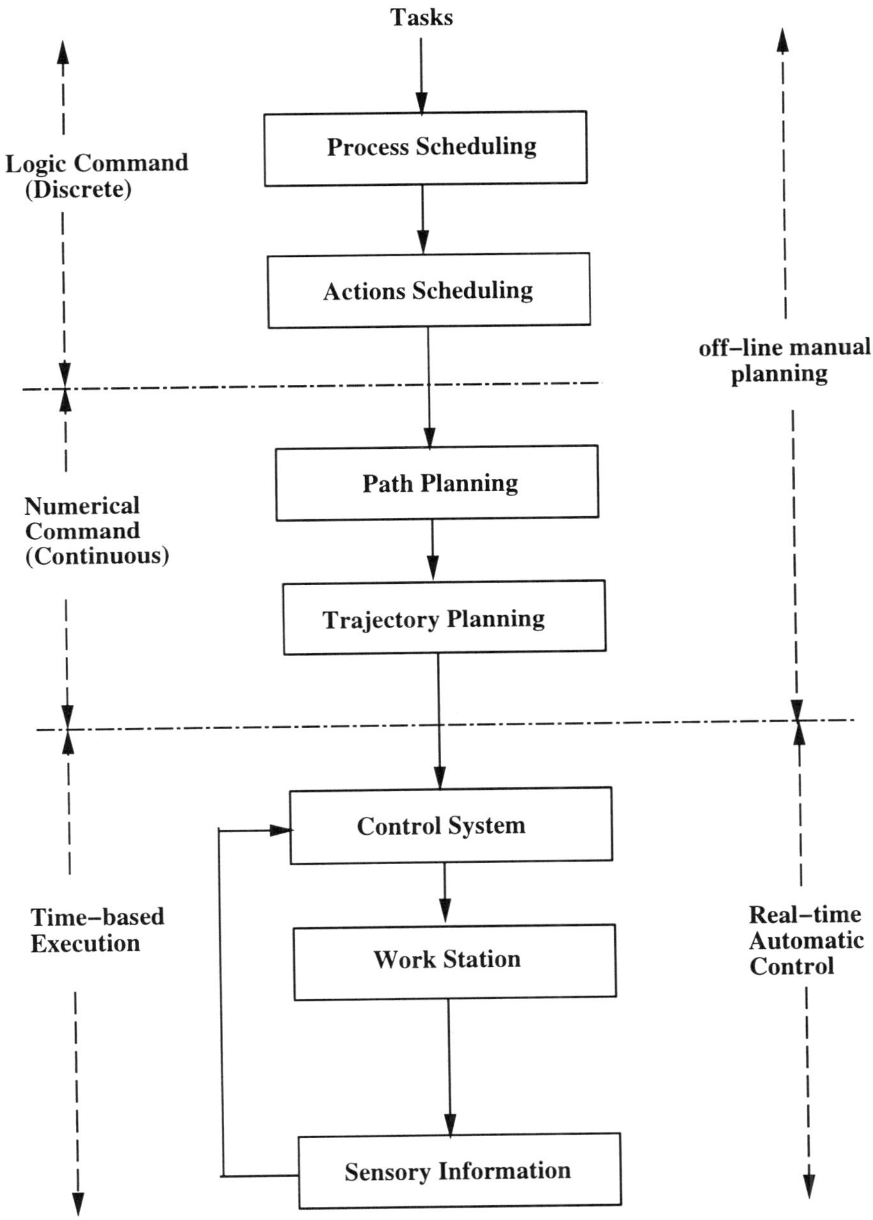

Fig. 1.1. Scheduling, planning and control

2. A Max-Plus Algebra Model

As shown in Fig. 2.1, a manufacturing work-cell considered in this chapter consists of a pair of robotic manipulators, a controlled disc conveyor, and a sensing system. Three-dimensional parts are distributed on the disc conveyor. In this manufacturing work-cell a task can be assembly, disassembly, or sorting etc.

Usually, each single task contains a sequence of the robotic operations and the disc operations. All of these operations could be considered as discrete events which are happened sequentially or concurrently. The Max-Plus algebra provides a mathematical tool such that we are able to model and analyze this kind of discrete event system easily. The most important step in modeling and design is to determine a task reference for the robotic operations and the disc operations. The task reference provides a reference frame for task scheduling and action planning. It should be directly related to the states of the execution of the task, which could be determined by the sensory information. Hence, task scheduling and action planning become processes of designing the relationship between a task or action with respect to the task reference.

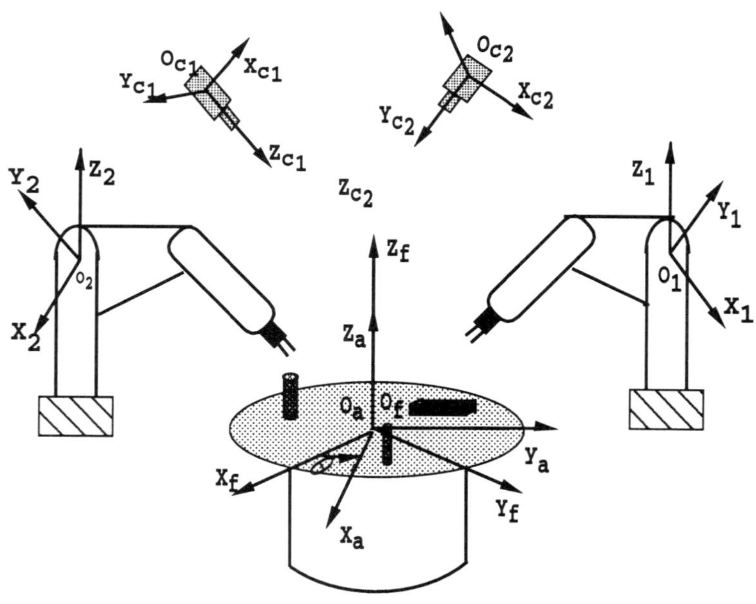

Fig. 2.1. Dual arm manufacturing work-cell

A simple parts inserting task is considered as an example in this dual robot manufacturing work-cell. The sequence of the robotic operations and the disc operations can be described as the following:

- The disc conveyor rotates an angle such that the robot r_1 can pick up a part p_1 at location ℓ_b on the disc;
- The disc conveyor rotates an angle such that the robot r_2 can pick up another part p_2 at location ℓ_d;
- Both robots move to the location ℓ_c and insert the parts, where we assume that the distance between the robots is close enough to be ignored;
- The robot r_1 lifts the finished product to location ℓ_a;
- The disc conveyor rotates an angle such that the robot r_2 can pick up a part again at location ℓ_d on the disc conveyor;
- The disc conveyor rotates an angle such that the robot r_1 can pick up another part at location ℓ_b on the disc conveyor.

It can be easily shown that there exists a definition of normalized location for any concurrent process of a robotic system such that the process can be represented by a two dimensional diagram, the so called timing diagram as Fig. 2.2. The variable t_i represents the nominal time required to complete the corresponding action.

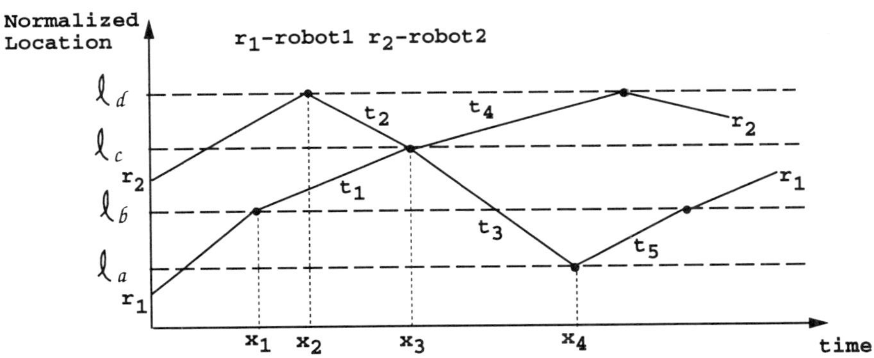

Fig. 2.2. Timing diagram of the robotic operations

Designing the task reference is the same as determining a sequence of events in Fig. 2.2, which happen sequentially, called a critical event sequence. The time to complete critical event sequence will determine the total time of execution of an assembly cycle. Therefore the task reference will represent the temporal relationship between different events. In addition, it is also required that the rest of the events in the manufacturing work-cell which happen concurrently with critical event sequences can be referenced to the events involved in the critical event sequence. This will represent the spatial relationship amongst different events. The task reference variable will be the combination of all action reference variables associated with the events in the critical event sequence.

A Max-Plus Algebra Model is proposed to analyze and determine the critical event sequence. Let

$x_i(k)$ – The time to the kth occurance of event i.

$u_i(k)$ – The time to the kth occurance of input (or disturbance) i.

$y_i(k)$ – A function (or observation) of $\{x_i(k) : i = 1, 2, \ldots, n\}$.

t_j – The time taken to complete a single segment of task (segment j).

A Max-Plus Algebra Model is

$$\begin{cases} X(k+1) = A \otimes X(k) \oplus B \otimes u(k) \\ Y(k) = C \otimes X(k) \end{cases} \tag{2.1}$$

where

$$X(k) \in \Re^n_{\max}, \quad Y(k) \in \Re^m_{\max}, \quad u(k) \in \Re^\ell_{\max},$$

and

$$\Re_{\max} = \Re \cup \{-\infty\}, \quad \oplus : \text{ Max operation}, \quad \otimes : \text{ Plus operation.}$$

It can be proved that $\{\Re_{\max} : \oplus, \otimes\}$ is an idempotent and commutative semi-field with zero element $\varepsilon = -\infty$ and identity element $e = 0$.

The system described by Fig. 2.2 can be modeled by (2.1) with:

$$X(k) = \begin{bmatrix} x_1(k) \\ x_2(k) \\ x_3(k) \\ x_4(k) \end{bmatrix}, \quad u(k) = \begin{bmatrix} u_1(k) \\ u_2(k) \end{bmatrix}, \quad Y(k) = \begin{bmatrix} y_1(k) \\ y_2(k) \\ y_3(k) \\ y_4(k) \end{bmatrix};$$

$$A = \begin{bmatrix} \varepsilon & \varepsilon & \varepsilon & t_5 \\ \varepsilon & \varepsilon & t_4 & \varepsilon \\ \varepsilon & \varepsilon & t_2 + t_4 & t_1 + t_5 \\ \varepsilon & \varepsilon & t_2 + t_3 + t_4 & t_1 + t_3 + t_5 \end{bmatrix},$$

$$B = \begin{bmatrix} e & \varepsilon \\ \varepsilon & e \\ t_1 & t_2 \\ t_1 + t_3 & t_2 + t_3 \end{bmatrix}, C = \begin{bmatrix} e & \varepsilon & \varepsilon & \varepsilon \\ \varepsilon & e & \varepsilon & \varepsilon \\ \varepsilon & \varepsilon & e & \varepsilon \\ \varepsilon & \varepsilon & \varepsilon & e \end{bmatrix},$$

where $x_i(k)$ is the time at which the event x_i in Fig. 2.2 happens for the kth time; $y_i(k)$ is the observation of $x_i(k)$. For instance, $x_1(k)$ is the time at which robot 2 picks up a part from location ℓ_b for the kth time. In this case, $y_1(k)$ is simply $x_1(k)$. $u_i(k)$ is the time at which certain outside input occurs for the kth time. $u_1(k)$ here means the time when the part for the robot 1 arrives the picking position. $u_2(k)$ is the time when the part arrives the picking position for the robot 2. The state transition relation in Fig. 2.2 can be written as

$$\begin{cases} x_1(k) &= \max\{x_4(k-1)+t_5,\ u_1(k-1)\} \\ x_2(k) &= \max\{x_3(k-1)+t_4,\ u_2(k-1)\} \\ x_3(k) &= \max\{x_4(k-1)+t_1+t_5,\ x_3(k-1)+t_2+t_4, \\ & \quad u_1(k-1)+t_1,\ u_2(k-1)+t_2\} \\ x_4(k) &= \max\{x_4(k-1)+t_1+t_3+t_5, \\ & \quad x_3(k-1)+t_2+t_3+t_4, \\ & \quad u_1(k-1)+t_1+t_3,\ u_2(k-1)+t_2+t_3\} \end{cases} \tag{2.2}$$

The system (2.1) has the solution [1]

$$X(k) = A^k \otimes X(0) \oplus \sum_{j=0}^{k-1} \left[A^{k-j-1} \otimes B \otimes u(j)\right].$$

From the order of $\{x_i(k)\}$ in the above solution, a sequence of critical events can be determined. Therefore, the task reference is obtained according to the critical event sequence.

In addition, the time period of the assembly process

$$\begin{aligned} T &= \max\{x_1(k)-x_1(k-1), x_2(k)-x_2(k-1), \\ &\quad \ldots x_4(k)-x_4(k-1)\} \\ &= [x_1(k)-x_1(k-1)] \oplus [x_2(k)-x_2(k-1)] \oplus \\ &\quad \cdots \oplus [x_4(k)-x_4(k-1)] \end{aligned}$$

can also be obtained from the system model (2.1).

Remark 2.1. It can be shown [1] that an eigenvalue λ and the corresponding eigenvector ν exist for the matrix A defined in (2.1). They can be defined by

$$A \otimes \nu = \lambda \otimes \nu.$$

It can further be shown that the cycle time of the manufacturing work-cell can be determined by the eigenvalue of A, λ. As a matter of fact, the throughput rate is $1/\lambda$.

Remark 2.2. The performance of the manufacturing work-cell is completely determined by matrices A, B, and C. They can be designed to optimize the system performance. In other words, the Max-Plus Algebra also provides an analytical model for design and optimization of manufacturing systems.

By the definition of $u_i(k)$, the input of the system (2.1) is a time of part arriving. In other words, when we solve the Max-Plus Algebra equation (2.1), it is necessary to have complete states of the parts, i.e. their position, orientation and velocity so that the time that parts arrive can be determined. A multi-sensor data fusion scheme is then developed for this purpose.

3. Centralized Multi-Sensor Data Fusion

In order to make a task schedule for a sequence of robotic operations in real time, the states of the parts in the robotic task space, such as their position, orientation and velocity, have to be determined by the sensing system. This sensing system, in the proposed manufacturing work-cell, includes two types of sensors, two uncalibrated CCD cameras and an encoder mounted in the disc conveyor. Since the position and orientation of the cameras are unknown, any individual sensor is not able to give three-dimensional information of the parts in the robotic task space. A centralized multi-sensor fusion scheme, which fuses the raw sensory data from both cameras and the encoder, has been developed to determine the complete states of the parts in the robotic task space, as shown in Fig. 3.1.

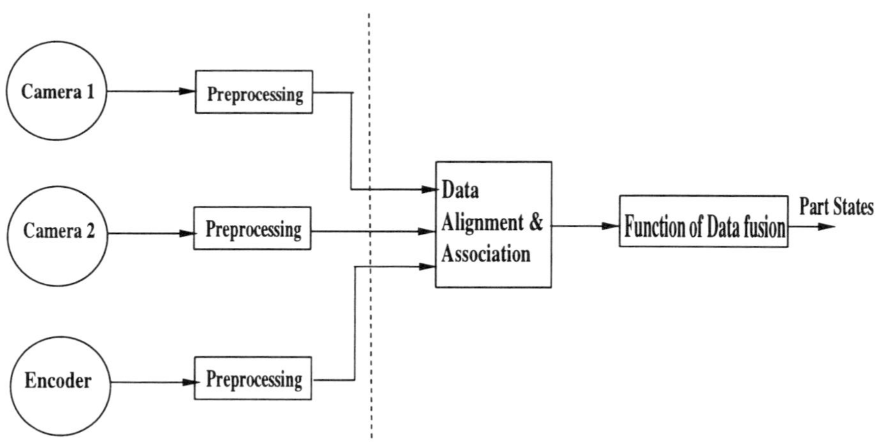

Fig. 3.1. Multi-sensor data fusion scheme

First of all, a set of coordinate frames has to be introduced. As shown in Fig. 2.1, O_{c_i} is the optical center of the camera coordinate frames $O_{c_i}X_{c_i}Y_{c_i}Z_{c_i}$, and $O_{c_i}Z_{c_i}$ is the optical axis of lens of ith camera ($i = 1, 2$). $O_fX_fY_fZ_f$ and $O_aX_aY_aZ_a$ are the fixed and the attached disc coordinate frame respectively. The $O_f Z_f$ and $O_a Z_a$ are coincident. The robots are uniformed by the frames $O_i X_i Y_i Z_i$ $(i = 1, 2)$.

It should be noted that, in this work, it is not necessary to know the exact relationship between camera coordinate frames or between the camera and the world frame, which implies that camera calibration is unnecessary. We have developed an calibration-free stereo vision algorithm [18], which has the ability of providing position information of any point in the disc attached frame $O_a X_a Y_a Z_a$. In other words, the data in the two camera frames are transformed into the frame $O_a X_a Y_a Z_a$ through the vision algorithm. This is the first step of our data fusion scheme.

Next, each part on the disc conveyor can be represented in the $O_a X_a Y_a Z_a$ by a set of finite points, whose coordinates have been provided by the visual sensors,

$$\{P_{a,i} | P_{a,i} = (x_{a,i}, y_{a,i}, z_{a,i})^T; i = 1, 2, ..., n\}.$$

The position of the centroid is recognized as the location of the part, which is calculated by

$$P_{a,cen} = \frac{1}{n} \sum_{i=1}^{i=n} P_{a,i}.$$

Assuming each geometrical point has unit mass, the inertia matrix can be expressed by

$$M = \begin{bmatrix} i_{xx} & i_{xy} & i_{xz} \\ i_{yx} & i_{yy} & i_{yz} \\ i_{zx} & i_{zy} & i_{zz} \end{bmatrix},$$

where $i_{\lambda\mu} = \sum_{i=1}^{i=n} (\lambda_{a,i} - \lambda_{a,cen})(\mu_{a,i} - \mu_{a,cen})$ and $\lambda, \mu = x, y, z$. The principal axis of maximum inertia I_a, then, is considered as the orientation of the part, which can be determined by the eigenvector of the maximum eigenvalue of the matrix M. Therefore, the configuration of the part in the frame $O_a X_a Y_a Z_a$ is obtained by knowing the pair $(P_{a,cen}, I_a)$.

Furthermore, the state of the part in the fixed disc coordinate frame $O_f X_f Y_f Z_f$ is given by the follows:

$$\begin{aligned} P_{f,cen} &= R_{Z_f}(\theta(t)) P_{a,cen} \\ I_f &= R_{Z_f}(\theta(t)) I_a, \end{aligned} \tag{3.1}$$

where $\theta(t)$ is the angle of the disc conveyor measured by the encoder, and $R_{Z_f}(\theta(t))$ is a rotational transformation around Z_f axis. Since all parts are stationary relative to the attached disc conveyor, linear velocity of the part in the fixed disc frame can be described as

$$\begin{aligned} \dot{P}_{f,cen} &= \dot{R}_{Z_f}(\theta(t)) \dot{\theta}(t) P_{a,cen} \\ \dot{I}_f &= \dot{R}_{Z_f}(\theta(t)) \dot{\theta}(t) I_a, \end{aligned} \tag{3.2}$$

where $\dot{\theta}(t)$ is the angular velocity read from the encoder. Equations (3.1) and (3.2) show how the absolute position, orientation and velocity of the part can be obtained from relative information in the attached disc frame through the usage of encoder measurement $\theta(t)$ and $\dot{\theta}(t)$.

Lastly, a transformation exists between the fixed disc coordinate frame and each of the robotic frames, which is given by the rotational transformation matrix R^i_{wf} and translational transformation matrix T^i_{wf} $(i = 1, 2)$. A fusion function is described as follows:

$$\begin{aligned} P^i_{w,cen} &= R^i_{wf} R_{Z_f}(\theta(t)) P_{a,cen} + T^i_{wf} \\ I^i_w &= R^i_{wf} R_{Z_f}(\theta(t)) I_a, \end{aligned} \tag{3.3}$$

and

$$\begin{aligned} \dot{P}^i_{w,cen} &= R^i_{wf} \dot{R}_{Z_f}(\theta(t)) \dot{\theta}(t) P_{a,cen} \\ \dot{I}^i_w &= R^i_{wf} \dot{R}_{Z_f}(\theta(t)) \dot{\theta}(t) I_a. \end{aligned} \tag{3.4}$$

Equations (3.3) and (3.4) show that, based on the data from different types of sensors, the coordinates of the parts have been transferred from relative coordinate frame to world frame. A complete information of the parts in robotic task space is, then, available for scheduling robotic operations.

4. Event-based Planning and Control

The robotic systems or the disc conveyor has its own special action reference. The tasks and actions of the robotic system and the disc conveyor should be synchronized and coordinated according to the given action reference. Traditionally, this reference is the time, which means that a task schedule or action plan is usually described with respect to the time. It is understandable to use time as the action reference since it is easy to obtain and be referenced by different entities of a system.

An event-based planning and control method has been successfully applied to a robotic operation with a single segment [19]. An intuitive idea of this method is to introduce a new action reference variable different from time. It is related to the measurements of system directly. Instead of using time as the action reference, the action plan or the desired system input is parameterized by using the new reference variable. This action reference can be designed to carry the sensory information, which is needed for a planner to adjust and modify the original plan to generate the desired system input. As a result, the desired system input becomes a function of the system output and the sensory measurements. This gives a real time planning process to adjust and modify the plan based on the system output and sensory information. The event-based planning and control method is shown in Fig. 4.1.

Figure 4.1 shows a slight difference from the traditional planning and control system block diagram. A block, which is called "action reference", is involved. The function of this block is to generate the action reference for the

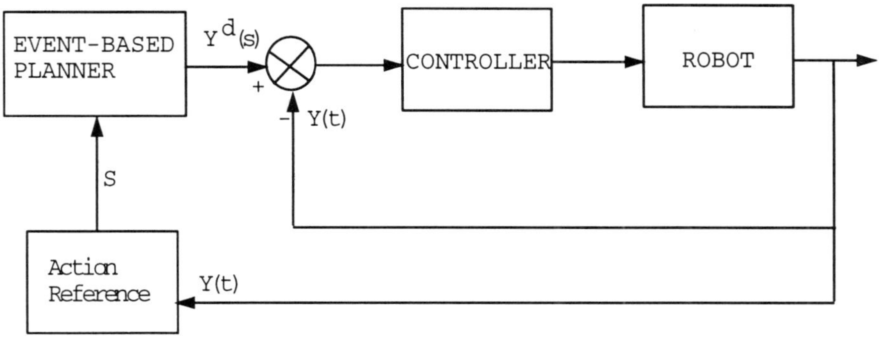

Fig. 4.1. Event-based planning and control

planner based on the system output measurement. The planner then gives a desired system input relying on this reference. From system point of view, here the planner becomes an investigation/decision agent. This agent gives the robotic system an ability to deal with unexpected or uncertain events. Based on the action reference variable s, the desired plan, y^d, for a single event can be planned as

$$y^d = Y^d(s).$$

Additionally, since the action reference is capable of being computed at the same rate as that in feedback control loop, a quick modification of the plan is allowed. In other words, the event-based planning and control scheme is able to deal with not only discrete unexpected events and uncertainties, but continuous unexpected events and uncertainties, such as system parameter drifting, modeling error etc.

As discussed in Sect. 2. the task in the manufacturing work-cell, usually contains a sequence of events, such as the robotic operations and the operations of the disc conveyor. The action references of these events are mostly different from each other. In order to extend the event-based planning and control method to handle multiple events, a unified action reference is created. This action reference is able to integrate the information of temporal, and spatial as well as logical connection and dependency of the different events. Now the function of the action reference block, as shown in Fig. 4.1, is improved to provide not only the action reference variable, but a logical command to switch to different action reference for different individual action. Hence, the schedule of these different actions is able to be implemented by switching the definition of the action reference with respect to the task reference determined by the Max-Plus Algebra model. This scheme is illustrated in Fig. 4.2.

Figure 4.2 shows that, based on the sensory measurements, the task reference block integrates two types of commands: continuous command, i.e. the

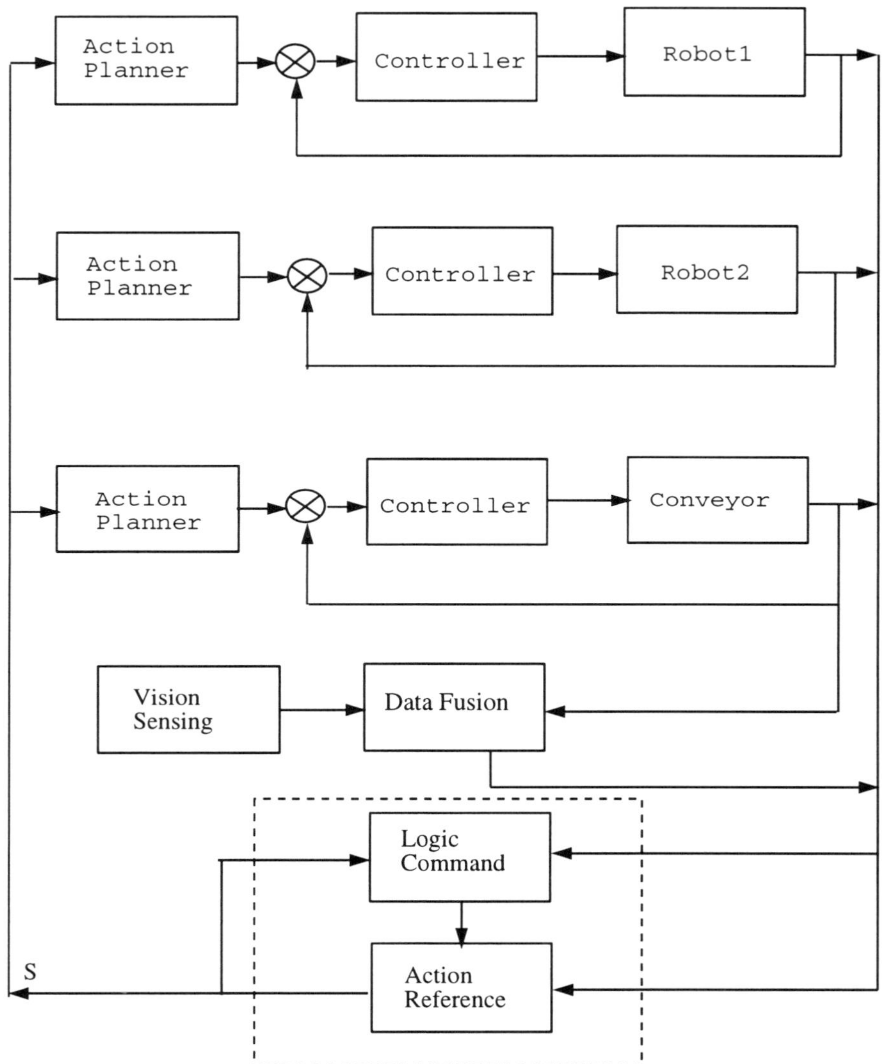

Fig. 4.2. Integrated sensing, planning and control

action reference variable, and discrete command, i.e. the logical command for switching action references. We can realize that the action planning for all concurrent actions in the overall system can be eventually described by function of the task reference variable. Hence, the synchronization of concurrent events is implemented through the task reference variable. Furthermore, the task reference is dependent on the sensor measurement of both robotic system and disc conveyor. Therefore, the sensing, the planning and control are well integrated through an unified action reference.

5. Experimental Results

An experimental system has been setup in the Center for Robotics and Automation at Washington University in St. Louis. It consists of two PUMA560 robot manipulators and a controllable disc conveyor. The diameter of the disc conveyor is 0.9 m. The maximum angular speed is 0.175 rad/s.

The vision system consists of two CCD cameras for stereo setup. The image resolution is 256 × 256. The Intelledex vision processor is based on a 16MHz Intel 80386 CPU. It interfaces to the main computer, an SGI IRIX 4D/340VGX. Visual measurements are sent to SGI by a parallel interface. The robot is controlled by a UMC controller that also interfaces to the SGI computer through memory mapping. The Planning and control algorithm of the robotic manipulator and the disc conveyor run in SGI. The parts used in the experiment are three bolts, which have different diameters (0.015 m, 0.012 m, and 0.008 m), and different height (0.11 m, 0.085 m, 0.045 m). Here, tasks assigned in this work-cell are to let robot pick up the parts in order of their height, from highest to lowest, and move each of the parts to three different specific locations.

The experimental results are shown in Figs. 5.1 and 5.2. In Fig. 5.1, the trajectories of the robot and the bolts in XYZ coordinates are shown. This result demonstrates that, through a unified action reference, the sensing, planning and control are well integrated, and the tasks and the actions in the work-cell are synchronized.

Figure 5.2 shows that the robot was stopped by an obstacle in the time period [14.923 s, 20.255 s] while approaching the second specific dropping place. Since the action plans for the robot and the disc conveyor are functions of the task reference variable which is determined by the output measurement of the robot and the disc conveyor, the task reference variable stops increasing as well. As a result, the disc conveyor also stops. Once the obstacle is removed, robot and disc conveyor start moving again. Without any rescheduling or replanning, the assembly process seamlessly resume. This experiment demonstrates that the method of the extended event-based planning and control is capable of coping with unexpected events occurring during the execution of an operation.

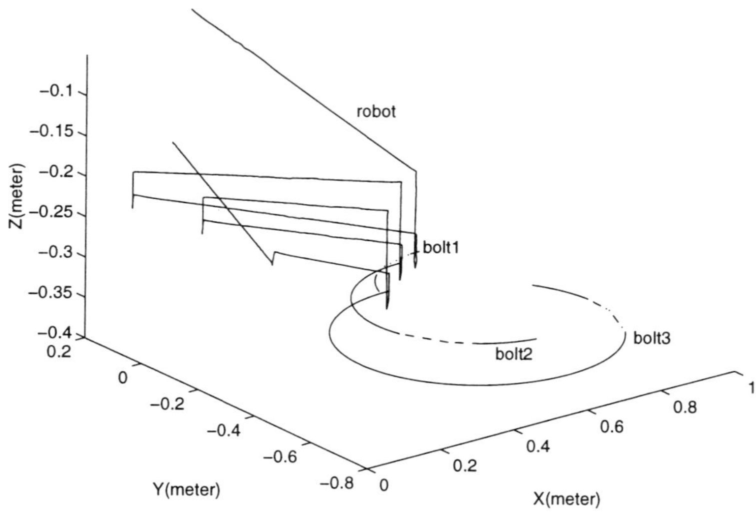

Fig. 5.1. Parts sorting operations

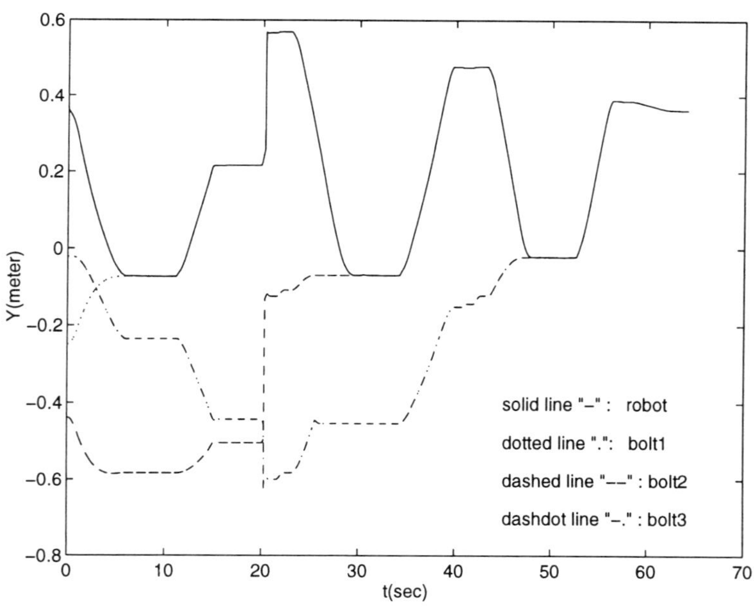

Fig. 5.2. Integrated sensing, planning and control

6. Conclusions

A new integrated sensing, planning and control method for robotic work-cell has been proposed. Based on the output of the centralized multi-sensor data fusion scheme, a Max-Plus Algebra model has a recursive solution such that a task reference can be determined, and an unified action reference variable can be obtained, which carries all the information obtained from sensors of the system. This enables the event-based planning and control method be extended to deal with multiple tasks. Therefore, all operations are synchronized by the unified action reference. Furthermore, the system is capable of coping with uncertainty and unexpected events. Contending with uncertainty and unexpected events is extremely important for achieving intelligent, robust and efficient performance for robotic systems.

References

[1] Baccelli F L, Cohen G, Quadrat J 1992 *Synchronization and Linearity: An Algebra for Discrete Event Systems.* Wiley, New York
[2] Bencze W, Franklin G 1995 A separation principle for hybrid control system design. *IEEE Contr Syst Mag.* 15
[3] Brandin B A, Wonham W M, Benhabib B 1992 Manufacturing cell supervisory control — A time discrete event system approach. In: *Proc 1992 IEEE Int Conf Robot Automat.* Nice, France, pp 931–936
[4] Brockett R 1993 Hybrid models for motion control systems. In: Trentelman H L, Willems J C (eds) *Essays on Control: Perspectives in the Theory and its Applications.* Birkhäuser, Boston, MA, pp 29–53
[5] Gollu A, Vavraiya P 1989 Hybrid dynamical system. In: *Proc 28th IEEE Conf Decision Contr.* Tampa, FL, pp 2708–2712
[6] Homem de Mello L S, Sanderson A C 1991 Representation of mechanical assembly sequences. *IEEE Trans Robot Automat.* 7:221–227
[7] Hsu H, Fu L H 1995 Fully automated robotic assembly cell: Scheduling and simulation. In: *Proc 1995 IEEE Int Conf Robot Automat.* Nagoya, Japan, pp 208–213
[8] Kanehara T, Suzuki T, Inaba A, Okuma S 1993 On algebraic and graph structural properties of assembly Petri net. In: *Proc 1993 IEEE Int Conf Robot Automat.* Atlanta, GA, vol 2, pp 507–514
[9] Kim G H, Lee C S G Genetic 1995 Reinforcement learning approach to the machine scheduling problem. In: *Proc 1995 IEEE Int Conf Robot Automat.* Nagoya, Japan, pp 196–201
[10] Latombe J 1991 *Robot Motion Planning.* Kluwer, Boston, MA
[11] Luh P B, Hoitomt D J 1993 Scheduling of manufacturing systems using the Lagrangian relaxation technique. *IEEE Trans Automat Contr.* 38:1066–1079
[12] McCarragher B, Asada H 1993 A discrete event approach to the control of robotic assembly tasks. In: *Proc 1993 IEEE Int Conf Robot Automat.* Atlanta, GA, vol 1, pp 331–336
[13] Nerode A, Kohn W 1992 An autonomous systems control theory: An overview. In: *Proc IEEE Symp Computer-Aided Contr Syst Des.* Napa, CA

[14] Saridis G N 1987 Knowledge implementation: Structure of intelligent control system. In: *Proc 1987 IEEE Int Symp Intel Contr*

[15] Shin K S, Zheng Q 1991 Scheduling job operations in an automatic assembly line. *IEEE Trans Robot Automat.* 7:333–341

[16] Sriskandrajah C, Ladet P, Germain R 1986 *Scheduling Methods for a Manufacturing System.* Elsevier, Amsterdam, The Netherlands

[17] Tarn T J, Bejczy A K, Xi N 1993 Intelligent motion planning and control for robot arms. In: *Proc 12th IFAC World Congr.* Sydney, Australia

[18] Tarn T J, Song M, Xi N, Ghosh B J 1996 Multi-sensor fusion scheme for calibration-free stereo vision in a manufacturing workcell. In: *Proc 1996 IEEE Int Conf Multisens Fusion Integr Intel Syst.* Washington, DC, pp 416–423

[19] Xi N, Tarn T J, Bejczy A K 1993 Event-based planning and control for multi-robot coordination. In: *Proc 1993 IEEE Int Conf Robot Automat.* Atlanta, GA, vol 1, pp 251–258

Advanced Air Traffic Automation: A Case Study in Distributed Decentralized Control

Claire J. Tomlin, George J. Pappas, Jana Košecká, John Lygeros, and Shankar S. Sastry

Department of Electrical Engineering and Computer Science, University of California at Berkeley, USA

In this survey chapter, we present some of the issues in designing algorithms for the control of distributed, multi-agent systems. The control of such systems is becoming an increasing issue in many areas owing to technological advances which make it possible to take "legacy" systems to new levels of functioning and efficiency. Of specific interest to us in this chapter is advanced air traffic management (ATM) to increase the efficiency and safety of air travel while accommodating the growing demand for air traffic. ATM systems will replace the completely centralized, ground-based air traffic control procedures. Within ATM, the concept of *free flight* allows each aircraft to plan four dimensional trajectories in real time, thus replacing the rigid and inefficient discrete airspace structure. These changes are feasible due to technological innovations such as advanced flight management systems with GPS. In this chapter, we propose a *decentralized* ATM architecture, in which some of the current air traffic control functionality is moved on board aircraft. Within this framework, we present the issues in hybrid systems verification and design for safe conflict resolution strategies between aircraft. Both cooperative and noncooperative conflict resolution strategies are presented along with verification methods based on Hamilton-Jacobi theory, automata theory, and the theory of games.

1. New Challenges: Intelligent Multi-agent Systems

To a large extent, control theory has investigated the very important paradigm of Central Control. In this paradigm, sensory information is collected from sensors observing a material process that may be distributed over space. This information is transmitted over a communication network to one center, where the commands that guide the process are calculated and transmitted back to the process actuators that implement those commands. In engineering practice, of course, as soon as the process becomes even moderately large, the Central Control paradigm breaks down. What we find instead is distributed control: a set of control stations, each of which receives some data and calculates some of the actions. Important examples of distributed control are air traffic management, the control system of an interconnected power grid, the telephone network, a chemical process control system. Although a Central Control paradigm no longer applies here, control engineers have with great success used its theories and its design and analysis tools to build and operate these distributed control systems. There are two reasons why the paradigm succeeded in practice, even when it failed in principle. First, in each case the complexity and scale of the material process grew incrementally and relatively slowly. Each new increment to the process was controlled using the paradigm, and adjustments were slowly made after extensive (but by no means exhaustive) testing to ensure that the new controller worked in relative harmony with the existing controllers. Second, the processes were operated with a considerable degree of "slack." That is, the process was operated well within its performance limits to permit errors in the extrapolation of test results to untested situations and to tolerate a small degree of disharmony among the controllers. However, in each system mentioned above, there were occasions when the material process was stressed to its limits and the disharmony became intolerable, leading to a spectacular loss of efficiency. For example, most air travelers have experienced delays as congestion in one part of the country is transmitted by the control system to other parts. The distributed control system of the interconnected power grid has sometimes failed to respond correctly and caused a small fault in one part of a grid to escalate into a system-wide blackout.

We are now attempting to build control systems for processes that are vastly more complex or that are to be operated much closer to their performance limits in order to achieve much greater efficiency of resource use. The attempt to use the central control paradigm cannot meet this challenge: the material process is already given and it is not practicable to approach its complexity in an incremental fashion as before. Moreover, the communication and computation costs in the central control paradigm would be prohibitive, especially if we insist that the control algorithms be fault-tolerant. What we need to meet the challenge of control design for a complex, high performance material process, is, we believe, a new paradigm for distributed control. It must distribute the control functions in a way that avoids the high communication and computation costs of central control, at the same time that it limits complexity. The distributed control must, nevertheless, permit centralized authority over those aspects of the material process that are necessary to achieve the high performance goals. We believe that such a challenge can be met by organizing the distributed control functions in a hierarchical architecture that makes those functions relatively autonomous (which permits using all the tools of central control), while introducing enough coordination and supervision to ensure the harmony of the distributed controllers necessary for high performance.

1.1 Analysis and Design of Multi-agent Hybrid Control Systems

One of the main incentives to move into the area of multi-agent large scale systems is economic. Preliminary studies indicate that automation can improve coordination in air traffic management systems, highway systems, chemical process control, power generation and distribution, etc. This in turn leads to performance improvement in terms of fuel consumption, safety, efficiency, and environmental impact. To deal with complex systems, engineers use a combination of continuous and discrete controllers. Continuous controllers are used primarily because interaction with the physical plant, through sensors and actuators, is essentially analog, and continuous models and design techniques have been developed, used, and validated extensively. An equally compelling case exists for discrete controllers: since discrete abstractions make it easier to manage system complexity, discrete models are easier to manipulate, and discrete abstractions more naturally accommodate linguistic and qualitative information in the controller design. We will use the term "hybrid systems" to describe systems that incorporate both continuous and discrete dynamics.

1.1.1 Multi-agent scarce resource systems. An important class of systems that are well suited for hybrid control are multi-agent, scarce resource systems. Their common characteristic is that many agents are trying to make use of a common, congestible resource. For example, in highway systems, the vehicles are agents competing for scarce highway space-time resources, while in air traffic management systems the aircraft compete for air space and runway space. To achieve the common optimum we should ideally have a centralized control scheme that computes the global optimum and commands the agents accordingly. A solution like this may be undesirable, however, for several reasons:

- it is likely to be very computationally intensive, as a large centralized computer is needed to make all the decisions;
- it may be less reliable, as the consequences may be catastrophic if the centralized controller is disabled;
- the information that needs to be exchanged may be too expensive; and
- the number of agents may be large and/or dynamically changing.

If the performance degradation of a completely decentralized solution is unacceptable and a completely centralized solution is prohibitively complex or expensive, a compromise will have to be found. Such a compromise will feature semi-autonomous agent operation. In this case, each agent is trying to optimize its own usage of the resource and coordinates with "neighboring" agents in case there is a conflict of objectives. It should be noted that *semi-autonomous agent control is naturally suited for hybrid designs*. At the continuous level, each agent chooses its own optimal strategy, while discrete coordination is used to resolve conflicts. Thus, the class of hybrid systems that we will be most interested in is multi-agent systems, where the hybrid dynamics arise from the interaction between continuous single agent "optimal" strategies and discrete conflict resolution or coordination protocols.

We have been involved in such a research program at Berkeley bringing to bear tools from control, robotics, and artificial intelligence into this framework. In addition to synthesizing diverse approaches and experiences into a unified paradigm, we will confirm or validate this new paradigm by using it for controlling our test processes. Thus, our program follows the classical pattern of scientific progress: the first phase of "induction" or the integration of approaches and experiences that go beyond the current practice into a new paradigm which subsumes the current one; and the second phase of "deduction" or the application of the new paradigm to concrete situations to test its validity. We have been guided in our choice of problems by a number of detailed case studies of large, complex systems with multiple agents arising in intelligent vehicle highway systems, air traffic management systems, and intelligent telemedicine.

In this chapter, we will give the details of the broad program discussed above in the context of *air traffic management systems*. This is an area of great commercial and technological importance which has unfortunately not yet received the level of attention that it deserves from the research community and exemplifies the broad issues discussed thus far. Section 2. gives a brief background of ATM. In Sect. 3. we discuss the architectural issues regarding ATM. Section 4. presents our view of ground and on-board air automation systems in the proposed distributed ATM system. In Sect. 5. we present hybrid system issues which arise in non-cooperative and cooperative conflict resolution. In particular, we discuss our approach to the design and verification of hybrid systems using Hamilton Jacobi theory, automata theory, and the theory of games. Some concluding remarks are in Sect. 6.

2. Introduction to Air Traffic Management

Air transportation systems are faced with soaring demands for air travel. According to the Federal Aviation Administration (FAA), the annual air traffic rate in the U.S. is expected to grow by 3 to 5 percent annually for at least the next 15 years [8]. The current National Airspace System (NAS) architecture and air traffic management will not be able to efficiently handle this increase because of several limiting factors including inefficient airspace utilization, increased Air Traffic Control (ATC) workload, and out of date technology. In view of the above problems and in an effort to meet the challenges of the next century, the aviation community is working towards an innovative concept called *Free Flight* [23]. Free Flight allows pilots to choose their own routes, altitude and speed and gives each aircraft the freedom to optimize their routes based on criteria such as fuel consumption, avoidance of bad weather and other factors, referred to as User Preferred Routing or UPR. Aircraft flexibility will be restricted only in congested airspace in order to ensure separation among aircraft, or to prevent unauthorized entry of special use airspace (such as military airspace).

The economic benefits of Free Flight are immediate. Direct great circle routes, optimal altitudes, optimal avoidance of developing weather hazards and utilization of favorable winds will result in fuel burn and flight time operating cost savings. NASA studies [4] estimate that in a free flight scenario, user preferred trajectories could have resulted in annual potential savings of $1.28 billion in 1995 and could result in $1.47 billion savings in 2005[1]. Free Flight is potentially feasible because of enabling technologies such as Global Positioning Systems (GPS), Datalink communications [9], Automatic Dependence Surveillance-Broadcast (ADS-B) [9], Traffic Alert and Collision Avoidance Systems (TCAS) [7] and powerful on-board computation. In addition, tools such as the Center-TRACON Automation System (CTAS) [6] will serve as decision support tools for ground controllers in an effort to reduce ATC workload and optimize capacity close to highly congested urban airports.

The technological advances will also enable air traffic controllers to accommodate future air traffic growth by restructuring NAS towards a more decentralized architecture. The current system is extremely centralized with ATC assuming most of the workload. Sophisticated on-board equipment allow aircraft to share some of the workload, such as navigation, weather prediction and aircraft separation, with ground controllers. In order to improve the current standards of safety in an unstructured, Free Flight environment, automatic conflict detection and resolution algorithms are vital. Sophisticated algorithms which predict and automatically resolve conflicts would be used either on the ground or on-board, either as advisories or as part of the Flight Vehicle Management System (FVMS) of each aircraft. The resulting air traffic management system requires coordination and control of a large number of semi-autonomous aircraft. The number of control decisions that have to be made and the complexity of the resulting decision process dictates a hierarchical, decentralized solution. Complexity management is achieved in a hierarchy by moving from detailed, decentralized models at the lower levels to abstract, centralized models at the higher levels. Coordination among the agents is usually in the form of communication protocols which are modeled by discrete event systems. Since the dynamics of individual agents is modeled by differential equations, we are left with a combination of interacting discrete event dynamical systems and differential equations, the so called *hybrid systems*. Hybrid systems also arise in the operation of a single aircraft because of *flight mode switching*. The use of discrete modes to describe phases of the aircraft operation is a common practice for pilots and autopilots and is dictated partly by the aircraft dynamics themselves. The modes may reflect, for example, changes in the outputs that the controller is asked to regulate: depending on the situation, the controller may try to achieve a certain air-

[1] Using forecasted air traffic demand for 2005.

speed, climb rate, angle of attack, etc. or combinations of those. We do not discuss these further in this chapter but refer the reader to [16].

3. A Distributed Decentralized ATM

One of the most important conceptual issues to be addressed in the architecture of large scale control systems is their degree of decentralization. Completely decentralized systems are inefficient and lead to conflict, while completely centralized ones are not tolerant of faults in the central controller, are computationally and conceptually complicated, and are slow to respond to emergencies.

The tradeoff between centralized and decentralized decision making raises a fundamental issue that has to be addressed by any proposed ATM. The current ATC system is primarily centralized; all safety critical decisions are taken centrally (at the ATC units) and distributed to the aircraft for execution. Because of the complexity of the problem and the limited computational power (provided primarily by the human operators in the current system) this practice may lead to inefficient operation.

A number of issues should be considered when deciding on the appropriate level of centralization. An obvious one is the *optimality* of the resulting design. Even though optimality criteria may be difficult to define for the air traffic problem it seems that, in principle, the higher the level of centralization the closer one can get to the globally optimal solution. However, the complexity of the problem also increases in the process; to implement a centralized design one has to solve a small number of complex problems as opposed to large number of simple ones. As a consequence the implementation of a centralized solution requires a greater effort on the part of the designer to produce control algorithms and greater computational power to execute them. One would ideally like to reach a compromise that leads to acceptable efficiency while keeping the problem tractable.

Another issue that needs to be considered is *reliability* and *scalability*. The greater the responsibility assigned to a central controller the more dramatic are likely to be the consequences if this controller fails. In this respect there seems to be a clear advantage in implementing a decentralized design: if a single aircraft's computer system fails, most of the ATM system is still intact and the affected aircraft may be guided by voice to the nearest airport. Similarly, a distributed system is better suited to handling increasing number of aircraft, since each new aircraft can easily be added to the system, its own computer contributing to the overall computational power. A centralized system on the other hand would require regular upgrades of the ATC computers. This may be an important feature given the current rate of increase of the demand for air travel.

Finally, the issue of *flexibility* should also be taken into account. A decentralized system will be more flexible from the point of view of the agents, in this case the pilots and airlines. This may be advantageous for example in avoiding turbulence or taking advantage of favorable winds, as the aircraft will not have to wait for clearance from ATC to change course in response to such transients or local phenomena. Improvements in performance may also be obtained by allowing aircraft to individually fine tune their trajectories making use of the detailed dynamical models contained in the autopilot. Finally, greater flexibility may be preferable to the airlines as it allows them to utilize their resources in the best way they see fit.

The focus of our research has been to strike a compromise in the form of partially decentralized control laws for guaranteeing *reliable, safe control of the individual agents* while providing *some measure of unblocked, fair, and optimum utilization of the scarce resource. In our design paradigm, agents have control laws which maintain their safe operation and try to optimize their own performance measures. They also coordinate with neighboring agents and a centralized controller to resolve conflicts as they arise and maintain efficient operation.* In the next section we present a control architecture that implements what we believe is a reasonable balance between complete centralization and complete decentralization.

4. Advanced Air Transportation Architectures

This section describes the balance between the ATM on the ground and in the air. Currently, ATC in the United States is organized hierarchically with a single *Air Traffic Control System Command Center (ATCSCC)* supervising the overall traffic flow management. This is supported by 20 *Air Traffic Control System Command Centers (ARTCCs)*, or simply Centers, organized by geographical area. Coastal Centers have jurisdiction over oceanic waters. For example, the Fremont (California) ARTCC has jurisdiction from roughly Eureka to Santa Barbara and from Japan in the West to the Sierra Nevada mountains in the East. In addition, around large urban airports there are *Terminal Radar Approach Control facilities (TRACONs)* numbering over 150. For instance, the Bay Area TRACON includes the San Francisco, Oakland and San Jose airports along with smaller airfields at Moffett Field, San Carlos, Fremont, etc. The TRACONs are supported by control towers at more than 400 airports. There are roughly 17,000 landing facilities in the United States serving nearly 220,000 aircraft. Of these the commercial aircraft number about 6,000 and the number of commercially used airstrips is roughly the 400 that have control towers. The overall system is referred to as *National Airspace System (NAS)* [11]. The main goal of both the ARTCCs and the TRACONs is to maintain safe separation between aircraft while guiding them to their destinations.

4.1 Automation on the Ground

In an effort to increase the runway throughput, airport capacity as well as reduce delays, fuel consumption and controller workload in the vicinity of highly congested urban airports, NASA has designed the Center-TRACON Automation System (CTAS) [6]. CTAS is a collection of planning and control functions which generate advisories to assist, but not replace, the controllers in handling traffic in the Center and TRACON areas. CTAS consists of three main components: the *Traffic Management Advisor (TMA)*, the *Descent Advisor (DA)* and the *Final Approach Spacing Tool (FAST)*. TMA and DA coexist and operate in Center airspace whereas FAST operates as a standalone in TRACON airspace. CTAS receives input from radar sensors which transmit the aircraft state; from Center and TRACON controllers who allocate runways and routes to particular aircraft as well as alter the capacity or acceptance rate of the TRACON, airport or runway; and finally from weather reports which include wind, temperature and pressure profiles. The main outputs of CTAS are arrival schedules which meet all the capacity, separation and flow rate constraints as well as advisories to Center or TRACON controllers. CTAS is currently being field tested at Denver and Dallas-Fort Worth. A similar ground system called User Request Evaluation Tool (URET) has been developed by MITRE Corp. [2] and is being field tested at Indianapolis.

In our proposed ATM system, we will assume that a ground system (either CTAS or URET) will have jurisdiction over highly congested TRACON airspace, that airspace structure exists inside the TRACON and that controllers have active control over aircraft in the TRACON, sending the aircraft heading, speed and altitude advisories. The advisories provide a suggested arrival schedule at the destination airport, which is designed to meet the announced arrival times while resolving conflicts. The schedule reflects compromises between airline schedules as well as possible negotiation between ATC and the aircraft.

4.2 Automation in the Air

In the less congested Center airspace, aircraft are allowed to choose their own routes in the spirit of Free Flight. In addition, aircraft may resolve potential conflicts by inter-aircraft coordination. The role of the ATC in Center airspace is limited to performing flow management, providing the aircraft with global information about en-route traffic and weather conditions, as well as providing advisories in case aircraft are unable to resolve conflicts on their own. Currently, nominal trajectories through the airspace are defined in terms of *waypoints*, which are fixed points in the airspace defined by VOR (VHF Omni-Directional Range) points on the ground. The waypoints are a necessary navigation tool for aircraft which are not equipped with GPS. Waypoints have resulted in a discrete airspace structure and an underutilization of airspace. On the other hand, they have resulted in a predictable environment which allows controllers to resolve conflicts in congested airspace. GPS and Free Flight will remove this structure which will lead to greater efficiency and airspace capacity. Aircraft may choose their own routes instead of following a sequence of waypoints. However, inside the crowded TRACONs, airspace structure will be necessary in order to simplify the controller's task of landing aircraft while resolving conflicts.

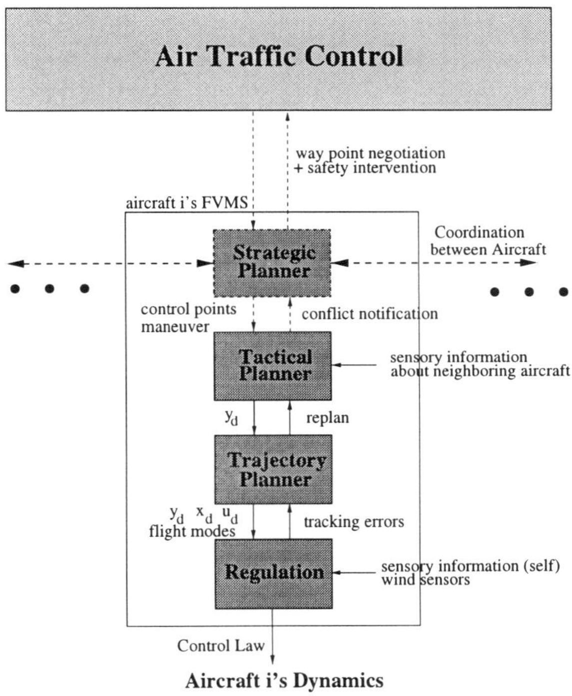

Fig. 4.1. Proposed ATM structure

In our proposed ATM structure, each aircraft is equipped with various planning and control algorithms. The aircraft will perform real time trajectory planning and tracking, conflict detection and resolution, as well as automatic mode switching. These smart aircraft will be extremely complex and each will be a large scale system in its own right. In order to reduce the resulting complexity and assist pilots in better performing their task, each aircraft is modeled using the hierarchical structure shown in Fig. 4.1. The levels of architecture below ATC reside on the aircraft and comprise what is known as the aircraft's *Flight Management System*, or FMS. The FMS consists of four layers, the strategic, tactical, and trajectory planners, and the regulation layer. Higher levels of the FMS architecture are associated with higher objectives and coarser models. Each layer of this architecture is described below.

4.2.1 Strategic planner. The main objectives of the strategic planner are to design a coarse trajectory for the aircraft and to resolve conflicts between aircraft. The trajectory is designed from origin to destination in some optimal sense, and is frequently redesigned in order to adapt to changes in the environment, such as weather patterns, potential conflicts and airport traffic. Inside TRACONs, the strategic planner simply accepts the advisories of the controllers. In Center airspace, the strategic planners of all aircraft involved in the potential conflict determine a sequence of maneuvers which will result in conflict-free trajectories, either using communication with each other through satellite datalink, or by calculating safe trajectories assuming the worst possible actions of the other aircraft [26]. Each strategic planner sends its most recently designed trajectory to the tactical planner in the form of a sequence of control points and/or a maneuver.

4.2.2 Tactical planner. The tactical planner refines the strategic plan by interpolating the control points with a smooth output trajectory, denoted by y_d in Fig. 4.1. The tactical planner uses a simple kinematic model of the aircraft for all trajectory calculations. Simple models are used at this stage since very detailed models may unnecessarily complicate the calculations, which are assumed to be approximate and have large safety margins. The output trajectory is then passed to the trajectory planner.

4.2.3 Trajectory planner. The trajectory planner uses a detailed dynamic model of the aircraft, sensory data about the wind magnitude and direction, and the tactical plan consisting of an output trajectory, to design full state and input trajectories for the aircraft, and a sequence of *flight modes* necessary to execute the dynamic plan. The flight modes represent different modes of operation of the aircraft and correspond to controlling different variables in the aircraft dynamics. A derivation of the flight mode logic necessary for safe operation of a CTOL (Conventional Take Off and Landing) aircraft is presented in [15].

The resulting trajectory, denoted y_d, x_d, and u_d in Fig. 4.1, is given to the regulation layer which directly controls the aircraft. The task of the trajectory planner is complicated by the presence of non-minimum phase dynamics [25, 27] and actuator saturation [21].

4.2.4 Regulation layer. Once a feasible dynamic trajectory has been determined, the regulation layer is asked to track it. Assuming that the aircraft dynamic model used by the trajectory planner is a good approximation of the true dynamics of the aircraft, tracking should be nearly perfect. In the presence of large external disturbances (such as wind shear or malfunctions), however, tracking can severely deteriorate. The regulation layer has access to sensory information about the actual state of the aircraft dynamics, and can calculate tracking errors. These errors are passed back to the trajectory planner, to facilitate replanning if necessary.

The structure of the proposed Flight Management System leads to various interesting questions regarding hierarchical systems. First, the convergence of the overall scheme to an acceptable and safe trajectory needs to be shown. Due to the complexity of the overall system and very nonlinear nature of the continuous dynamics it is unlikely that purely continuous or purely discrete techniques alone will be adequate in this setting. More elaborate hybrid techniques are needed. In addition, higher levels of the hierarchy use coarser system models or coarser abstractions. This raises the interesting notions of consistent abstractions or implementability, which is the ability of a lower level system to execute the commands of a higher level system. Preliminary work along this direction may be found in [22].

5. Conflict Resolution

The operation of the proposed ATM involves the interaction of continuous and discrete dynamics. Such *hybrid* phenomena arise, for example, from the coordination between aircraft at the strategic level when resolving a potential conflict. The conflict resolution maneuvers are implemented in the form of discrete communication protocols. These maneuvers appear to the (primarily continuous) tactical planner as discrete resets of the desired waypoints. One would like to determine the effect of these discrete changes on the continuous dynamics (and vice versa) and ultimately obtain guarantees on the minimum aircraft separation possible under the proposed control scheme.

Research in the area of conflict detection and resolution for air traffic has been centered on predicting conflict and deriving maneuvers assuming that the intent of each aircraft is known to all other aircraft involved in the conflict, for both deterministic [13, 28, 24], and probabilistic [14, 20] models.

In our research, we differentiate between two types of conflict resolution: *noncooperative* and *cooperative* [26]. In noncooperative conflict resolution, each aircraft involved in the conflict derives a safe avoidance maneuver without coordinating with the other aircraft. Such a situation occurs when there is an emergency and there is not enough time to establish communication with other aircraft, as was encountered by Air Force I with a United Parcel Service aircraft over the coast of Ireland in June 1997. The safest action that this aircraft can take is to choose a strategy which resolves the conflict for any possible action, within bounds, of the other aircraft. We formulate the noncooperative conflict resolution strategy as a zero sum dynamical game of the pursuit-evasion style [10]. The aircraft are treated as players, aware only of the *set* of possible actions of the other agents. These actions are modeled as disturbances, assumed to lie within a known set but with their particular values unknown, and the aircraft solves the game for the worst possible disturbance. The performance index for the game is the relative distance between the aircraft, required to be above a certain threshold (the Federal Aviation Administration requires a 5 mile horizontal separation in en-route airspace).

Assuming that a saddle solution to the game exists, the saddle solution is *safe* if the performance index evaluated at the saddle solution is above the required threshold. The sets of *safe states* and *safe control actions* for each aircraft may be calculated: the saddle solution defines the boundaries of these sets. The aircraft may choose any trajectory in its set of safe states, and a control policy from its set of safe control actions; coordination with the other aircraft is unnecessary.

The model used is a relative kinematic model for two aircraft, aircraft 1 and aircraft 2, which describes the motion of aircraft 2 with respect to aircraft 1:

$$\begin{aligned} \dot{x}_r &= -v_1 + v_2 \cos\phi_r + \omega_1 y_r \\ \dot{y}_r &= v_2 \sin\phi_r - \omega_1 x_r \\ \dot{\phi}_r &= \omega_2 - \omega_1 \end{aligned} \tag{5.1}$$

in which (x_r, y_r, ϕ_r) is the relative position and orientation of aircraft 2 with respect to aircraft 1, and v_i and ω_i are the linear and angular velocities of each aircraft.

In cooperative conflict resolution, safety is ensured by full coordination among the aircraft. The aircraft follow predefined maneuvers, inspired by robot collision avoidance maneuvers, which are proven to be safe. The class of maneuvers constructed to resolve conflicts must be rich enough to cover all possible conflict scenarios. In this case, the predefined resolution protocols dictate a hybrid nature in the overall system.

We will discuss these two scenarios in some detail now.

5.1 Noncooperative Conflict Resolution

First, we describe our noncooperative conflict resolution design philosophy on a general relative configuration model in $\mathbb{R}^n$. Consider the system

$$\dot{x} = f(x, u, d) \qquad x(t) = x \tag{5.2}$$

where $x \in \mathbb{R}^n$ describes the relative configuration of one of the aircraft with respect to the other, $u \in \mathcal{U}$ is the control input of one agent, and $d \in \mathcal{D}$ is the control of the other agent. We assume that the system starts at state x at initial time t. Both $\mathcal{U}$ and $\mathcal{D}$ are known sets, but whereas the control input u may be chosen by the designer, the disturbance d is unknown.

The goal is to maintain safe operation of the system (5.2), meaning that the system trajectories do not enter a prespecified unsafe region of the state space, called the Target set and denoted T with boundary ∂T. We assume that there exists a differentiable function $l(x)$ so that $T = \{x \in \mathbb{R}^n \mid l(x) \leq 0\}$ and $\partial T = \{x \in \mathbb{R}^n \mid l(x) = 0\}$. In this chapter, T represents the protected zone around the aircraft at the origin of the relative axis frame (Fig. 5.1).

Suppose that the two aircraft are conflict-prone, and they cannot cooperate to resolve conflict due to any one of the reasons mentioned in the previous

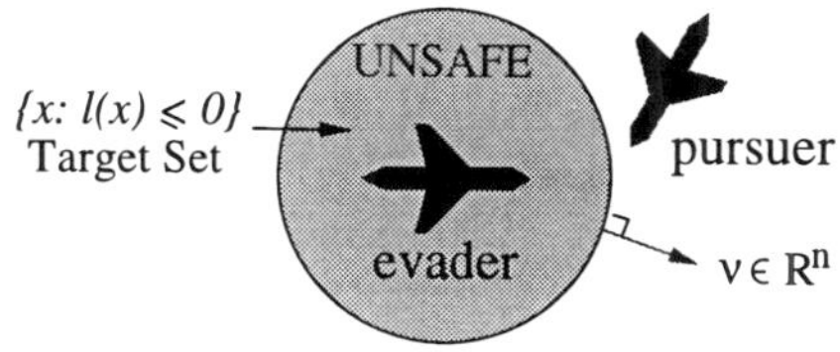

Fig. 5.1. The evader and pursuer, with Target set and its outward pointing normal ν

section. Then the safest possible strategy of each aircraft is to fly a trajectory which guarantees that the minimum allowable separation with the other aircraft is maintained, *regardless* of the actions of the other aircraft. Since the intent of each aircraft is unknown to the other, then this strategy must be safe for the *worst possible actions* of the other aircraft. We formulate this problem as a two-person, zero-sum dynamical game of the pursuer-evader variety. Call the aircraft at the origin of the relative frame the *evader* with control input u, and the other aircraft the *pursuer* with control input d; the goal of the evader is to drive the system outside T whereas the worst possible action of the disturbance is to try to drive the system into T. We solve the dynamical game for system (5.2) over the time interval $[t, t_f]$, where t_f is defined as

$$t_f = \inf\{\tau \in I\!R^+ \mid x(\tau) \in T\} \tag{5.3}$$

with initial state x at time t. If $t_f = \infty$, then for all possible control actions and disturbances the trajectory never enters T. The game is a variational problem without a running cost, or Lagrangian: we are interested only in whether or not the state enters T. The cost $J_1(x, t, u, d)$ is therefore defined as a function (only) of the terminal state:

$$J_1(x, t, u, d) = l(x(t_f)) \tag{5.4}$$

Given $J_1(x, t, u, d)$, we first characterize the *unsafe* portion of ∂T, defined as those states $x \in \partial T$ for which there exists some disturbance $d \in \mathcal{D}$ such that for all inputs $u \in \mathcal{U}$ the vector field points into T; the *safe* portion of ∂T consists of the states $x \in \partial T$ for which there is some input $u \in \mathcal{U}$ such that for all disturbances $d \in \mathcal{D}$, the vector field points outward from T. More formally, we denote the outward pointing normal to T as

$$\nu = \frac{\partial l}{\partial x}(x(t_f)) \tag{5.5}$$

as in Fig. 5.1 which allows us to define

$$\begin{array}{ll} \text{Safe portion of } \partial T & \{x \in \partial T : \exists u \forall d \quad \nu^T f(x, u, d) \geq 0\} \\ \text{Unsafe portion of } \partial T & \{x \in \partial T : \forall u \exists d \quad \nu^T f(x, u, d) < 0\} \end{array} \tag{5.6}$$

Given the above anatomy of ∂T, the game is won by the pursuer if the terminal state $x(t_f)$ belongs in the unsafe portion of the boundary, and is won by the evader otherwise. It is clear that the optimal control $u^* \in \mathcal{U}$ is the one which maximizes $J_1(x,t,u,d)$, and the worst disturbance $d^* \in \mathcal{D}$ is the one which minimizes $J_1(x,t,u,d)$:

$$\begin{aligned} u^* &= \arg\max_{u\in\mathcal{U}} J_1(x,t,u,d) && (5.7)\\ d^* &= \arg\min_{d\in\mathcal{D}} J_1(x,t,u,d) && (5.8)\end{aligned}$$

The game is said to have a saddle solution if the cost $J_1^*(x,t)$ does not depend on the order in which the maximization and minimization is performed:

$$J_1^*(x,t) = \max_{u\in\mathcal{U}} \min_{d\in\mathcal{D}} J_1(x,t,u,d) = \min_{d\in\mathcal{D}} \max_{u\in\mathcal{U}} J_1(x,t,u,d) \tag{5.9}$$

The concept of a saddle solution is key to our computation of the safe regions of operation of the aircraft, since a solution of (5.2) with $u = u^*$ and $d = d^*$ represents an optimal trajectory for *each* player under the assumption that the other player plays its optimal strategy.

Safety is maintained by operating within the *safe set* of states V_1, which is the largest subset of $\mathbb{R}^n \backslash T$ which can be rendered invariant using inputs $u \in \mathcal{U}$ regardless of the disturbance $d \in \mathcal{D}$. We formally define V_1 as

$$\begin{aligned} V_1 &= \{x \in \mathbb{R}^n \backslash T \mid \exists u \in \mathcal{U}, J_1(x,t,u,d) \geq 0, \forall d \in \mathcal{D}\} && (5.10)\\ &= \{x \in \mathbb{R}^n \backslash T \mid \exists u \in \mathcal{U}, J_1(x,t,u,d^*) \geq 0\} && (5.11)\end{aligned}$$

Let ∂V_1 denote the boundary of V_1. At any instant t, the set $\{x \in \mathbb{R}^n \backslash T \mid J_1^*(x,t) \geq 0\}$ defines the set of safe states starting from time t. We would like to calculate the "steady state" safe set, or the safe set of states for all $t \in (-\infty, t_f]$. For this purpose we construct the Hamilton-Jacobi (Isaacs) equation for this system and attempt to calculate its steady state solution. Define the Hamiltonian $H(x,p,u,d) = p^T f(x,u,d)$ where $p \in T^*\mathbb{R}^n$ is the costate. The optimal Hamiltonian is given by:

$$H^*(x,p) = \max_{u\in U} \min_{d\in D} H(x,p,u,d) = H(x,p,u^*,d^*) \tag{5.12}$$

and satisfies Hamilton's equations (provided $H^*(x,p)$ is smooth in x and p):

$$\begin{aligned} \dot{x} &= \frac{\partial H^*}{\partial p}(x,p)\\ \dot{p} &= -\frac{\partial H^*}{\partial x}(x,p)\end{aligned} \tag{5.13}$$

with the boundary conditions $p(t_f) = \partial l(x(t_f))/\partial x$ and $x(t_f) \in \partial T$. If $J_1^*(x,t)$ is a smooth function of x and t, then $J_1^*(x,t)$ satisfies the Hamilton-Jacobi equation:

$$\frac{\partial J_1^*(x,t)}{\partial t} = -H^*(x, \frac{\partial J_1^*(x,t)}{\partial x}) \tag{5.14}$$

with boundary condition $J_1^*(x, t_f) = l(x(t_f))$. Our goal is to compute the safe set $V_1 = \{x \in I\!\!R^n \backslash T \mid J_1^*(x, -\infty) \geq 0\}$ where $J_1^*(x, -\infty)$ is the steady state solution of Eq. 5.14. However, it is difficult to guarantee that the PDE (Eq. 5.14) has solutions for all $t \leq 0$, due to the occurrence of "shocks", i.e. discontinuities in J as a function of x. If there are no shocks in the solution of Eq. 5.14, we may compute $J_1^*(x, -\infty)$ by setting the left hand side of the Hamilton-Jacobi equation to zero, thus $H^*(x, \frac{\partial J_1^*(x,-\infty)}{\partial x}) = 0$ which implies that $\frac{\partial J_1^*(x,-\infty)}{\partial x}$ is normal to the vector field $f(x, u^*, d^*)$.

To compute the safe set, we can propagate the boundaries of the safe set (those points for which $l(x) = 0$ and $H^*(x, p) = 0$) backwards in time, using the Hamilton-Jacobi (Isaacs) equation (Eq. 5.14) to determine the safe and unsafe sets over the state space $I\!\!R^n$ (see Fig. 5.2).

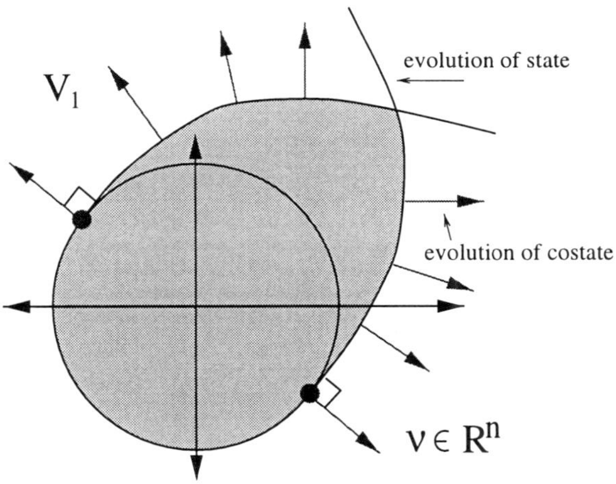

Fig. 5.2. The unsafe set of states (shaded) and its complement (the safe set V_1)

The set V_1 defines the *least restrictive control scheme* for safety. If the pursuer is inside V_1, any control input may be safely applied by the evader, whereas on the boundary, the only input which may be safely applied to ensure safety is u^*. If the pursuer is inside the unsafe set, it will eventually end up in the target set regardless of the actions of the evader. The safe set of control inputs associated with each state $x \in V_1$ is

$$\mathcal{U}_1(x) = \{u \in \mathcal{U} \mid J_1(x, t, u, d) \geq 0, \forall d \in \mathcal{D}\} \qquad (5.15)$$

Since all $u \in \mathcal{U}_1$ guarantee safety from state x, it is advantageous to find the optimal control policy $u \in \mathcal{U}_1$, for example the one that minimizes deviation from the nominal trajectory, which is encoded by a second cost function J_2, usually a quadratic function of the tracking error. To do this, we solve the optimal control problem which is *nested inside* the differential game calculation:

$$\min_{u \in \mathcal{U}_1} J_2 \tag{5.16}$$

subject to the original differential equations (5.2) which describe the aircraft motion in absolute coordinates. Additional system requirements, such as *passenger comfort*, can now be incorporated by extending the above nested chain of games and optimal control problems following the multiobjective design methodology of [15].

5.2 Resolution by Angular Velocity

Let us first consider the case in which the linear velocities of both aircrafts are fixed, $v_1, v_2 \in \mathbb{R}$, and the aircrafts avoid conflict solely by using their angular velocities, thus $u = \omega_1$ and $d = \omega_2$, and the model (5.1) becomes:

$$\begin{aligned} \dot{x}_r &= -v_1 + v_2 \cos\phi_r + u y_r \\ \dot{y}_r &= v_2 \sin\phi_r - u x_r \\ \dot{\phi}_r &= d - u \end{aligned} \tag{5.17}$$

with state variables $x_r, y_r \in \mathbb{R}$, $\phi_r \in [-\pi, \pi)$, and control and disturbance inputs $u \in \mathcal{U} = [\underline{\omega}_1, \overline{\omega}_1] \subset \mathbb{R}$, $d \in \mathcal{D} = [\underline{\omega}_2, \overline{\omega}_2] \subset \mathbb{R}$. Without loss of generality (we scale the coefficients of u and d if this is not met), assume that $\underline{\omega}_i = -1$ and $\overline{\omega}_i = 1$, for $i = 1, 2$.

The target set T is the protected zone of the evader:

$$T = \{(x_r, y_r) \in \mathbb{R}^2, \phi_r \in [-\pi, \pi) \mid x_r^2 + y_r^2 \leq 5^2\} \tag{5.18}$$

which is a 5-mile-radius cylinder in the (x_r, y_r, ϕ_r) space. Thus the function $l(x)$ may be defined as

$$l(x) = x_r^2 + y_r^2 - 5^2 \tag{5.19}$$

The optimal Hamiltonian is

$$H^*(x,p) = \max_{u \in U} \min_{d \in D} [-p_1 v_1 + p_1 v_2 \cos\phi_r + p_2 v_2 \sin\phi_r + (p_1 y_r - p_2 x_r - p_3) u + p_3 d] \tag{5.20}$$

Defining the *switching functions* $s_1(t)$ and $s_2(t)$, as

$$\begin{aligned} s_1(t) &= p_1(t) y_r(t) - p_2(t) x_r(t) - p_3(t) \\ s_2(t) &= p_3(t) \end{aligned} \tag{5.21}$$

the saddle solution u^*, d^* exists when $s_1 \neq 0$ and $s_2 \neq 0$ and are calculated as

$$\begin{aligned} u^* &= \operatorname{sgn}(s_1) \\ d^* &= -\operatorname{sgn}(s_2) \end{aligned} \tag{5.22}$$

The equations for $\dot{p}$ are obtained through Eq. 5.13 and are

$$\begin{aligned} \dot{p}_1 &= u^* p_2 \\ \dot{p}_2 &= -u^* p_1 \\ \dot{p}_3 &= p_1 v_2 \sin\phi_r - p_2 v_2 \cos\phi_r \end{aligned} \tag{5.23}$$

with $p(t_f) = (x_r, y_r, 0)^T = \nu$, the outward pointing normal to ∂T at any point (x_r, y_r, ϕ_r) on ∂T.

The safe and unsafe portions of ∂T are calculated using Eqs. 5.6 with $\nu = (x_r, y_r, 0)^T$. Thus, those (x_r, y_r, ϕ_r) on ∂T for which

$$-v_1 x_r + v_2(x_r \cos\phi_r + y_r \sin\phi_r) < 0 \tag{5.24}$$

constitute the unsafe portion, and those (x_r, y_r, ϕ_r) on ∂T for which

$$-v_1 x_r + v_2(x_r \cos\phi_r + y_r \sin\phi_r) = 0 \tag{5.25}$$

are the final state conditions for the boundary of the safe set V_1. To solve for $p(t)$ and $x(t)$ along this boundary for $t < t_f$, we must first determine $u^*(t_f)$ and $d^*(t_f)$. Equations 5.22 are not defined at $t = t_f$, since $s_1 = s_2 = 0$ on ∂T, giving rise to "abnormal extremals" (meaning that the optimal Hamiltonian loses dependence on u and d at these points). Analogously to [1] (Chapter 8), we use an indirect method to calculate $u^*(t_f)$ and $d^*(t_f)$: at any point (x_r, y_r, ϕ_r) on ∂T, the derivatives of the switching functions s_1 and s_2 are

$$\dot{s}_1 = y_r v_1 \tag{5.26}$$

$$\dot{s}_2 = x_r v_2 \sin\phi_r - y_r v_2 \cos\phi_r \tag{5.27}$$

For example, for points $(x_r, y_r, \phi_r) \in \partial T$, such that $\phi_r \in (0, \pi)$, it is straightforward to show that $\dot{s}_1 > 0$ and $\dot{s}_2 > 0$, meaning that for values of t slightly less than t_f, $s_1 < 0$ and $s_2 < 0$. Thus for this range of points along ∂T, $u^*(t_f) = -1$ and $d^*(t_f) = 1$. These values for u^* and d^* remain valid for $t < t_f$ as long as $s_1(t) < 0$ and $s_2(t) < 0$. When $s_1(t) = 0$ and $s_2(t) = 0$, the saddle solution switches and the computation of the boundary continues with the new values of u^* and d^*, thus introducing "kinks" into the safe set boundary. These points correspond to the shocks in the Hamilton-Jacobi (Isaacs) equation discussed above. Figure 5.3 displays the resulting boundary of the safe set V_1, for $t < t_f$ until the first time that either $s_1(t)$ or $s_2(t)$ switches.

The automaton illustrating the *least restrictive control scheme* for safety is shown in Fig. 5.4.

The computation of the boundary of V_1 is in general difficult. For certain ranges of $\mathcal{U}$ and $\mathcal{D}$, the surfaces shown in Fig. 5.3 intersect. At the intersection, it is not clear that u^* is the unique safe input.

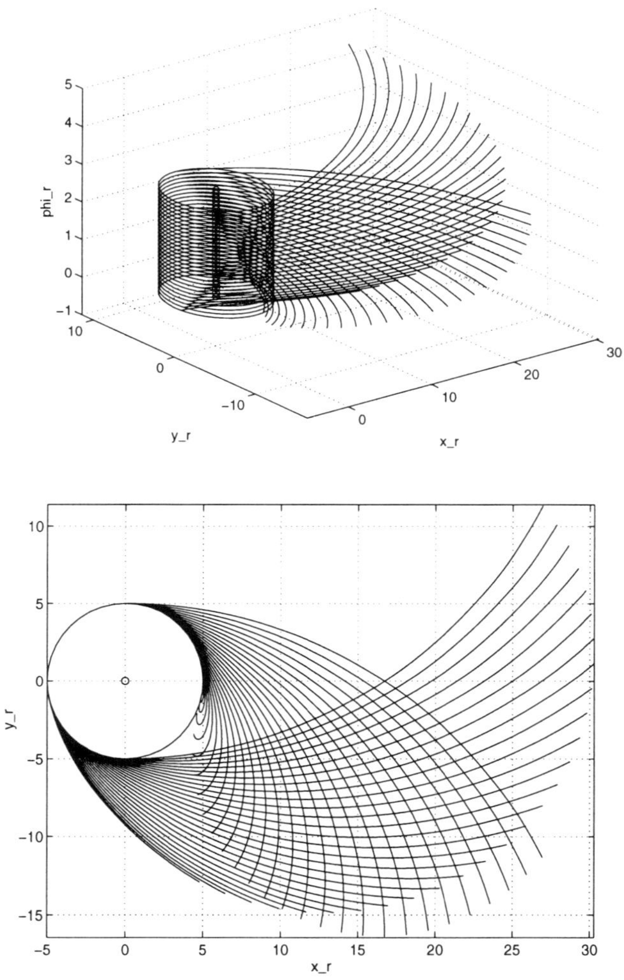

Fig. 5.3. The Target set $T = \{(x_r, y_r), \phi_r \in (0, \pi) \mid x_r^2 + y_r^2 \leq 5^2\}$ (cylinder) and the boundary of the safe set V_1 for $t \leq t_f$ until the first switch in either $s_1(t)$ or $s_2(t)$. The unsafe set is enclosed by the boundary. The second picture is a top view of the first

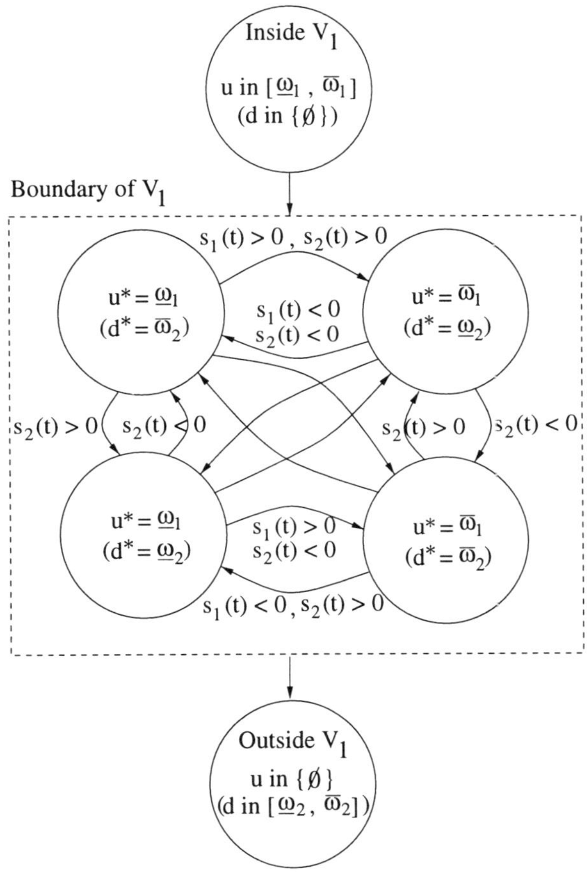

Fig. 5.4. Switching law governing the two aircraft system with angular velocity control inputs. The law is least restrictive in that the control u is not restricted when the state is inside V_1. The diagonal transitions in the automaton for the boundary of V_1 are not labeled for legibility. This automaton may be thought of as the composition of two switching automata, one each for the pursuer and evader. The individual switching automaton for u is easily derived by neglecting transitions for d, and conversely

5.3 Resolution by Linear Velocity

We now consider the case in which the angular velocities of the two aircraft are zero, and collision is avoided by altering the velocity profile of the trajectories. Thus, $u = v_1$, $d = v_2$, and model (5.1) reduces to:

$$\begin{aligned} \dot{x}_r &= -u + d\cos\phi_r \\ \dot{y}_r &= d\sin\phi_r \\ \dot{\phi}_r &= 0 \end{aligned} \tag{5.28}$$

The input and disturbance lie in closed subsets of the positive real line $u \in \mathcal{U} = [\underline{v}_1, \overline{v}_1] \subset \mathbb{R}^+$, $d \in \mathcal{D} = [\underline{v}_1, \overline{v}_2] \subset \mathbb{R}^+$.

The Target set T and function $l(x)$ are defined as in the previous example. In this example, it is straightforward to calculate the saddle solution (u^*, d^*) directly, by integrating Eqs. 5.28 for piecewise constant u and d, and substituting the solutions into the cost function (Eq. 5.4). To do this we first define the switching functions s_1 and s_2 as

$$\begin{aligned} s_1(t) &= x_r \\ s_2(t) &= x_r\cos\phi_r + y_r\sin\phi_r \end{aligned} \tag{5.29}$$

Proposition 5.1. [Saddle Solution for Linear Velocity Controls] *The global saddle solution (u^*, d^*) to the game described by system (Eq. 5.28) for the cost $J_1(x,t,u,d)$ given by equation (Eq. 5.4) is*

$$u^* = \begin{cases} \underline{v}_1 & \textit{if}\, \operatorname{sgn}(s_1) > 0 \\ \overline{v}_1 & \textit{if}\, \operatorname{sgn}(s_1) < 0 \end{cases} \tag{5.30}$$

$$d^* = \begin{cases} \underline{v}_2 & \textit{if}\, \operatorname{sgn}(s_2) > 0 \\ \overline{v}_2 & \textit{if}\, \operatorname{sgn}(s_2) < 0 \end{cases} \tag{5.31}$$

As can be seen from equation (5.30), the optimal speed of the evader depends on the position of the pursuer relative to the evader. If the pursuer is ahead of the evader in the relative axis frame, then u^* is at its lower limit, if the pursuer is behind the evader in the relative axis frame then u^* is at its upper limit. If the pursuer is heading towards the evader, then d^* is at its upper limit, if the pursuer is heading away from the evader, d^* is at its lower limit. The bang-bang nature of the saddle solution allows us to abstract the system behavior by the hybrid automaton shown in Fig. 5.5, which describes the least restrictive control scheme for safety.

The unsafe sets of states are illustrated in Fig. 5.6 for various values of ϕ_r, and speed ranges as illustrated.

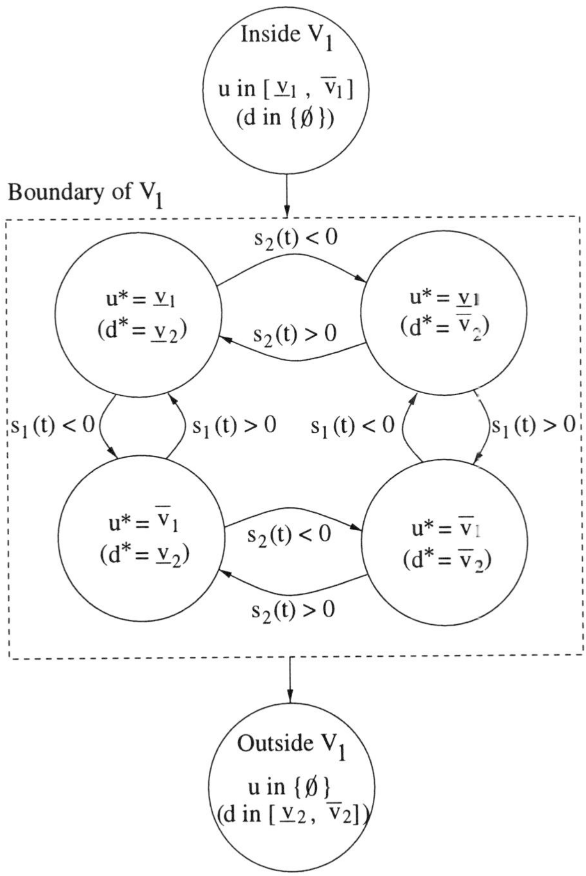

Fig. 5.5. Switching law governing the two aircraft system with linear velocity control inputs. The note in the caption of the automaton of the previous example, about the composition of automata for u and d, applies here also

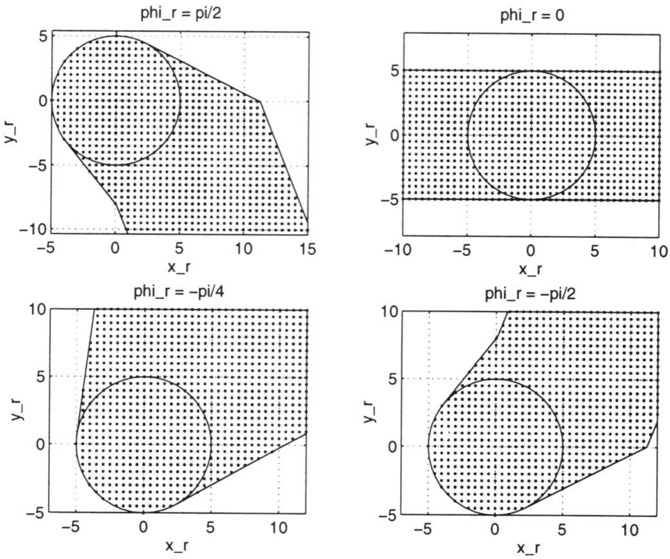

Fig. 5.6. Unsafe sets (x_r, y_r) for $[\underline{v}_1, \overline{v}_1] = [2, 4]$, $[\underline{v}_2, \overline{v}_2] = [1, 5]$ and $\phi_r = \pi/2$, 0, $-\pi/4$, and $-\pi/2$

5.4 Cooperative Conflict Resolution

In *cooperative conflict resolution* the aircraft coordinate amongst themselves if trajectory conflicts occur, and perform predefined, *a-priori safe* maneuvers in order to resolve the conflict. The class of maneuvers constructed to resolve conflicts must be rich enough to cover most possible conflict scenarios. Examples of *head-on* and *overtake* maneuvers can be seen in Figs. 5.7 and 5.8.

In order to construct a complete set of collision avoidance maneuvers which cover general conflict scenarios involving more than two agents, there is a need to classify all kinds of possible collisions, and the maneuvers to resolve these. To do this, we use the strategy outlined in Fig. 5.9. First we employ the automated method of potential field based motion planning to generate "prototype maneuvers", which inspire the actual collision avoidance maneuver. We believe that distributed on-line motion planning techniques and their application to ATM can be inspirational for deriving a set of possible maneuvers for collision avoidance between aircraft. In spite of the fact that the feasibility of the individual trajectories can be asserted by simulations, the proof of the safety of the maneuvers for dynamic models of aircraft remains a challenging problem. It is for this reason that we wish to construct the simplest possible maneuvers from the prototypes, those made up of straight lines. This discretized version of the maneuvers can be modeled as a hybrid

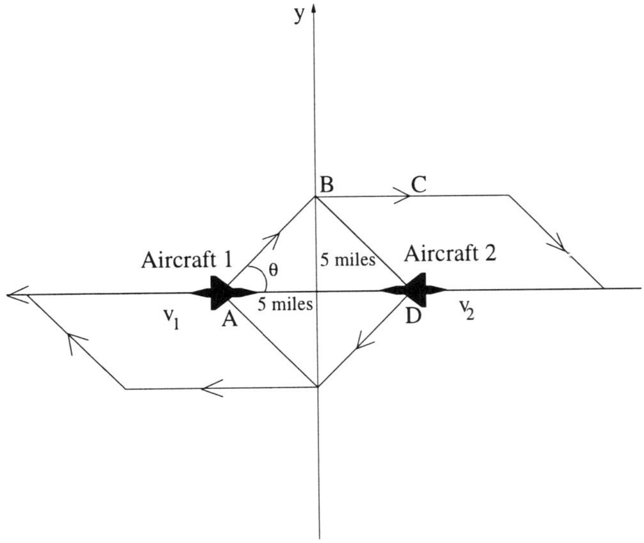

Fig. 5.7. Generalized *head-on* conflict

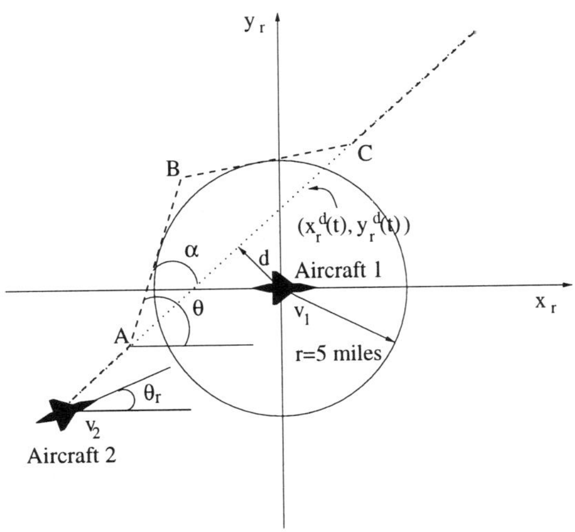

Fig. 5.8. Generalized *overtake* conflict

system and can be proven to be safe using hybrid verification techniques. The verification step is crucial for building an off-line database of safe conflict resolution maneuvers.

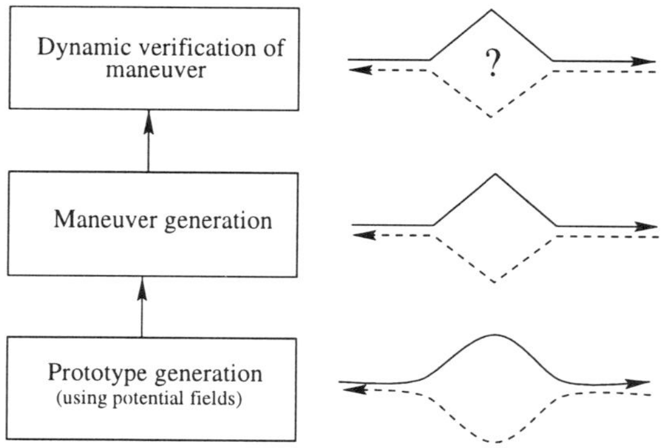

Fig. 5.9. Generation of potential field inspired maneuvers

5.4.1 Robot collision avoidance. There is a large number of theoretical studies in the classical motion planning literature regarding the multiple robot planning problem. Algorithms embedded in time-extended configuration space [5] prove to be computationally hard, and with additional velocity bounds the multi-agent motion planning problem has been shown to be NP-complete [3]. In applications the scenarios considered most often involve navigation in the presence of other moving agents and obstacles [18]. The proposed solutions are geared towards distributed settings, in which only the local information about the state of the environment and the other agents in the vicinity is available to each agent. These techniques are based on potential and vortex field methods [19] and the complexity is proportional to the number of agents. An attempt to guarantee that the agents achieve their goals without colliding with each other has been proposed by Masoud [17]. In spite of the fact that collision avoidance is an integral part of agents' navigation capabilities, the requirements for safety and optimality have not been addressed to any great extent. This is partly due to the fact that the agent velocities have traditionally been relatively small and safety issues not so prominent: low velocity collisions occasionally occur, but various recovery strategies allow the agents to further pursue their tasks. In path planning for more than two agents, prioritized schemes have been used to fix the order in which the conflicts are resolved.

In our air traffic collision avoidance problem, we use the approach based on potential and vortex fields to generate *prototype* maneuvers from which we derive the *qualitative* properties of the actual maneuver that each aircraft follows, and then classify these prototypes into discrete sets of avoidance maneuvers which we prove to be safe.

5.4.2 Maneuver generation. We adopt the potential and vortex field approach for distributed motion planning proposed in [17]. In ATM the absence of stationary obstacles and the approximation of individual agents by circles with a specified radius constitute reasonable assumptions prior to formulating the collision avoidance strategy[2]. We consider the *planar* collision avoidance problem with multiple moving agents.

The planner is obtained by the superposition of several vector fields representing qualitatively different steering actions of each agent. Suppose we have m agents, with the i-th agent represented by a circle with radius r_i and its configuration denoted by $\boldsymbol{x}_i = (x_i, y_i)$. The desired destination of the i-th agent $\boldsymbol{x}_{di} = (x_{di}, y_{di})$ is represented by an attractive potential function:

$$U_a(\boldsymbol{x}_i, \boldsymbol{x}_{di}) = \frac{1}{2}(\boldsymbol{x}_{di} - \boldsymbol{x}_i)^2$$

In order to achieve the desired destination a force proportional to the negative gradient of the U_a needs to be exerted:

$$F_a(\boldsymbol{x}_i, \boldsymbol{x}_{di}) = -\nabla U_a(\boldsymbol{x}_i, \boldsymbol{x}_{di}) = -(\boldsymbol{x}_i - \boldsymbol{x}_{di})$$

To prevent collisions between agents i and j, the following spherically symmetric repulsive field $U_r(\boldsymbol{x}_i, \boldsymbol{x}_j)$ is associated with each agent:

$$U_r(\boldsymbol{x}_i, \boldsymbol{x}_j) = \begin{cases} -\dfrac{(\boldsymbol{r}_{ij} - (\boldsymbol{r}_j + \delta_{rj}))^2}{2\delta_{rj}} & \text{if } \boldsymbol{r}_j \le \boldsymbol{r}_{ij} \le \boldsymbol{r}_j + \delta_{rj} \\ 0 & \text{otherwise} \end{cases}$$

where $\boldsymbol{r}_{ij} = \sqrt{(x_i - x_j)^2 + (y_i - y_j)^2}$ is the distance between the i-th and jth agent, $\boldsymbol{r}_j$ is the radius of jth agent and δ_{rj} is the influence zone of its repulsive field. The repulsive force associated with this field is:

$$F_r(\boldsymbol{x}_i, \boldsymbol{x}_j) = \nabla U_r(\boldsymbol{x}_i, \boldsymbol{x}_j)$$

A vortex field, used to ensure that all agents turn in the same direction when encountering a conflict, is constructed around each agent tangential to the repulsive field $U_r(\boldsymbol{x}_i, \boldsymbol{x}_j)$:

$$F_v(\boldsymbol{x}_i, \boldsymbol{x}_j) = \pm \begin{bmatrix} \dfrac{\partial U_r(\boldsymbol{x}_i, \boldsymbol{x}_j)}{\partial y} \\ -\dfrac{\partial U_r(\boldsymbol{x}_i, \boldsymbol{x}_j)}{\partial x} \end{bmatrix}$$

Note that by the choice of the sign in the above vortex field expression one can determine the direction of the circulating field. Setting the direction to

[2] The aircraft is considered to be a "hockey puck" of a specified safety radius representing the desired clearance from the other aircraft.

a particular sign for all agents corresponds essentially to a "rule of the road" which specifies the direction of the avoidance for conflict maneuvers.

The dynamic planner for a single agent in the presence of multiple agents is obtained by superposition of participating potential and vector fields and becomes:

$$\dot{\boldsymbol{x}}_i = \frac{F_a(\boldsymbol{x}_i, \boldsymbol{x}_{di})}{\|F_a(\boldsymbol{x}_i, \boldsymbol{x}_{di})\|} + \sum_j (k_{ri}\, F_r(\boldsymbol{x}_i, \boldsymbol{x}_j) + k_{vi}\, F_v(\boldsymbol{x}_i, \boldsymbol{x}_j))$$

where $j = 1, ..., m, \quad i \neq j$. The contributions from repulsive and vortex fields range between [0, 1], increasing as the agent approaches the boundary of another agent. Normalization of the attractive field component makes its contribution comparable to the magnitudes of the repulsive and vortex fields. The strength of the field then becomes independent of the distance to the goal, capturing merely the heading to the goal. The individual contributions are then weighted by k_{ri}, k_{vi} and the resulting vector is again normalized and scaled by k_{di}, a constant proportional to the desired velocity of the i-th agent. The velocity of i-th agent is then:

$$\boldsymbol{v}_i = k_{di} \frac{\dot{\boldsymbol{x}}_i}{\|\dot{\boldsymbol{x}}_i\|}$$

In the following paragraph we demonstrate the capability of the planner to generate trajectories for general classes of collision avoidance maneuvers.

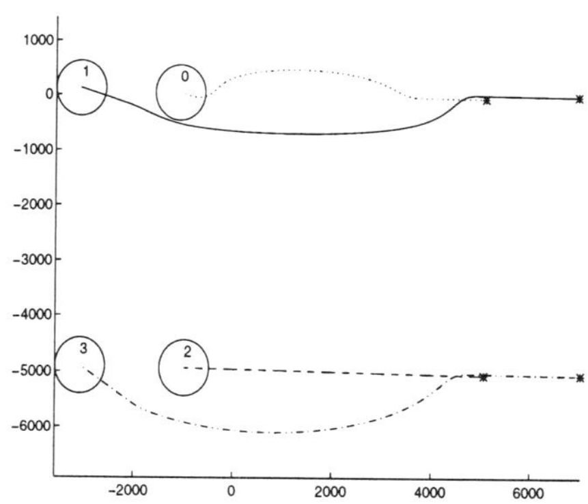

Fig. 5.10. *Overtake* maneuvers. Top: $1.5k_{d0} = k_{d1}$, $k_{r0} = k_{r1} = k_{v0} = k_{v1} = 1.0$. Bottom: $1.5k_{d2} = k_{d3}$, $k_{r2} = k_{v2} = 0.0$; $k_{r3} = k_{v3} = 1.0$

5.4.3 Overtake conflicts. The *overtake* conflict can be resolved by the planner in several qualitatively different ways obtained by adjusting the parameters in the individual contributions of the participating vector fields. In the outlined experiments two agents having different velocities participate in conflict resolution. In Fig. 5.10 agent 1 is 1.5 times faster than agent 0. In the top maneuver, agent 1 overtakes agent 0 and agent 0 moves away from agent 1, resulting in smaller deviations from the original trajectory for both agents. The strength of contribution from the repulsive and vortex fields is the same for both agents. The willingness of the slower agent to cooperate in the overtake maneuver can be modeled by the strength of the agent's repulsive and vortex fields: in the bottom maneuver of Fig. 5.10 the contributions of agent 2's vortex and repulsive fields are set to zero and agent 2 does not deviate from its original trajectory. Figure 5.11 demonstrates generalized overtake maneuvers

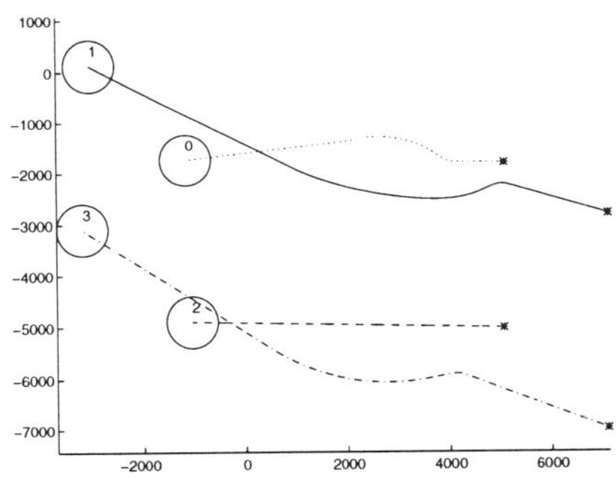

Fig. 5.11. Generalized *overtake* maneuver. In the conflict at the top both agents participate in the maneuver while at the bottom the conflict is resolved solely by agent 3

5.4.4 Head-on conflicts. Similarly, several qualitatively different *head-on* maneuvers can be generated by changing the contributions of individual fields. In Fig. 5.12 there are two head-on maneuvers where both agents actively cooperate in resolving the conflict, i.e. the strength of the repulsive and vortex fields is the same for each agent.

Figure 5.13 demonstrates generalized head-on maneuvers where the agents are not initially aligned.

5.4.5 Multiple aircraft conflicts. When multiple aircraft are involved in conflict the vector field based planner is very instructional: the direction of the vortex field contribution serves as a coordination element between the

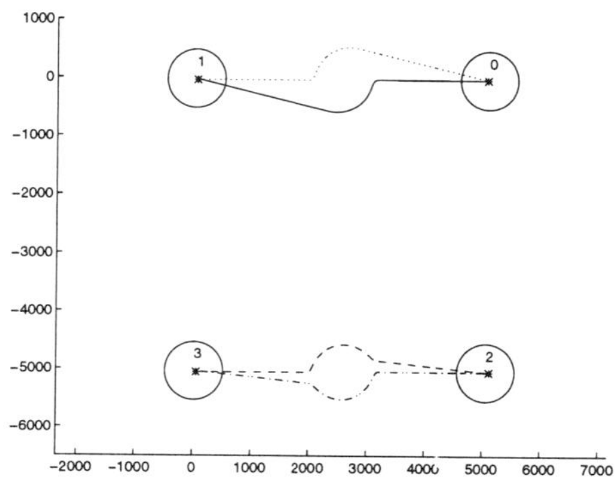

Fig. 5.12. *Head-on* maneuvers. Top: symmetric head-on with all the parameters the same. Bottom: $k_{d2} = k_{d3} = 10.0$, $k_{r2} = k_{r3} = 0.3$, $k_{v2} = k_{v3} = 5.0$ with the influence of the vortex field emphasized

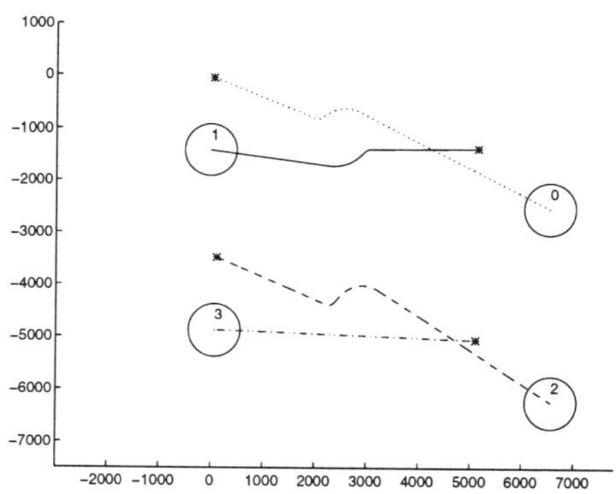

Fig. 5.13. Generalized *head-on* maneuver. Top: the velocities of the agents are the same and both agents participate in the maneuver. Bottom: agent 3 does not participate in the maneuver since $k_{r3} = k_{v3} = 0$

aircraft. Figure 5.14 depicts a symmetric *roundabout* maneuver similar to the one proposed in [26]. The agents involved in the resolution of the conflict are homogeneous, having the same velocities, willing to participate equally in the maneuver (the strength of the repulsive and vortex fields is the same for all agents). Figure 5.15 demonstrates a scenario where agent 0 does not participate in the coordination (k_{v0} and k_{r0} are 0) and is willing only to adjust its velocity slightly. This particular conflict can be still resolved and the resulting trajectories are flyable.

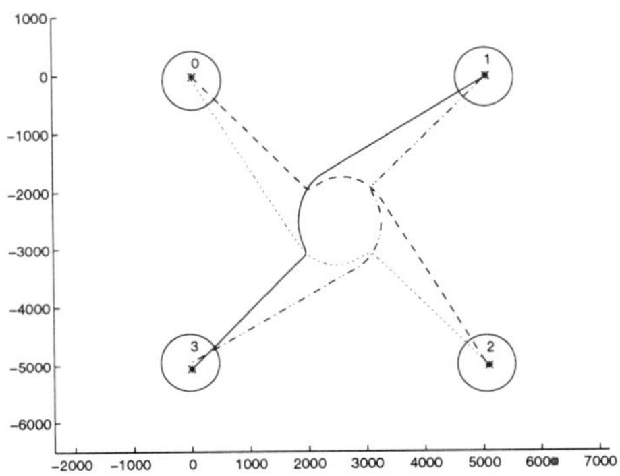

Fig. 5.14. Symmetric *roundabout*, gain factors for individual agents are the same

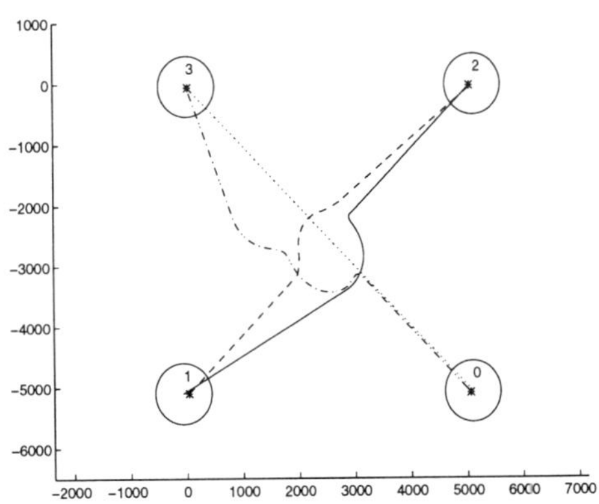

Fig. 5.15. Partial *roundabout*, $k_{vi} = k_{ri} = 1.0$ and $k_{d0} = 0.5k_{di}$ for i = 1,2,3 with the maximal velocity of agent 0 reduced by a factor of 2 and $k_{r0} = k_{v0} = 0$

5.4.6 Observations. The presented planner has the capability of changing the spatial behavior of individual agents and always resolved the conflict if the agents were homogeneous and there were no restrictions on the temporal profiles of the agents' paths. Given particular constraints on agents' velocities certain conflicts may result in "loss of separation" or trajectories which are not flyable, due to the violation of the limits on turn angles (Fig. 5.16). In such cases the shape of the path can be affected by changing parameters of contributing vector fields. The adjustment of influence zones δ_{ri} and δ_{vi} as well as the relative strength of the repulsive and vortex vector fields, k_{ri} and k_{vi}, can affect the turn angle and maximal deviation from the original trajectory needed to resolve the conflict. The change of the temporal profiles of the path by adjusting the velocities of individual agents (k_{di}) has the most profound affect on the capability of resolving general conflict scenarios. In Fig. 5.17 the unflyable trajectory from Fig. 5.16 can be changed by adjusting the velocity of agent 2 resulting in a flyable trajectory.

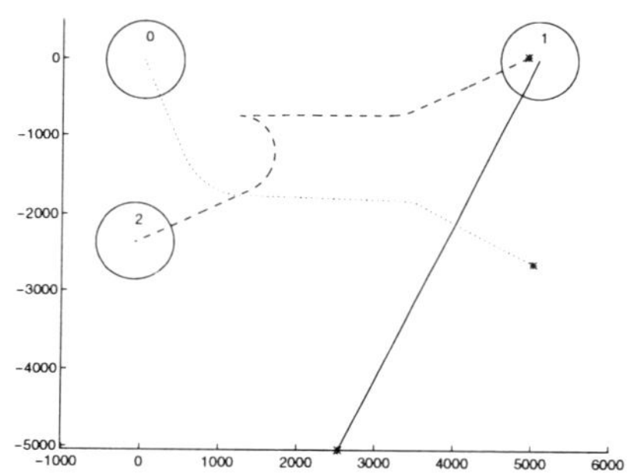

Fig. 5.16. General conflict scenario. Trajectory of agent 2 is not flyable

5.4.7 Maneuver approximation and verification. The discretization of the prototype maneuver is motivated by techniques currently performed by air traffic controllers which resolve conflicts by "vectoring" the aircraft in the airspace. This is partly due to the current status of the communication technology between the air traffic control center and the aircraft as well as the state of current avionics (autopilot) on board the aircraft which operate in a set-point mode. We consider two types of approximations: turning point and offset.

The individual approximation can be obtained from the trajectories generated by the dynamic planner by recursive least squares linear fit (see Fig. 5.18).

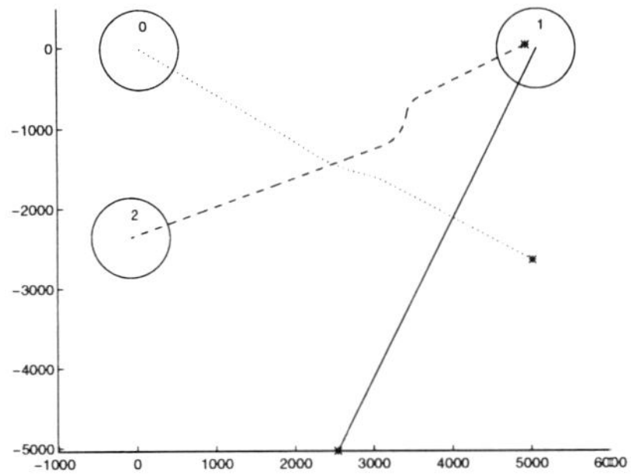

Fig. 5.17. Velocity profile agent 2 is adjusted resulting in a flyable trajectory

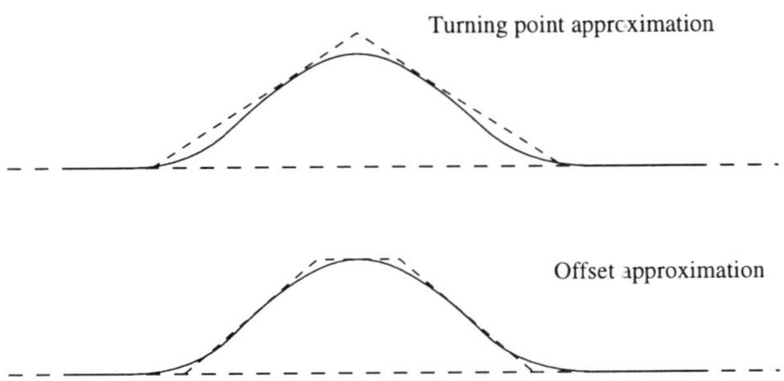

Fig. 5.18. Turning point and offset approximation

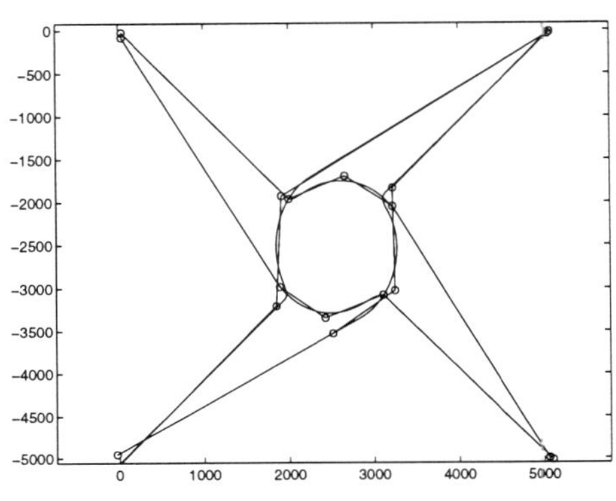

Fig. 5.19. Discretized *roundabout* maneuver

5.5 Verification of the Maneuvers

The approximation phase is followed by the verification of the obtained maneuvers. The purpose of the verification step is to prove the safety of the maneuver by taking into account the velocity bounds and sets of initial conditions of individual aircraft. The collision avoidance problem lends itself to a hybrid system description: the continuous modes of the hybrid model correspond to individual parts of the maneuver (e.g. straight, turn right θ_1 degrees, turn left θ_2 degrees) and the transitions between modes correspond to switching between individual modes of the maneuver. Within each mode the speed of each aircraft can be specified in terms of lower and upper bounds. This suitable simplification of the problem allows us to model the collision avoidance maneuver in terms of hybrid automata. Each aircraft is modeled by a hybrid automaton, and an additional controller automaton implements the discrete avoidance maneuver strategy. The verification results can assert that the maneuver is safe for given velocity bounds and given set of initial conditions. To relate the verification of the cooperative schemes to the use of the Hamilton-Jacobi equation of the previous section, we only mention that this approach can be used to compute the safe set of initial conditions in the iterations required to verify the safety of the maneuver. Further details may be found in [12].

The previously presented simulation results suggest that the generalized *overtake* and generalized *head-on* maneuvers may be used to solve all possible two-aircraft conflicts. This allows us to classify two-aircraft maneuvers by the angle at which the aircraft approach each other, and to design simple deviation maneuvers as sequences of straight line segments which approximate the trajectories derived from the potential and vortex field algorithm. For more than two aircraft the obtained discretized version of the *roundabout* maneuver is proposed. For this maneuver, the radius of a circular path around the conflict point is proportional to the influence zones of the aircrafts' repulsive and vortex fields. We propose this methodology as a suitable step of automation of conflict resolution in ATM given currently available technology. The complete classification of a library of reasonably complete conflict scenarios and maneuvers remains a challenging problem.

6. Conclusions

The technological advances that make free flight feasible include on-board GPS, satellite datalink, and powerful on-board computation such as the Traffic Collision and Avoidance System (TCAS), currently certified by the FAA to provide warnings of ground, traffic, and weather proximity. Navigation systems use GPS which provides each aircraft with its four dimensional coordinates with extreme precision. For conflict detection, current radar systems are adequate. Conflict prediction and resolution, however, require information regarding the position, velocity and intent of other aircraft in the vicinity. This will be accomplished by the proposed ADS-B broadcast information system. These advances will be economically feasible only for commercial aviation aircraft: how to merge the proposed architecture with general aviation aircraft (considered disturbances in the system in this chapter) is a critical issue. Furthermore, the transition from the current to the proposed system must be smooth and gradual. Above all, the algorithms must be verified for correctness and safety before the implementation stage. This is one of the main challenges facing the systems and verification community. The accent in this chapter has been on "safety" proofs for hybrid systems. In fact there are other properties of hybrid system such as *non-blockage of time*, *fairness*, etc. which are so-called liveness properties which also need to verified. Techniques for studying these are in their infancy except for very simple classes of hybrid system models.

Another important area of investigation in large scale systems design (such as the ATMS just described) is the *global or emergent* characteristics of the system. We have discussed how conflict resolution can provide autonomy for aircraft to decide how to plan their trajectories in the airspace between TRACONs, and for air traffic controllers to implement conflict resolution inside the TRACONs. The study of the composite automated system frequently reveals some surprising characteristics. For example, it was found from the implementation of CTAS at Dallas Fort Worth and UPR in the flight sector from Dallas to Washington that all aircraft tended to prefer the same route resulting in congestion at specific times in the Dallas TRACON. Another phenomenon associated with UPR is the formation of "convoys" of aircraft in the Asian airspace en-route from South East Asia to Europe. This latter phenomenon has spurred the study of the benefits of explicitly convoying aircraft in groups to their destination. Theoretical tools for the study of aggregate behavior arising from protocols for individual groups of agents are necessary to be able to assess the economic impact of air traffic automation strategies.

Acknowledgement. This research is supported by NASA under grant NAG 2-1039, and by ARO under grants DAAH 04-95-1-0588 and DAAH 04-96-1-0341.

References

[1] Başar T, Olsder G J 1995 *Dynamic Non-cooperative Game Theory.* 2nd ed, Academic Press, New York

[2] Brudnicki D J, McFarland A L 1997 User request evaluation tool (URET) conflict probe performance and benefits assessment. In: *Proc USA/Europe ATM Seminar.* Eurocontrol, Paris, France

[3] Canny J, Reif J 1987 New lower bound techniques for robot motion planning problems. In: *Proc 28th Annual IEEE Symp Found Comp Science.* pp 49–60
[4] Couluris G J, Dorsky S 1995 Advanced air transportation technologies (AATT) potential benefits analysis. Tech Rep NASA Contract NAS2-13767, Seagull Technology Inc, Cupertino, CA
[5] Erdman M, Lozano-Perez T 1987 On multiple moving objects. *Algorithmica.* 2:477–595
[6] Erzberger H 1992 CTAS: Computer intelligence for air traffic control in the terminal area. Tech Rep NASA TM-103959, NASA Ames Research Center, Moffett Field, CA
[7] Harman W H 1989 TCAS: A system for preventing midair collisions. *Lincoln Lab J.* 2:437–457
[8] Honeywell Inc 1996 Markets Report. Tech Rep NASA Contract NAS2-114279
[9] Honeywell Inc 1996 Technology and Procedures Report. Tech Rep NASA Contract NAS2-114279
[10] Isaacs R 1967 *Differential Games.* Wiley, New York
[11] Kahne S, Frolow I 1996 Air traffic management: Evolution with technology. *IEEE Contr Syst Mag.* 16(4):12–21
[12] Košecká J, Tomlin C, Pappas G, Sastry S 1997 Verification of cooperative conflict resolution maneuvers. Submitted to: *Hybrid Systems V*
[13] Krozel J, Mueller T, Hunter G 1996 Free flight conflict detection and resolution analysis. In: *Proc AIAA Guid Navig Contr Conf.* paper AIAA-96-3763
[14] Kuchar J K 1995 A unified methodology for the evaluation of hazard alerting systems. PhD thesis, Massachussets Institute of Technology
[15] Lygeros J, Tomlin C, Sastry S 1996 Multiobjective hybrid controller synthesis. In: *Proc Int Work Hybrid Real-Time Syst.* Grenoble, France, pp 109–123
[16] Lygeros J, Tomlin C, Sastry S 1997 Multi-objective hybrid controller synthesis. In: Maler O (ed) *Proc HART97.* Springer-Verlag, Berlin, Germany, pp 109–123
[17] Masoud A 1996 Using hybrid vector-harmonic potential fields for multi-robot, multi-target navigation in stationary environment. In: *Proc 1996 IEEE Int Conf Robot Automat.* Minneapolis, MN, pp 3564–3571
[18] Mataric M J 1993 Issues and approaches in the design of collective autonomous agents. *Robot Autonom Syst.* 16:321–331
[19] Medio C D, Oriolo G 1991 Robot obstacle avoidance using vortex fields. In: Stifter S, Lenarčič (eds) *Advances in Robot Kinematics.* Kluwer, Dordrecht, The Netherlands, pp 227–235
[20] Paielli R A, Erzberger H 1996 Conflict probability estimation and resolution for free flight. NASA Ames Research Center, preprint
[21] Pappas G J, Lygeros J, Godbole D N 1995 Stabilization and tracking of feedback linearizable systems under input constraints. In: *Proc 34th IEEE Conf Decision Contr.* New Orleans, LA, pp 596–601
[22] Pappas G J, Sastry S 1997 Towards continuous abstractions of dynamical and control systems In: Antsaklis P, Kohn W, Nerode A, Sastry S (eds) *Hybrid Systems IV*, Springer-Verlag, New York
[23] Radio Technical Commission for Aeronautics 1995 Final report of RTCA task force 3: Free flight implementation. Tech Rep, Washington, DC
[24] Shewchun M, Feron E 1997 Linear matrix inequalities for analysis of free flight conflict problems. In: *36th IEEE Conf Decision Contr.* San Diego, CA
[25] Tomlin C, Lygeros J, Benvenuti L, Sastry S 1995 Output tracking for a nonminimum phase dynamic CTOL aircraft model. In: *Proc 34th IEEE Conf Decision Contr.* New Orleans, LA

[26] Tomlin C, Pappas G, Sastry S 1997 Conflict resolution for air traffic management: A case study in multi-agent hybrid systems. Tech Rep UCB/ERL M97/33, University of California at Berkeley, to appear in *IEEE Trans Automat Contr.*

[27] Tomlin C, Sastry S 1995 Bounded tracking for nonminimum phase nonlinear systems with fast zero dynamics. In: *Proc 35th IEEE Conf Decision Contr.* Kobe, Japan, pp 2058–2063

[28] Zhao Y, Schultz R 1997 Deterministic resolution of two aircraft conflict in free flight In: *Proc AIAA Guid Navig Contr Conf.* New Orleans, LA, paper AIAA-97-3547

Lecture Notes in Control and Information Sciences

Edited by M. Thoma

1993–1997 Published Titles:

Vol. 186: Sreenath, N.
Systems Representation of Global Climate Change Models. Foundation for a Systems Science Approach.
288 pp. 1993 [3-540-19824-5]

Vol. 187: Morecki, A.; Bianchi, G.; Jaworeck, K. (Eds)
RoManSy 9: Proceedings of the Ninth CISM-IFToMM Symposium on Theory and Practice of Robots and Manipulators.
476 pp. 1993 [3-540-19834-2]

Vol. 188: Naidu, D. Subbaram
Aeroassisted Orbital Transfer: Guidance and Control Strategies
192 pp. 1993 [3-540-19819-9]

Vol. 189: Ilchmann, A.
Non-Identifier-Based High-Gain Adaptive Control
220 pp. 1993 [3-540-19845-8]

Vol. 190: Chatila, R.; Hirzinger, G. (Eds)
Experimental Robotics II: The 2nd International Symposium, Toulouse, France, June 25-27 1991
580 pp. 1993 [3-540-19851-2]

Vol. 191: Blondel, V.
Simultaneous Stabilization of Linear Systems
212 pp. 1993 [3-540-19862-8]

Vol. 192: Smith, R.S.; Dahleh, M. (Eds)
The Modeling of Uncertainty in Control Systems
412 pp. 1993 [3-540-19870-9]

Vol. 193: Zinober, A.S.I. (Ed.)
Variable Structure and Lyapunov Control
428 pp. 1993 [3-540-19869-5]

Vol. 194: Cao, Xi-Ren
Realization Probabilities: The Dynamics of Queuing Systems
336 pp. 1993 [3-540-19872-5]

Vol. 195: Liu, D.; Michel, A.N.
Dynamical Systems with Saturation Nonlinearities: Analysis and Design
212 pp. 1994 [3-540-19888-1]

Vol. 196: Battilotti, S.
Noninteracting Control with Stability for Nonlinear Systems
196 pp. 1994 [3-540-19891-1]

Vol. 197: Henry, J.; Yvon, J.P. (Eds)
System Modelling and Optimization
975 pp approx. 1994 [3-540-19893-8]

Vol. 198: Winter, H.; Nüßer, H.-G. (Eds)
Advanced Technologies for Air Traffic Flow Management
225 pp approx. 1994 [3-540-19895-4]

Vol. 199: Cohen G.; Quadrat, J.-P. (Eds)
11th International Conference on Analysis and Optimization of Systems – Discrete Event Systems: Sophia-Antipolis, June 15–16–17, 1994
648 pp. 1994 [3-540-19896-2]

Vol. 200: Yoshikawa, T.; Miyazaki, F. (Eds)
Experimental Robotics III: The 3rd International Symposium, Kyoto, Japan, October 28-30, 1993
624 pp. 1994 [3-540-19905-5]

Vol. 201: Kogan, J.
Robust Stability and Convexity
192 pp. 1994 [3-540-19919-5]

Vol. 202: Francis, B.A.; Tannenbaum, A.R. (Eds)
Feedback Control, Nonlinear Systems, and Complexity
288 pp. 1995 [3-540-19943-8]

Vol. 203: Popkov, Y.S.
Macrosystems Theory and its Applications: Equilibrium Models
344 pp. 1995 [3-540-19955-1]

Vol. 204: Takahashi, S.; Takahara, Y.
Logical Approach to Systems Theory
192 pp. 1995 [3-540-19956-X]

Vol. 205: Kotta, U.
Inversion Method in the Discrete-time Nonlinear Control Systems Synthesis Problems
168 pp. 1995 [3-540-19966-7]

Vol. 206: Aganovic, Z.;.Gajic, Z.
Linear Optimal Control of Bilinear Systems with Applications to Singular Perturbations and Weak Coupling
133 pp. 1995 [3-540-19976-4]

Vol. 207: Gabasov, R.; Kirillova, F.M.; Prischepova, S.V.
Optimal Feedback Control
224 pp. 1995 [3-540-19991-8]

Vol. 208: Khalil, H.K.; Chow, J.H.; Ioannou, P.A. (Eds)
Proceedings of Workshop on Advances inControl and its Applications
300 pp. 1995 [3-540-19993-4]

Vol. 209: Foias, C.; Özbay, H.; Tannenbaum, A.
Robust Control of Infinite Dimensional Systems: Frequency Domain Methods
230 pp. 1995 [3-540-19994-2]

Vol. 210: De Wilde, P.
Neural Network Models: An Analysis
164 pp. 1996 [3-540-19995-0]

Vol. 211: Gawronski, W.
Balanced Control of Flexible Structures
280 pp. 1996 [3-540-76017-2]

Vol. 212: Sanchez, A.
Formal Specification and Synthesis of Procedural Controllers for Process Systems
248 pp. 1996 [3-540-76021-0]

Vol. 213: Patra, A.; Rao, G.P.
General Hybrid Orthogonal Functions and their Applications in Systems and Control
144 pp. 1996 [3-540-76039-3]

Vol. 214: Yin, G.; Zhang, Q. (Eds)
Recent Advances in Control and Optimization of Manufacturing Systems
240 pp. 1996 [3-540-76055-5]

Vol. 215: Bonivento, C.; Marro, G.; Zanasi, R. (Eds)
Colloquium on Automatic Control
240 pp. 1996 [3-540-76060-1]

Vol. 216: Kulhavý, R.
Recursive Nonlinear Estimation: A Geometric Approach
244 pp. 1996 [3-540-76063-6]

Vol. 217: Garofalo, F.; Glielmo, L. (Eds)
Robust Control via Variable Structure and Lyapunov Techniques
336 pp. 1996 [3-540-76067-9]

Vol. 218: van der Schaft, A.
L_2 Gain and Passivity Techniques in Nonlinear Control
176 pp. 1996 [3-540-76074-1]

Vol. 219: Berger, M.-O.; Deriche, R.; Herlin, I.; Jaffré, J.; Morel, J.-M. (Eds)
ICAOS '96: 12th International Conference on Analysis and Optimization of Systems - Images, Wavelets and PDEs:
Paris, June 26-28 1996
378 pp. 1996 [3-540-76076-8]

Vol. 220: Brogliato, B.
Nonsmooth Impact Mechanics: Models, Dynamics and Control
420 pp. 1996 [3-540-76079-2]

Vol. 221: Kelkar, A.; Joshi, S.
Control of Nonlinear Multibody Flexible Space Structures
160 pp. 1996 [3-540-76093-8]

Vol. 222: Morse, A.S.
Control Using Logic-Based Switching
288 pp. 1997 [3-540-76097-0]

Vol. 223: Khatib, O.; Salisbury, J.K.
Experimental Robotics IV: The 4th International Symposium, Stanford, California,
June 30 - July 2, 1995
596 pp. 1997 [3-540-76133-0]

Vol. 224: Magni, J.-F.; Bennani, S.; Terlouw, J. (Eds)
Robust Flight Control: A Design Challenge
664 pp. 1997 [3-540-76151-9]

Vol. 225: Poznyak, A.S.; Najim, K.
Learning Automata and Stochastic Optimization
219 pp. 1997 [3-540-76154-3]

Vol. 226: Cooperman, G.; Michler, G.; Vinck, H. (Eds)
Workshop on High Performance Computing and Gigabit Local Area Networks
248 pp. 1997 [3-540-76169-1]

Vol. 227: Tarbouriech, S.; Garcia, G. (Eds)
Control of Uncertain Systems with Bounded Inputs
203 pp. 1997 [3-540-76183-7]

Vol. 228: Dugard, L.; Verriest, E.I. (Eds)
Stability and Control of Time-delay Systems
344 pp. 1998 [3-540-76193-4]

Vol. 229: Laumond, J.-P. (Ed.)
Robot Motion Planning and Control
360 pp. 1998 [3-540-76219-1]